A Problem-Solving Approach to Multi-Variable Calculus

Farzin Asadi

A Problem-Solving Approach to Multi-Variable Calculus

Volume II

Springer

Farzin Asadi
Department of Computer Engineering
OSTIM Technical University
Ankara, Türkiye

ISBN 978-3-032-21114-9 ISBN 978-3-032-21115-6 (eBook)
https://doi.org/10.1007/978-3-032-21115-6

This Springer imprint is published by the registered company Springer Nature Switzerland AG
The registered company address is: Gewerbestrasse 11, 6330 Cham, Switzerland

Preface

Multivariable calculus is the natural evolution of mathematical modeling, extending the principles of change to the complex, multi-dimensional world we inhabit. While single-variable calculus examines a single cause and its effect, multivariable calculus provides the framework for students in the basic sciences, engineering, and economics to analyze systems where numerous variables interact simultaneously. It adapts our two indispensable tools for a higher-dimensional reality: the partial derivative, which quantifies how a system changes as one specific factor varies among many, and the multiple integral, which calculates total accumulation over surfaces and volumes.

These concepts are critical for mastering modern quantitative disciplines. For example, engineers move beyond linear motion to analyze stress distributions in solid structures and the flow of fluids in three dimensions; scientists apply multivariable models to describe electromagnetic fields, heat diffusion, and the complex gradients found in atmospheric weather patterns; and economists utilize these tools to solve optimization problems involving multiple constraints, such as maximizing utility across a diverse basket of goods or optimizing global supply chains. By moving beyond the single line, multivariable calculus equips us to map and master the intricate landscapes of the real world.

To truly grasp the concepts and techniques of multivariable calculus, practice is paramount. This book is your comprehensive companion—a three-volume set filled with meticulously solved, step-by-step problems designed to build your skills and deepen your understanding. With these resources, you will gain the confidence to approach any multivariable calculus problem and transform challenges into achievements. A solid foundation in single-variable calculus is the only prerequisite.

This book is designed as a valuable supplement to any standard multivariable calculus textbook. Each chapter begins with a concise review of key concepts, followed by a collection of fully worked examples presented in a clear, step-by-step format. You are encouraged to attempt each problem on your own before consulting the detailed solutions.

Because computational methods are central to modern science and engineering, this book adopts a dual approach to problem-solving. It first guides you through traditional hand calculations and then demonstrates how to apply the same concepts using MATLAB®, bridging theoretical understanding with practical computational skills.

I hope this book will serve as a reliable companion throughout your studies. By embracing this problem-solving approach, you will not only build a strong understanding of multivariable calculus but also develop the confidence to tackle new and challenging problems. I wish you a productive and rewarding journey as you master this fascinating subject.

Ankara, Türkiye Farzin Asadi

Competing Interests The author has no competing interests to declare that are relevant to the content of this manuscript.

Contents

Notation Used in This Book

$\sin^{-1}(x)$ denotes the arcsine of x.

$\cos^{-1}(x)$ denotes the arccosine of x.

$\tan^{-1}(x)$ denotes the arctangent of x.

$\cot^{-1}(x)$ denotes the arccotangent of x.

$\sec^{-1}(x)$ denotes the arcsecant of x.

$\csc^{-1}(x)$ denotes the arccosecant of x.

$\sec(x)$ denotes the secant of x ($\sec(x) = \frac{1}{\cos(x)} = (\cos(x))^{-1}$).

$csc(x)$ denotes the cosecant of x ($\csc(x) = \frac{1}{\sin(x)} = (\sin(x))^{-1}$).

i and j are the imaginary unit, where $i = j = \sqrt{-1}$.

Note that, for example, $\cos^{-1}(x)$ denotes the inverse cosine function (also written as $arccos(x)$, meaning the function that undoes cosine), while $(\cos(x))^{-1}$ represents the reciprocal of cosine, which equal $\frac{1}{\cos(x)}$.

The notation $|A|$ denotes the determinant when A is a matrix, and the absolute value when A is a scalar expression.

Chapter 1
Double Integrals

1.1 Introduction

While a single integral allows us to calculate the area under a curve, the double integral extends this concept into three dimensions. It is the primary tool used to calculate the volume under a surface defined by a function of two variables, $z = f(x, y)$. By summing up infinitesimal rectangular prisms across a region in the xy-plane, double integrals provide a powerful way to analyze physical properties such as mass, center of gravity, and surface area.

1.1.1 Summary of Integration Formulas

A strong grasp of single-variable calculus is essential for computing double integrals. To assist you, a review of fundamental integration formulas is provided here for reference.

1. $\int k_1 dx = k_1 x + C$
2. $\int x^n dx = \frac{1}{n+1} x^{n+1} + C \; n \in \mathbb{R}, n \neq -1$
3. $\int x^{-1} dx = \int \frac{1}{x} dx = \ln(|x|) + C$
4. $\int \frac{u(x)'}{u(x)} dx = \ln(|u(x)|) + C$
5. $\int \frac{1}{ax+b} dx = \frac{1}{a} \ln(|ax+b|) + C \quad a, b \in \mathbb{R}, a \neq 0$
6. $\int (ax+b)^n dx = \frac{1}{a} \frac{1}{n+1} (ax+b)^{n+1} + C \quad a, b, n \in \mathbb{R}, a \neq 0, n \neq -1$
7. $\int \frac{u'(x)}{(u(x))^2 + a^2} dx = \frac{1}{a} \tan^{-1}\left(\frac{u(x)}{a}\right) + C$
8. $\int \frac{1}{x^2+a^2} dx = \frac{1}{a} \tan^{-1}\left(\frac{x}{a}\right) + C$
9. $\int \frac{u'(x)}{\sqrt{a^2 - u(x)^2}} dx = \sin^{-1}\left(\frac{u(x)}{a}\right) + C$
10. $\int \frac{u'(x)}{\sqrt{1 - u(x)^2}} dx = \sin^{-1}(u(x)) + C$

F. Asadi, *A Problem-Solving Approach to Multi-Variable Calculus*,
https://doi.org/10.1007/978-3-032-21115-6_1

11. $\int \frac{u'(x)}{\sqrt{a^2-u(x)^2}}dx = -\cos^{-1}\left(\frac{u(x)}{a}\right) + C$
12. $\int \frac{u'(x)}{\sqrt{1-u(x)^2}}dx = -\cos^{-1}(u(x)) + C$
13. $\int \frac{u'(x)}{a^2+u(x)^2}dx = \frac{1}{a}\tan^{-1}\left(\frac{u(x)}{a}\right) + C$
14. $\int \frac{u'(x)}{1+u(x)^2}dx = \tan^{-1}(u(x)) + C$
15. $\int \frac{u'(x)}{a^2+u(x)^2}dx = -\frac{1}{a}\cot^{-1}\left(\frac{u(x)}{a}\right) + C$
16. $\int \frac{u'(x)}{1+u(x)^2}dx = -\cot^{-1}(u(x)) + C$
17. $\int \frac{u'(x)}{1-u(x)^2}dx = \frac{1}{2}\ln\left(\left|\frac{1+u(x)}{1-u(x)}\right|\right) + C$
18. $\int \frac{u'(x)}{|u(x)|\sqrt{(u(x))^2-a^2}}dx = \frac{1}{a}\sec^{-1}\left(\left|\frac{u(x)}{a}\right|\right) + C$
19. $\int \frac{u'(x)}{|u(x)|\sqrt{u^2(x)-a^2}}dx = -\frac{1}{a}\csc^{-1}\left(\left|\frac{u(x)}{a}\right|\right) + C$
20. $\int u'(x)\sin(u(x))dx = -\cos(u(x)) + C$
21. $\int u'(x)\cos(u(x))dx = \sin(u(x)) + C$
22. $\int u'(x)\big(1+\tan^2(u(x))\big)dx = \tan(u(x)) + C$
23. $\int u'(x)\big(1+\cot^2(u(x))\big)dx = -\cot(u(x)) + C$
24. $\int \sin(ax+b)dx = -\frac{1}{a}\cos(ax+b) + C \quad a, b \in \mathbb{R}, a \neq 0$
25. $\int \cos(ax+b)dx = \frac{1}{a}\sin(ax+b) + C \quad a, b \in \mathbb{R}, a \neq 0$
26. $\int \big(1+\tan^2(ax+b)\big)dx = \frac{1}{a}\tan(ax+b) + C \quad a, b \in \mathbb{R}, a \neq 0$
27. $\int \big(1+\cot^2(ax+b)\big)dx = -\frac{1}{a}\cot(ax+b) + C \quad a, b \in \mathbb{R}, a \neq 0$
28. $\int u'(x)e^{u(x)}dx = e^{u(x)} + C$
29. $\int u'(x)(u(x))^n dx = \frac{1}{n+1}u^{n+1}(x) + C$
30. $\int e^{ax}dx = \frac{1}{a}e^{ax} + C$

1.1.2 Wallis' Formula

Wallis' Formula is a highly efficient shortcut in calculus used to evaluate definite integrals of the form $\int_0^{\frac{\pi}{2}} \sin^n(x)dx$ or $\int_0^{\frac{\pi}{2}} \cos^n(x)dx$, where n is a positive integer.

For $n = 1$, $\int_0^{\frac{\pi}{2}} \sin(x)dx = \int_0^{\frac{\pi}{2}} \cos(x)dx = 1$.

For an integer $n \geq 2$, the formula is:

$$\int_0^{\frac{\pi}{2}} \sin^n(x)dx = \int_0^{\frac{\pi}{2}} \cos^n(x)dx = \begin{cases} \frac{n-1}{n} \times \frac{n-3}{n-2} \times \cdots \times \frac{2}{3} \times 1 & \textit{if n is odd} \\ \frac{n-1}{n} \times \frac{n-3}{n-2} \times \cdots \times \frac{1}{2} \times \frac{\pi}{2} & \textit{if n is even} \end{cases}$$

The "rule of thumb" to remember is to start with $n-1$ in the numerator and n in the denominator, then subtract 2 from both until you can't go any further without hitting zero or negative numbers.

If *n* is odd: The sequence ends at $\frac{2}{3}$ and is multiplied by 1.

If *n* is even: The sequence ends at $\frac{1}{2}$ and is multiplied by $\frac{\pi}{2}$.

For example,

$$\int_0^{\frac{\pi}{2}} \sin^5(x)dx = \frac{4}{5} \times \frac{2}{3} \times 1 = \frac{8}{15}$$

$$\int_0^{\frac{\pi}{2}} \cos^6(x)dx = \frac{5}{6} \times \frac{3}{4} \times \frac{1}{2} \times \frac{\pi}{2} = \frac{15\pi}{96} = \frac{5\pi}{32}$$

These formulas simplify the calculation of integrals on the interval $[0, \pi]$:

$$\forall n \in \mathbb{N} \Rightarrow \int_0^{\pi} \sin^n(x)dx = 2\int_0^{\frac{\pi}{2}} \sin^n(x)dx$$

$$\forall n \in \mathbb{N} \Rightarrow \int_0^{\pi} \cos^n(x)dx = \begin{cases} 2\int_0^{\frac{\pi}{2}} \cos^n(x)dx & \textit{if } n \textit{ is even} \\ 0 & \textit{if } n \textit{ is odd} \end{cases}$$

1.1.3 The Separability Rule

If $f(x, y) = g(x)h(y)$ and the region of integration is a rectangle $R = \{(x, y)a \leq x \leq b, c \leq y \leq d\}$, then:

$$\int_c^d \int_a^b f(x, y)dxdy = \int_c^d \int_a^b g(x)h(y)dxdy = \int_a^b g(x)dx \int_c^d h(y)dy$$

1.1.4 The Jacobian Matrix and Change of Variables in Multiple Integrals

The Jacobian matrix, J, is defined below for transformations involving two and three variables:

$$J = \frac{\partial(f_1, f_2)}{\partial(x_1, x_2)} = \begin{bmatrix} \frac{\partial f_1}{\partial x_1} & \frac{\partial f_1}{\partial x_2} \\ \frac{\partial f_2}{\partial x_1} & \frac{\partial f_2}{\partial x_2} \end{bmatrix}$$

$$J = \frac{\partial(f_1, f_2, f_3)}{\partial(x_1, x_2, x_3)} = \begin{bmatrix} \frac{\partial f_1}{\partial x_1} & \frac{\partial f_1}{\partial x_2} & \frac{\partial f_1}{\partial x_3} \\ \frac{\partial f_2}{\partial x_1} & \frac{\partial f_2}{\partial x_2} & \frac{\partial f_2}{\partial x_3} \\ \frac{\partial f_3}{\partial x_1} & \frac{\partial f_3}{\partial x_2} & \frac{\partial f_3}{\partial x_3} \end{bmatrix}$$

The change of variables method for double integrals allows you to transform a complex integral over an irregular region R in the xy-plane into a simpler integral over a new, typically rectangular or circular region S in the uv-plane. This is achieved by defining a one-to-one transformation, T, where $x = x(u, v)$ and $y = y(u, v)$. The general formula is:

$$\iint_R f(x, y)dA = \iint_S f(x(u, v), y(u, v))\left|\frac{\partial(x, y)}{\partial(u, v)}\right|dudv$$

The crucial part of this transformation is the factor $\left|\frac{\partial(x,y)}{\partial(u,v)}\right| = \left|\det\left(\begin{bmatrix} \frac{\partial x}{\partial u} & \frac{\partial x}{\partial v} \\ \frac{\partial y}{\partial u} & \frac{\partial y}{\partial v} \end{bmatrix}\right)\right|$, which is the absolute value of the Jacobian matrix determinant. This factor accounts for how the transformation stretches or shrinks the differential area element $dA = dxdy = dydx$ as it moves from the uv-plane to the xy-plane, ensuring the integral's value remains the same.

When u and v are given in terms of x and y, computing $\left|\det\left(\frac{\partial(x,y)}{\partial(u,v)}\right)\right|$ directly may be difficult. In such cases, you can use the identity:

$$\left|\det\left(\frac{\partial(x, y)}{\partial(u, v)}\right)\right| = \frac{1}{\left|\det\left(\frac{\partial(u,v)}{\partial(x,y)}\right)\right|} = \frac{1}{\left|\det\left(\begin{bmatrix} \frac{\partial u}{\partial x} & \frac{\partial u}{\partial y} \\ \frac{\partial v}{\partial x} & \frac{\partial v}{\partial y} \end{bmatrix}\right)\right|}$$

1.2 Solved Problems

Example 1.1 Calculate $\int_1^2 \int_0^y \frac{1}{(2x+3y)^5} dxdy$.

Solution To calculate the double integral $I = \int_1^2 \int_0^y \frac{1}{(2x+3y)^5} dxdy$, we must evaluate the inner integral with respect to x first, and then the outer integral with respect to y.

$$\int_0^y \frac{1}{(2x + 3y)^5} dx = \int_0^y (2x + 3y)^{-5} dx = \frac{1}{2} \int_0^y 2(2x + 3y)^{-5} dx$$

$$= \frac{1}{2} \frac{(2x + 3y)^{-5+1}}{-5 + 1}\Bigg|_{x=0}^{x=y}$$

$$= \frac{(2x+3y)^{-4}}{-8}\Big|_{x=0}^{x=y} = \frac{(2x+3y)^{-4}}{8}\Big|_{x=y}^{x=0}$$
$$= \frac{(2\times 0+3y)^{-4}}{8} - \frac{(2\times y+3y)^{-4}}{8}$$
$$= \frac{(3y)^{-4}-(5y)^{-4}}{8} = \frac{2^{-4}-5^{-4}}{8}y^{-4}$$

$$\int_1^2\int_0^y \frac{1}{(2x+3y)^5}dxdy = \int_1^2\left(\frac{3^{-4}-5^{-4}}{8}y^{-4}\right)dy = \frac{3^{-4}-5^{-4}}{8}\int_1^2 y^{-4}dy$$
$$= \frac{3^{-4}-5^{-4}}{8}\frac{y^{-4+1}}{-4+1}\Big|_0^2 = \frac{3^{-4}-5^{-4}}{8}\frac{y^{-3}}{-3}\Big|_1^2$$
$$= \frac{3^{-4}-5^{-4}}{-24}y^{-3}\Big|_1^2 = \frac{3^{-4}-5^{-4}}{24}y^{-3}\Big|_2^1$$
$$= \frac{3^{-4}-5^{-4}}{24}1^{-3} - \frac{3^{-4}-5^{-4}}{24}2^{-3}$$
$$= \frac{3^{-4}-5^{-4}}{24}\left(1^{-3}-2^{-3}\right) = \frac{5^4-3^4}{24\times 5^4\times 3^4}\times\frac{7}{8}$$
$$= \frac{3808}{9720000} \approx 0.00039177$$

Example 1.2 Calculate $\int_0^{\frac{\pi}{2}}\int_1^2 x^2y\sin\left(y^2\right)dxdy$.

Solution If you are integrating a function $f(x, y)$ over a rectangular region $R = [a, b]\times[c, d]$, and the function $f(x, y)$ is separable (meaning it can be written as the product of a function of x only, $g(x)$, and a function of y only, $h(y)$), then you can separate the double integral into the product of two single integrals.

$$\int_0^{\frac{\pi}{2}}\int_1^2 yx^2\sin\left(y^2\right)dxdy = \int_1^2 x^2dx\times\int_0^{\frac{\pi}{2}} y\sin\left(y^2\right)dy = \int_1^2 x^2dx\times\frac{1}{2}\int_0^{\frac{\pi}{2}} 2y\sin\left(y^2\right)dy$$
$$= \frac{x^3}{3}\Big|_1^2\times\frac{1}{2}\times-\cos\left(y^2\right)\Big|_0^{\frac{\pi}{2}} = \frac{x^3}{3}\Big|_1^2\times\frac{1}{2}\times\cos\left(y^2\right)\Big|_{\frac{\pi}{2}}^0$$
$$= \left(\frac{2^3}{3}-\frac{1^3}{3}\right)\times\frac{1}{2}\times\left(\cos(0^2)-\cos\left(\frac{\pi^2}{4}\right)\right)$$
$$= \frac{7\left(1-\cos\left(\frac{\pi^2}{4}\right)\right)}{6} \approx 2.0781$$

Example 1.3 Calculate $\int_0^{\frac{\pi}{2}}\int_1^2 xsin\left(x^2\right)e^{\cos\left(x^2\right)-5y}dxdy$.

Solution Because the integrand is separable—meaning it can be written as $g(x)h(y)$—the double integral can be calculated as the product of two single integrals.

$$\int_0^{\frac{\pi}{2}}\int_1^2 x\sin(x^2)e^{\cos(x^2)-5y}dxdy = \int_1^2 x\sin(x^2)e^{\cos(x^2)}dx \times \int_0^{\frac{\pi}{2}} e^{-5y}dy$$

$$= \frac{1}{-2}\int_1^2 -2x\sin(x^2)e^{\cos(x^2)}dx \times \frac{1}{-5}\int_0^{\frac{\pi}{2}} -5e^{-5y}dy$$

$$= \frac{1}{10} \times \int_1^2 -2x\sin(x^2)e^{\cos(x^2)}dx \times \int_0^{\frac{\pi}{2}} -5e^{-5y}dy$$

$$= \frac{1}{10} \times e^{\cos(x^2)}\Big|_1^2 \times e^{-5y}\Big|_0^{\frac{\pi}{2}}$$

$$= \frac{1}{10} \times \left(e^{\cos(4)} - e^{\cos(1)}\right) \times \left(e^{-\frac{5\pi}{2}} - 1\right)$$

$$\approx \frac{1}{10} \times -1.1964 \times -0.9996 = 0.1196$$

Example 1.4 Calculate $\int_1^{e^4}\int_0^y \frac{1}{(x^2+y^2)}dxdy$.

Solution Recall that $\int \frac{1}{x^2+a^2}dx = \frac{1}{a}\tan^{-1}\left(\frac{x}{a}\right) + C$. Therefore, $\int_0^y \frac{1}{x^2+y^2}dx = \frac{1}{y}\tan^{-1}\left(\frac{x}{y}\right)\Big|_0^y = \frac{1}{y}\tan^{-1}\left(\frac{y}{y}\right) - \frac{1}{y}\tan^{-1}\left(\frac{0}{y}\right) = \frac{1}{y}\tan^{-1}(1) - \frac{1}{y}\tan^{-1}(0) = \frac{1}{y} \times \frac{\pi}{4} - \frac{1}{y} \times 0 = \frac{\pi}{4y}$.

$$\int_1^{e^4}\int_0^y \frac{1}{(x^2+y^2)}dxdy = \int_1^{e^4} \frac{\pi}{4y}dy = \frac{\pi}{4}\ln(y)\Big|_1^{e^4}$$

$$= \frac{\pi}{4}\ln(e^4) - \frac{\pi}{4}\ln(1) = \frac{\pi}{4} \times 4 - \frac{\pi}{4} \times 0 = \pi$$

Example 1.5 Calculate $\int_{-2}^2\int_{-2}^{-\sqrt{3}}(x^2+y^2)dxdy$.

Solution

$$\int_{-2}^{-\sqrt{3}}(x^2+y^2)dx = \left(\frac{x^3}{3} + y^2x\right)\Big|_{-2}^{-\sqrt{3}} = \left(-\sqrt{3} - \sqrt{3}y^2\right) - \left(\frac{-8}{3} - 2y^2\right)$$

$$= \left(2 - \sqrt{3}\right)y^2 + \frac{8}{3} - \sqrt{3}$$

$$\int_{-2}^{2}\int_{-2}^{-\sqrt{3}}(x^2+y^2)dxdy = \int_{-2}^{2}\left(\left(2-\sqrt{3}\right)y^2+\frac{8}{3}-\sqrt{3}\right)dy$$

$$= 2\int_{0}^{2}\left(\left(2-\sqrt{3}\right)y^2+\frac{8}{3}-\sqrt{3}\right)dy$$

$$= 2\left(\frac{\left(2-\sqrt{3}\right)y^3}{3}+\left(\frac{8}{3}-\sqrt{3}\right)y\Bigg|_0^2\right)$$

$$= 2\left(\frac{\left(2-\sqrt{3}\right)\times 8}{3}+\left(\frac{8}{3}-\sqrt{3}\right)\times 2-0\right)$$

$$= \frac{16}{3}\left(4-\sqrt{3}\right)-4\sqrt{3} = \frac{64}{3}-\frac{28}{3}\sqrt{3} = \frac{64-28\sqrt{3}}{3}$$

$$\approx 5.1675$$

Example 1.6 Calculate $\int_0^1 \int_{\sqrt[4]{y}}^{y} 1dxdy$.

Solution $\int_{\sqrt[4]{y}}^{y} 1dx = x\Big|_{\sqrt[4]{y}}^{y} = y-\sqrt[4]{y}$

$$\int_0^1\int_{\sqrt[4]{y}}^{y} 1dxdy = \int_0^1 \left(y-\sqrt[4]{y}\right)dy = \frac{y^2}{2}-\frac{y^{\frac{1}{4}+1}}{\frac{1}{4}+1}\Bigg|_0^1$$

$$= \frac{y^2}{2}-\frac{4y^{\frac{5}{4}}}{5}\Bigg|_0^1 = \frac{1}{2}-\frac{4}{5}-0 = \frac{-3}{10} = -0.3$$

Example 1.7 Calculate $\int_{-\sqrt{3}}^{\sqrt{3}} \int_{-2}^{x^2-1}(x^2+y^2)dydx$.

Solution

$$\int_{-2}^{x^2-1}(x^2+y^2)dy = \left(x^2y+\frac{y^3}{3}\right)\Bigg|_{-2}^{x^2-1}$$

$$= \left(x^2(x^2-1)+\frac{\left(x^2-1\right)^3}{3}\right)-\left(-2x^2-\frac{8}{3}\right)$$

$$= x^4-x^2+\frac{\left(x^2-1\right)^3}{3}+2x^2+\frac{8}{3}$$

$$= x^4 + x^2 + \frac{\left(x^2 - 1\right)^3}{3} + \frac{8}{3}$$

$$\int_{-\sqrt{3}}^{\sqrt{3}} \int_{-2}^{x^2-1} (x^2 + y^2) dy dx = \int_{-\sqrt{3}}^{\sqrt{3}} \left(x^4 + x^2 + \frac{\left(x^2 - 1\right)^3}{3} + \frac{8}{3} \right) dx$$

$$= 2 \int_{0}^{\sqrt{3}} \left(x^4 + x^2 + \frac{\left(x^2 - 1\right)^3}{3} + \frac{8}{3} \right) dx$$

$$= 2 \int_{0}^{\sqrt{3}} \left(x^4 + x^2 + \frac{x^6 - 3x^4 + 3x^2 - 1}{3} + \frac{8}{3} \right) dx$$

$$= 2 \int_{0}^{\sqrt{3}} \left(\frac{x^6}{3} + 2x^2 + \frac{7}{3} \right) dx$$

$$= 2 \left(\frac{x^7}{21} + \frac{2x^3}{3} + \frac{7}{3} x \right) \Bigg|_{0}^{\sqrt{3}}$$

$$= 2 \left(\frac{27\sqrt{3}}{21} + 2\sqrt{3} + \frac{7}{\sqrt{3}} \right) - 0$$

$$= \frac{18\sqrt{3}}{7} + 4\sqrt{3} + \frac{14}{\sqrt{3}}$$

$$\approx 19.4650$$

Example 1.8 Calculate $\int_{-2}^{2} \int_{\sqrt{3}}^{2} (x^2 + y^2) dx dy$.

Solution $\int\limits_{\sqrt{3}}^{2} (x^2 + y^2) dx = \left(\frac{x^3}{3} + y^2 x \right) \Big|_{\sqrt{3}}^{2} = \frac{8}{3} + 2y^2 - \left(\sqrt{3} + \sqrt{3} y^2 \right) = \left(2 - \sqrt{3}\right) y^2 + \frac{8}{3} - \sqrt{3}$

$$\int_{-2}^{2} \int_{\sqrt{3}}^{2} (x^2 + y^2) dx dy = \int_{-2}^{2} \left(\left(2 - \sqrt{3}\right) y^2 + \frac{8}{3} - \sqrt{3} \right) dy$$

$$= 2 \int_{0}^{2} \left(\left(2 - \sqrt{3}\right) y^2 + \frac{8}{3} - \sqrt{3} \right) dy$$

$$= 2\left(\frac{\left(2-\sqrt{3}\right)y^3}{3} + \left(\frac{8}{3}-\sqrt{3}\right)y\Bigg|_0^2\right)$$

$$= 2\left(\frac{\left(2-\sqrt{3}\right)\times 8}{3} + \left(\frac{8}{3}-\sqrt{3}\right)\times 2 - 0\right)$$

$$= \frac{16}{3}\left(4-\sqrt{3}\right) - 4\sqrt{3} = \frac{64}{3} - \frac{28}{3}\sqrt{3}$$

$$= \frac{64-28\sqrt{3}}{3} \approx 5.1675$$

Example 1.9 Calculate $\int_{\frac{\pi}{2}}^{\frac{3\pi}{2}} \int_0^{3(1+\cos(y))} xdxdy$.

Solution $\int_0^{3(1+\cos(y))} xdx = \frac{x^2}{2}\Big|_0^{3(1+\cos(y))} = \frac{(3(1+\cos(y)))^2}{2} - \frac{0^2}{2} = \frac{9}{2} + 9\cos(y) + \frac{9}{2}\cos^2(y)$

$$\int_{\frac{\pi}{2}}^{\frac{3\pi}{2}} \int_0^{3(1+\cos(y))} xdxdy = \int_{\frac{\pi}{2}}^{\frac{3\pi}{2}} \left(\frac{9}{2} + 9\cos(y) + \frac{9}{2}\cos^2(y)\right)dy$$

$$= \int_{\frac{\pi}{2}}^{\frac{3\pi}{2}} \left(\frac{9}{2} + 9\cos(y) + \frac{9}{2}\times\frac{1+\cos(2y)}{2}\right)dy$$

$$= \int_{\frac{\pi}{2}}^{\frac{3\pi}{2}} \left(\frac{27}{4} + 9\cos(y) + \frac{9}{4}\cos(2y)\right)dy$$

$$= \left(\frac{27}{4}y + 9\sin(y) + \frac{9}{8}\sin(2y)\right)\Bigg|_{\frac{\pi}{2}}^{\frac{3\pi}{2}}$$

$$= \left(\frac{27}{4}\times\frac{3\pi}{2} + 9\sin\left(\frac{3\pi}{2}\right) + \frac{9}{8}\sin\left(2\times\frac{3\pi}{2}\right)\right)$$
$$- \left(\frac{27}{4}\times\frac{\pi}{2} + 9\sin\left(\frac{\pi}{2}\right) + \frac{9}{8}\sin\left(2\times\frac{\pi}{2}\right)\right)$$

$$= \left(\frac{81\pi}{8} - 9 + 0\right) - \left(\frac{27\pi}{8} + 9 + 0\right)$$

$$= \frac{54\pi}{8} - 18 = \frac{27\pi}{4} - 18 \approx 3.2058$$

Example 1.10 Calculate $\int_0^2 \int_0^x ydydx$.

Solution $\int_0^x y dy = \frac{y^2}{2}\Big|_0^x = \frac{x^2}{2} - \frac{0^2}{2} = \frac{x^2}{2}$

$$\int_0^2 \int_0^x y dy dx = \int_0^2 \frac{x^2}{2} dx = \frac{x^3}{6}\Big|_0^2 = \frac{2^3}{6} - 0 = \frac{8}{6} = \frac{4}{3} \approx 1.3333$$

Example 1.11 Evaluate the double integral $\iint_R y dx dy$ over the region R depicted in Fig. 1.1.

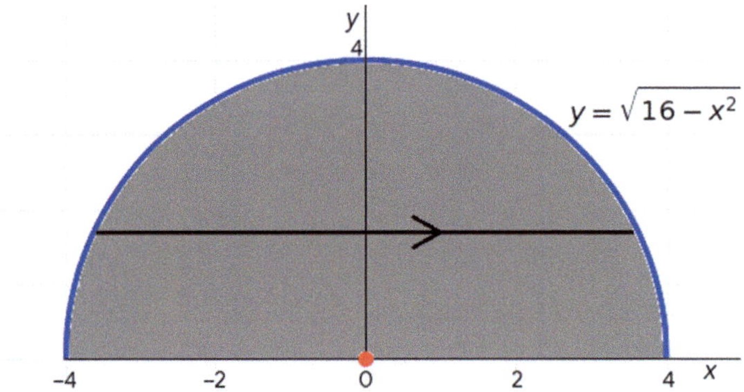

Fig. 1.1 Region R for Example 1.11

Solution $y = \sqrt{16 - x^2} \Rightarrow y^2 = 16 - x^2 \Rightarrow x^2 = 16 - y^2 \Rightarrow x = \pm\sqrt{16 - y^2}$.

From Fig. 1.2 we see that $R = \left\{ (x, y) | 0 \le y \le 4, -\sqrt{16 - y^2} \le x \le \sqrt{16 - y^2} \right\}$. Therefore,

$$\iint_R y dx dy = \int_0^4 \int_{-\sqrt{16-y^2}}^{\sqrt{16-y^2}} y dx dy$$

$$\int_{-\sqrt{16-y^2}}^{\sqrt{16-y^2}} y dx = yx \Big|_{-\sqrt{16-y^2}}^{\sqrt{16-y^2}} = y\sqrt{16 - y^2} - \left(-y\sqrt{16 - y^2}\right)$$

$$= 2y\sqrt{16 - y^2}$$

$$\int_0^4 \int_{-\sqrt{16-y^2}}^{\sqrt{16-y^2}} y dx dy = \int_0^4 2y\sqrt{16 - y^2} dy = -\int_0^4 -2y\sqrt{16 - y^2} dy$$

$$= -\frac{(16-y^2)^{\frac{1}{2}+1}}{\frac{1}{2}+1}\Bigg|_0^4 = -\frac{2(16-y^2)^{\frac{3}{2}}}{3}\Bigg|_0^4$$

$$= \frac{2(16-y^2)^{\frac{3}{2}}}{3}\Bigg|_4^0 = \frac{2(16-0^2)^{\frac{3}{2}}}{3}$$

$$-\frac{2(16-4^2)^{\frac{3}{2}}}{3} = \frac{2\times 16^{\frac{3}{2}}}{3} - 0$$

$$= \frac{2\times(4^2)^{\frac{3}{2}}}{3} = \frac{2\times 4^3}{3} = \frac{128}{3} \approx 42.6667$$

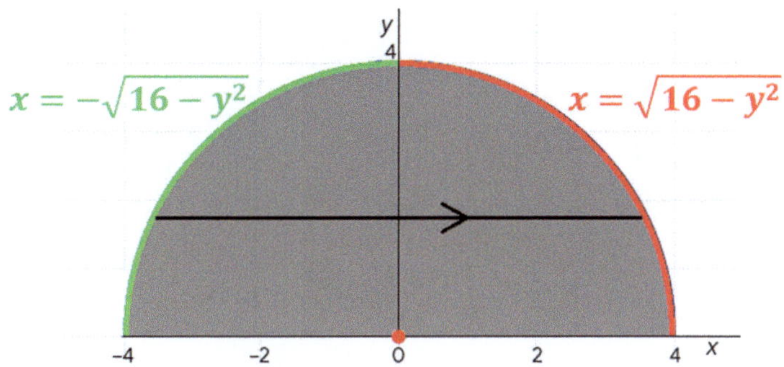

Fig. 1.2 Boundary equations for the region R

Example 1.12 Evaluate the double integral $\iint_R y\,dy\,dx$ over the region R depicted in Fig. 1.3.

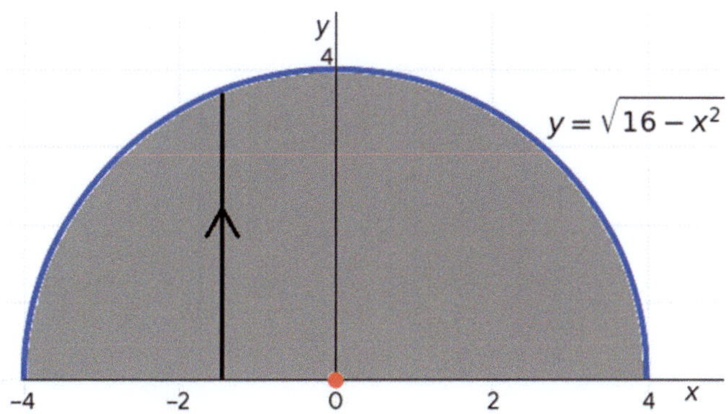

Fig. 1.3 Region R for Example 1.12

Solution From Fig. 1.3 we see that $R = \left\{ (x, y) | -4 \le x \le 4, 0 \le y \le \sqrt{16-y^2} \right\}$. Therefore,

$$\iint_R y\,dy\,dx = \int_{-4}^{4}\int_{0}^{\sqrt{16-x^2}} y\,dy\,dx$$

$$\int_{0}^{\sqrt{16-x^2}} y\,dy = \frac{y^2}{2}\Bigg|_{0}^{\sqrt{16-x^2}} = \frac{\left(\sqrt{16-x^2}\right)^2}{2} - \frac{0^2}{2}$$

$$= \frac{16-x^2}{2} - 0 = \frac{16-x^2}{2}$$

$$\int_{-4}^{4}\int_{0}^{\sqrt{16-x^2}} y\,dy\,dx = \int_{-4}^{4} \frac{16-x^2}{2}dx$$

The integrand is an even function, therefore,

$$\int_{-4}^{4}\int_{0}^{\sqrt{16-x^2}} y\,dy\,dx = \int_{-4}^{4} \frac{16-x^2}{2}dx = 2\int_{0}^{4} \frac{16-x^2}{2}dx$$

$$= \int_{0}^{4} (16-x^2)dx = 16x - \frac{x^3}{3}\Bigg|_{0}^{4}$$

$$= 16 \times 4 - \frac{4^3}{3} - \left(16 \times 0 - \frac{0^3}{3}\right)$$

$$= 64 - \frac{64}{3} - 0 = \frac{128}{3} \approx 42.6667$$

Example 1.13 Evaluate the double integral $\iint_R e^{\frac{x}{2y}}\,dA$ over the region R depicted in Fig. 1.4.

Fig. 1.4 Region R for Example 1.13

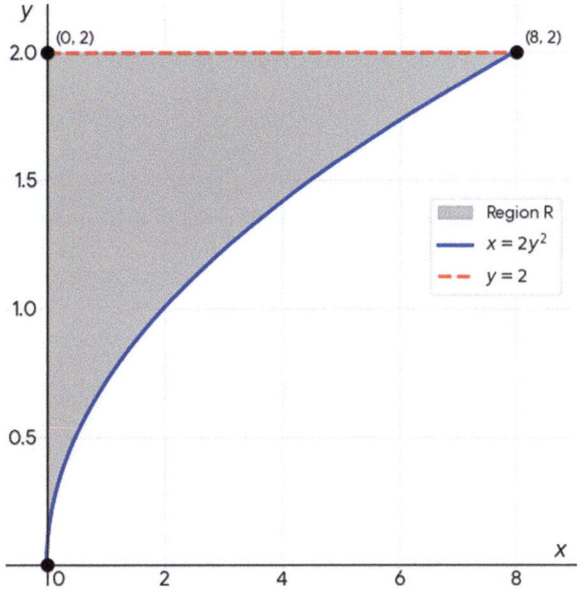

Solution Given that the problem statement does not dictate the sequence, we have the option to integrate in either order ($\iint_R e^{\frac{x}{2y}}\,dydx$ or $\iint_R e^{\frac{x}{2y}}\,dxdy$). We will attempt the calculation using the dy integral first, followed by the dx integral (Fig. 1.5).

Fig. 1.5 Evaluation of the double integral using the order of integration $dydx$

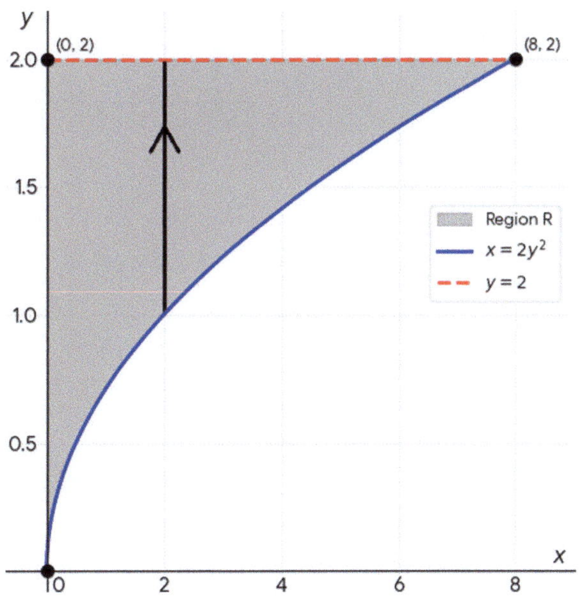

Based on Fig. 1.5,

$$\iint_R e^{\frac{x}{2y}}\,dA = \iint_R e^{\frac{x}{2y}}\,dydx = \int_0^8 \int_{\sqrt{\frac{x}{2}}}^{2} e^{\frac{x}{2y}}\,dydx$$

However, the inner integral, $\int_{\sqrt{\frac{x}{2}}}^{2} e^{\frac{x}{2y}}\,dy$, has no elementary antiderivative and we cannot proceed. Let's attempt the calculation using the dx integral first, followed by the dy integral (Fig. 1.6).

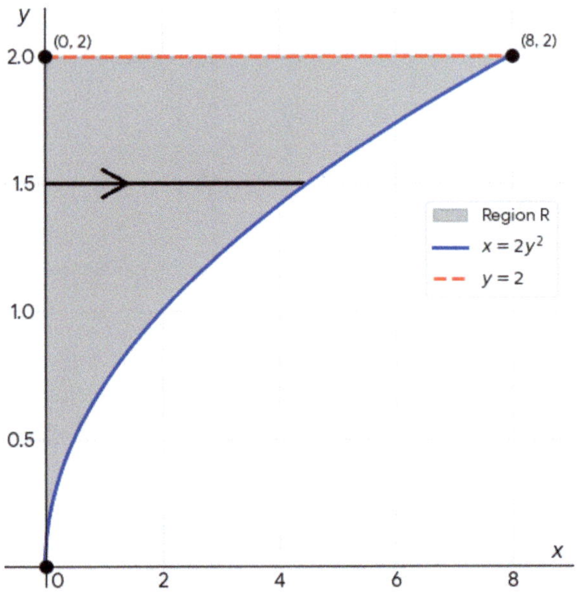

Fig. 1.6 Evaluation of the double integral using the order of integration *dxdy*

Based on Fig. 1.6,

$$\iint_R e^{\frac{x}{2y}}\,dA = \iint_R e^{\frac{x}{2y}}\,dxdy = \int_0^2 \int_0^{2y^2} e^{\frac{x}{2y}}\,dxdy$$

$$\int_0^{2y^2} e^{\frac{x}{2y}}\,dx = 2y \int_0^{2y^2} \frac{1}{2y} e^{\frac{x}{2y}}\,dx = 2ye^{\frac{x}{2y}} \Big|_0^{2y^2}$$

$$= 2ye^{\frac{2y^2}{2y}} - 2ye^{\frac{0}{2y}} = 2ye^{y} - 2y = 2y\left(e^{y} - 1\right)$$

$$\int_0^2 \int_0^{2y^2} e^{\frac{x}{2y}}\,dxdy = \int_0^2 2y\left(e^{y} - 1\right)dy = 2\left(\int_0^2 ye^{y}dy - \int_0^2 ydy\right)$$

Let's calculate each integral individually. The Tabular Integration (Integration by Parts) method used to evaluate $\int_0^2 ye^y dy$ is shown in Fig. 1.7.

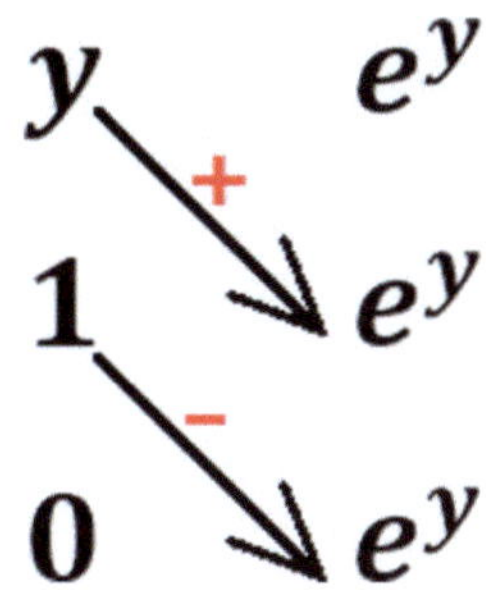

Fig. 1.7 Evaluation of $\int ye^y dy$

Based on Fig. 1.7,

$$\int ye^y dy = ye^y - e^y + C$$

$$\int_0^2 ye^y dy = ye^y - e^y \Big|_0^2 = 2e^2 - e^2 - \left(0 \times e^0 - e^0\right) = e^2 + 1$$

and

$$\int_0^2 ydy = \frac{y^2}{2}\Big|_0^2 = \frac{2^2}{2} - \frac{0^2}{2} = 2 - 0 = 2$$

Therefore,

$$\int_0^2 \int_0^{2y^2} e^{\frac{x}{2y}} dxdy = \int_0^2 2y(e^y - 1)dy = 2\left(\int_0^2 ye^y dy - \int_0^2 ydy\right)$$
$$= 2\left(e^2 + 1 - 2\right) = 2\left(e^2 - 1\right) \approx 12.7781$$

Although the integral $\int_0^2 \int_0^{2y^2} e^{\frac{x}{2y}} dxdy$ is an improper double integral due to the integrand being undefined at the boundary point (0, 0), its evaluation confirmed that the integral converges.

Example 1.14 Evaluate the double integral $\iint_R x^2 \ln(y)dA$ over the region R depicted in Fig. 1.8.

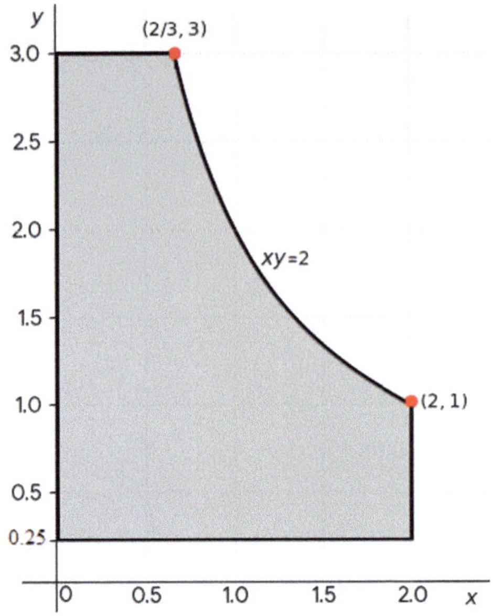

Fig. 1.8 Region R for Example 1.14

Solution Given that the problem statement does not dictate the sequence, we have the option to integrate in either order ($\iint_R x^2 \ln(y)dxdy$ or $\iint_R x^2 \ln(y)dydx$). We will attempt the calculation using the dx integral first, followed by the dy integral (Fig. 1.9). If this order leads to an unsolvable integral, we must change the order of integration.

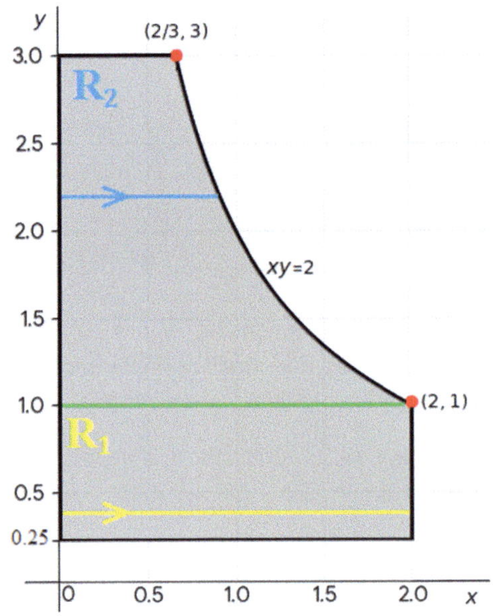

Fig. 1.9 Region R divided into R_1 and R_2

To facilitate the integration, we decompose the domain R into two non-overlapping regions, R_1 and R_2. The green line shown in Fig. 1.9 marks the common interface (or boundary) between R_1 and R_2.

$$\iint_R x^2 \ln(y)dA = \iint_{R_1} x^2 \ln(y)dA + \iint_{R_2} x^2 \ln(y)dA$$

$$= \int_{0.25}^{1}\int_{0}^{2} x^2 \ln(y)dxdy + \int_{1}^{3}\int_{0}^{\frac{2}{y}} x^2 \ln(y)dxdy$$

We will calculate each integral separately.

$$\begin{aligned}
\int_{0.25}^{1}\int_{0}^{2} x^2 \ln(y)dxdy &= \int_{0}^{2} x^2 dx \times \int_{0.25}^{1} \ln(y)dy \\
&= \left(\frac{x^3}{3}\bigg|_0^2\right) \times \left(y\ln(y) - y\bigg|_{0.25}^{1}\right) \\
&= \left(\frac{8}{3} - 0\right) \times (-1 - 0.25\ln(0.25) + 0.25) \\
&= -\frac{8}{3} \times \left(\frac{3}{4} + \frac{1}{4}\ln(0.25)\right) \\
&= -\frac{2}{3}(3 + \ln(0.25)) = -2 - \frac{2}{3}\ln(0.25) \\
&= -2 + \frac{2}{3}\ln\left(0.25^{-1}\right) = \frac{2}{3}\ln(4) - 2 \\
&\approx -1.0758
\end{aligned}$$

The double integral $\int_1^3 \int_0^{\frac{2}{y}} x^2 \ln(y)dxdy$ is calculated by evaluating the inner and outer integrals sequentially:

$$\int_{0}^{\frac{2}{y}} x^2 \ln(y)dx = \frac{x^3 \ln(y)}{3}\bigg|_0^{\frac{2}{y}} = \frac{\left(\frac{2}{y}\right)^3 \ln(y)}{3} - \frac{0^3 \ln(y)}{3} = \frac{8\ln(y)}{3y^3}$$

$$\int_{1}^{3}\int_{0}^{\frac{2}{y}} x^2 \ln(y)dxdy = \int_{1}^{3} \frac{8\ln(y)}{3y^3}dy = \frac{8}{3}\int_{1}^{3} y^{-3}\ln(y)dy$$

The integral $\int_1^3 y^{-3}\ln(y)dy$ can be calculated using the Tabular Integration method (Integration by Parts) (Fig. 1.10).

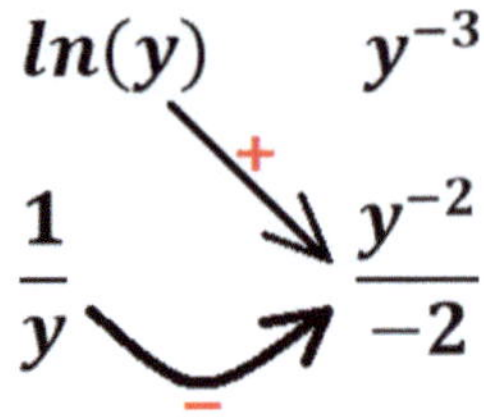

Fig. 1.10 Evaluation of $\int y^{-3}\ln(y)dy$

Based on Fig. 1.10,

$$\int y^{-3}\ln(y)dy = \ln(y)\times\frac{y^{-2}}{-2} - \int \frac{1}{y}\times\frac{y^{-2}}{-2}dy$$
$$\Rightarrow \int y^{-3}\ln(y)dy = \frac{\ln(y)y^{-2}}{-2} + \frac{1}{2}\int y^{-3}dy$$
$$\Rightarrow \int y^{-3}\ln(y)dy = \frac{\ln(y)y^{-2}}{-2} + \frac{1}{2}\frac{y^{-3+1}}{-3+1}$$
$$\Rightarrow \int y^{-3}\ln(y)dy = -\frac{\ln(y)}{2y^2} - \frac{1}{4y^2} + C$$

$$\int_1^3 y^{-3}\ln(y)dy = -\frac{\ln(y)}{2y^2} - \frac{1}{4y^2}\bigg|_1^3 = \frac{\ln(y)}{2y^2} + \frac{1}{4y^2}\bigg|_3^1$$
$$= \left(0+\frac{1}{4}\right) - \left(\frac{\ln(3)}{18} + \frac{1}{36}\right) = \frac{2}{9} - \frac{1}{18}\ln(3)$$
$$\approx 0.1612$$

$$\int_1^3 y^{-3}\ln(y)dy = \frac{2}{9} - \frac{1}{18}\ln(3)$$
$$\Rightarrow \frac{8}{3}\int_1^3 y^{-3}\ln(y)dy = \frac{16}{27} - \frac{4}{27}\ln(3)$$
$$\approx 0.4298$$

Therefore,

$$\iint_R x^2\ln(y)dA = \int_{0.25}^{1}\int_0^2 x^2\ln(y)dxdy + \int_1^3\int_0^{\frac{2}{y}} x^2\ln(y)dxdy$$
$$= \left(\frac{2}{3}\ln(4) - 2\right) + \left(\frac{16}{27} - \frac{4}{27}\ln(3)\right)$$

$$= \frac{2}{3}\ln(4) - \frac{4}{27}\ln(3) - \frac{38}{27} \approx -0.6460$$

Example 1.15 Evaluate the double integral $\iint_R x^2 \ln(y) dA$ over the region R depicted in Fig. 1.11.

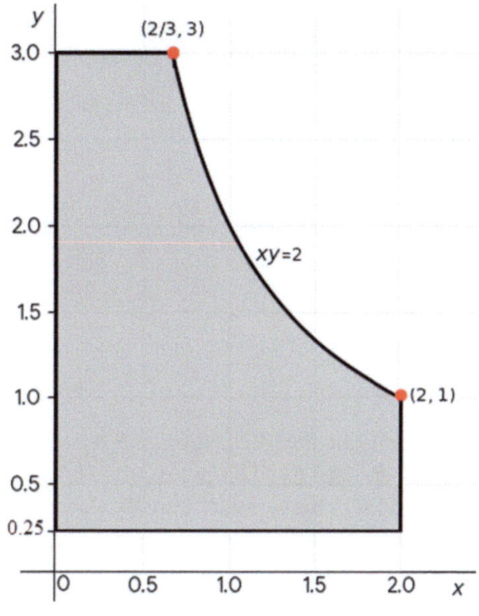

Fig. 1.11 Region R for Example 1.15

Solution This problem was solved in the previous example. Here we will re-evaluate it by reversing the order of integration (integrating with respect to y first, then x) (Fig. 1.12).

Fig. 1.12 Region R divided into R_1 and R_2

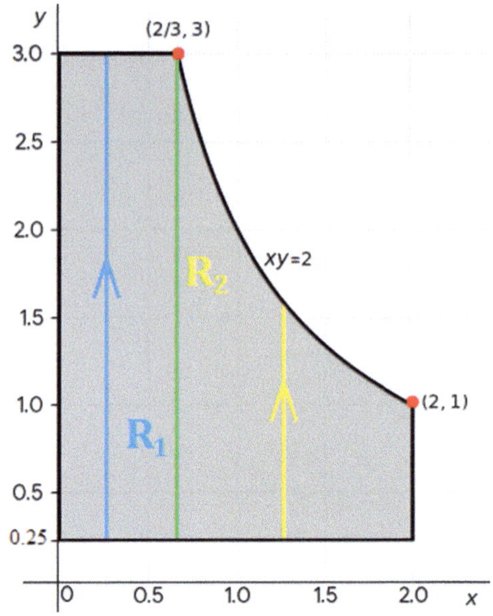

To facilitate the integration, we decompose the domain R into two non-overlapping regions, R_1 and R_2. The green line shown in Fig. 1.12 marks the common interface (or boundary) between R_1 and R_2.

$$\iint_R x^2 \ln(y)dA = \iint_{R_1} x^2 \ln(y)dA + \iint_{R_2} x^2 \ln(y)dA$$
$$= \int_0^{\frac{2}{3}} \int_{0.25}^{3} x^2 \ln(y)dydx + \int_{\frac{2}{3}}^{2} \int_{0.25}^{\frac{2}{x}} x^2 \ln(y)dydx$$

We will calculate each integral separately.

$$\int_0^{\frac{2}{3}} \int_{0.25}^{3} x^2 \ln(y)dydx = \int_0^{\frac{2}{3}} x^2 dx \times \int_{0.25}^{3} \ln(y)dy = \left(\frac{x^3}{3}\bigg|_0^{\frac{2}{3}}\right) \times \left(y\ln(y) - y\bigg|_{0.25}^{3}\right)$$
$$= \left(\frac{8}{81} - 0\right) \times (3\ln(3) - 3 - (0.25\ln(0.25) - 0.25))$$
$$= \frac{8}{81} \times (3\ln(3) - 3 - 0.25\ln(0.25) + 0.25)$$
$$= \frac{8}{81} \times \left(3\ln(3) + 0.25\ln\left(0.25^{-1}\right) - 2.75\right)$$
$$= \frac{8}{27}\ln(3) + \frac{2}{81}\ln(4) - \frac{22}{81} \approx 0.0881$$

The double integral $\int_{\frac{2}{3}}^{2}\int_{0.25}^{\frac{2}{x}} x^2 \ln(y)dydx$ is calculated by evaluating the inner and outer integrals sequentially:

$$\begin{aligned}
\int_{0.25}^{\frac{2}{x}} x^2 \ln(y)dy &= x^2 \int_{0.25}^{\frac{2}{x}} \ln(y)dy = x^2(y\ln(y) - y)\Big|_{0.25}^{\frac{2}{x}} \\
&= x^2\left(\frac{2}{x}\ln\left(\frac{2}{x}\right) - \frac{2}{x}\right) - \left(x^2(0.25\ln(0.25) - 0.25)\right) \\
&= 2x\ln\left(\frac{2}{x}\right) - 2x + (0.25 - 0.25\ln(0.25))x^2 \\
&= 2x\ln\left(\frac{2}{x}\right) - 2x + \left(0.25 + 0.25\ln\left(0.25^{-1}\right)\right)x^2 \\
&= 2x\ln\left(\frac{2}{x}\right) - 2x + \frac{1}{4}(1 + \ln(4))x^2
\end{aligned}$$

$$\begin{aligned}
\int_{\frac{2}{3}}^{2}\int_{0.25}^{\frac{2}{x}} x^2 \ln(y)dydx &= \int_{\frac{2}{3}}^{2}\left(2x\ln\left(\frac{2}{x}\right) - 2x + \frac{1}{4}(1 + \ln(4))x^2\right)dx \\
&= 2\int_{\frac{2}{3}}^{2} x\ln\left(\frac{2}{x}\right)dx - 2\int_{\frac{2}{3}}^{2} xdx + \frac{1}{4}(1 + \ln(4))\int_{\frac{2}{3}}^{2} x^2dx
\end{aligned}$$

We will calculate each integral separately. The second and third integrals are straightforward to calculate:

$$\int_{\frac{2}{3}}^{2} xdx = \frac{x^2}{2}\Big|_{\frac{2}{3}}^{2} = \frac{2^2}{2} - \frac{\left(\frac{2}{3}\right)^2}{2} = 2 - \frac{2}{9} = \frac{16}{9} \approx 1.7778$$

$$\int_{\frac{2}{3}}^{2} x^2dx = \frac{x^3}{3}\Big|_{\frac{2}{3}}^{2} = \frac{2^3}{3} - \frac{\left(\frac{2}{3}\right)^3}{3} = \frac{8}{3} - \frac{8}{81} = \frac{216 - 8}{81} = \frac{208}{81} \approx 2.5679$$

The integral $\int_{\frac{2}{3}}^{2} x\ln\left(\frac{2}{x}\right)dx$ can be calculated using the Tabular Integration method (Integration by Parts) (Fig. 1.13).

Fig. 1.13 Evaluation of $\int x \ln\left(\frac{2}{x}\right) dx$

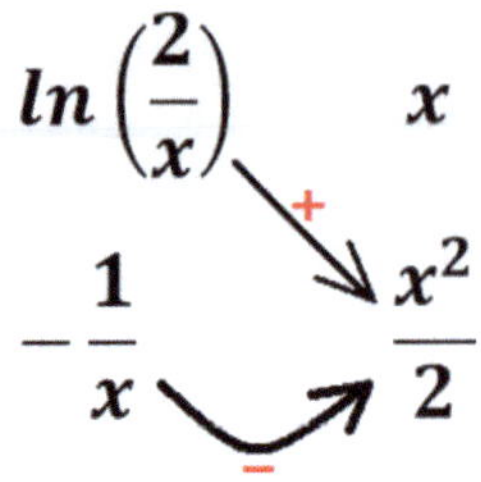

Based on Fig. 1.13,

$$\int x \ln\left(\frac{2}{x}\right) dx = \ln\left(\frac{2}{x}\right) \times \frac{x^2}{2} - \int -\frac{1}{x} \times \frac{x^2}{2} dx$$

$$\Rightarrow \int x \ln\left(\frac{2}{x}\right) dx = \frac{x^2}{2} \ln\left(\frac{2}{x}\right) + \frac{1}{2} \int x dx$$

$$\Rightarrow \int x \ln\left(\frac{2}{x}\right) dx = \frac{x^2}{2} \ln\left(\frac{2}{x}\right) + \frac{x^2}{4}$$

$$\int_{\frac{2}{3}}^{2} x \ln\left(\frac{2}{x}\right) dx = \left(\frac{x^2}{2} \ln\left(\frac{2}{x}\right) + \frac{x^2}{4}\right)\Bigg|_{\frac{2}{3}}^{2}$$

$$= \frac{2^2}{2} \ln\left(\frac{2}{2}\right) + \frac{2^2}{4} - \left(\frac{\left(\frac{2}{3}\right)^2}{2} \ln\left(\frac{2}{\frac{2}{3}}\right) + \frac{\left(\frac{2}{3}\right)^2}{4}\right)$$

$$= 0 + 1 - \left(\frac{2}{9} \ln(3) + \frac{1}{9}\right) = 1 - \frac{2}{9} \ln(3) - \frac{1}{9}$$

$$= \frac{8}{9} - \frac{2}{9} \ln(3) = \frac{8 - 2\ln(3)}{9} \approx 0.6448$$

Having completed the calculations for all three integrals, we now proceed to combine their values to find the final result.

$$\int_{\frac{2}{3}}^{2} \int_{0.25}^{\frac{2}{x}} x^2 \ln(y) dy dx = \int_{\frac{2}{3}}^{2} \left(2x \ln\left(\frac{2}{x}\right) - 2x + \frac{1}{4}(1 + \ln(4))x^2\right) dx$$

$$= 2 \int_{\frac{2}{3}}^{2} x \ln\left(\frac{2}{x}\right) dx - 2 \int_{\frac{2}{3}}^{2} x dx + \frac{1}{4}(1 + \ln(4)) \int_{\frac{2}{3}}^{2} x^2 dx$$

$$= 2\left(\frac{8 - 2\ln(3)}{9}\right) - 2\left(\frac{16}{9}\right) + \frac{1}{4}(1 + \ln(4))\left(\frac{208}{81}\right)$$

$$= \frac{52}{81} \ln(4) - \frac{4}{9} \ln(3) - \frac{92}{81}$$

$$\approx -0.7341$$

Therefore, the value of the requested integral is:

$$\iint_R x^2 \ln(y) dA = \iint_{R_1} x^2 \ln(y) dA + \iint_{R_2} x^2 \ln(y) dA$$
$$= \int_0^{\frac{2}{3}} \int_{0.25}^{3} x^2 \ln(y) dy dx + \int_{\frac{2}{3}}^{2} \int_{0.25}^{\frac{2}{x}} x^2 \ln(y) dy dx$$
$$= \left(\frac{8}{27}\ln(3) + \frac{2}{81}\ln(4) - \frac{22}{81}\right) + \left(\frac{52}{81}\ln(4) - \frac{4}{9}\ln(3) - \frac{92}{81}\right)$$
$$= \frac{54}{81}\ln(4) - \frac{4}{27}\ln(3) - \frac{114}{81} \approx -0.6460$$

The value obtained is identical to the result from the preceding problem.

Example 1.16 Evaluate the double integral $\iint_R xy dA$ over the region R shown in Fig. 1.14.

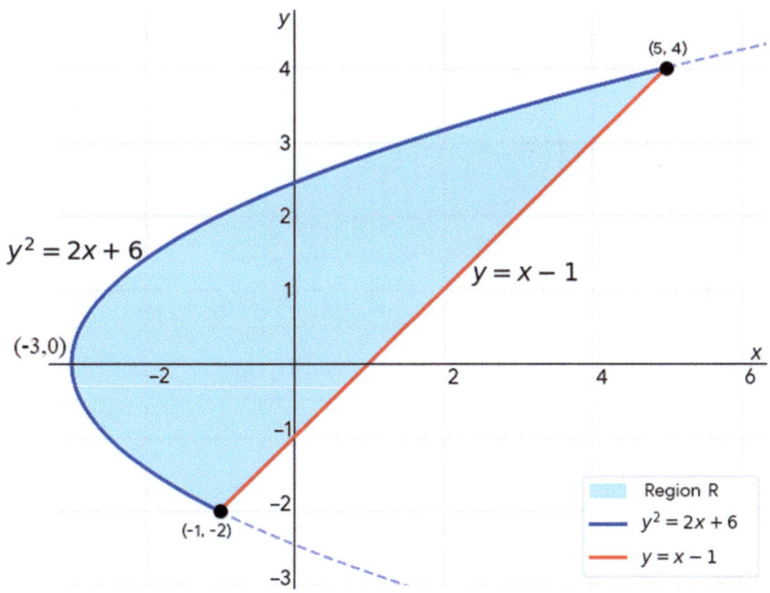

Fig. 1.14 Region R for Example 1.16

Solution Based on Fig. 1.15, $\iint_R xy dA = \int_{-2}^{4} \int_{\frac{y^2-6}{2}}^{y+1} xy dx dy$.

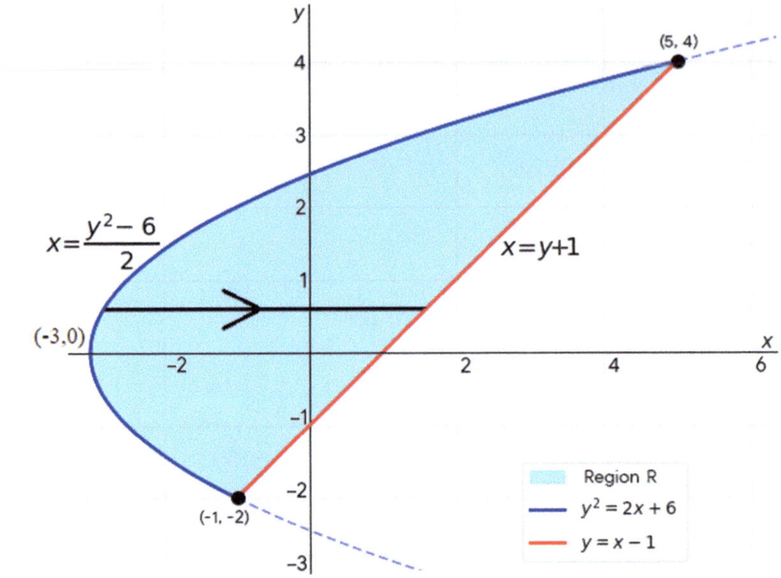

Fig. 1.15 Evaluation of the double integral using the order of integration *dxdy*

The double integral $\int_{-2}^{4}\int_{\frac{y^2-6}{2}}^{y+1} xydxdy$ is calculated by evaluating the inner and outer integrals sequentially:

$$\int_{\frac{y^2-6}{2}}^{y+1} xydx = y\int_{\frac{y^2-6}{2}}^{y+1} xdx = y\frac{x^2}{2}\Bigg|_{\frac{y^2-6}{2}}^{y+1} = y\frac{(y+1)^2}{2} - y\frac{\left(\frac{y^2-6}{2}\right)^2}{2}$$

$$= y\frac{y^2+2y+1}{2} - y\frac{y^4-12y^2+36}{8}$$

$$= \frac{4y^3+8y^2+4y}{8} - \frac{y^5-12y^3+36y}{8}$$

$$= \frac{4y^3+8y^2+4y-y^5+12y^3-36y}{8}$$

$$= \frac{16y^3+8y^2-y^5-32y}{8}$$

$$= -4y+y^2+2y^3-\frac{1}{8}y^5$$

$$\int_{-2}^{4}\int_{\frac{y^2-6}{2}}^{y+1} xydxdy = \int_{-2}^{4}\left(-4y+y^2+2y^3-\frac{1}{8}y^5\right)dy = -2y^2+\frac{y^3}{3}+\frac{y^4}{2}-\frac{y^6}{48}\Bigg|_{-2}^{4}$$

$$= \left(-2 \times 4^2 + \frac{4^3}{3} + \frac{4^4}{2} - \frac{4^6}{48}\right)$$
$$- \left(-2(-2)^2 + \frac{(-2)^3}{3} + \frac{(-2)^4}{2} - \frac{(-2)^6}{48}\right)$$
$$= \left(-32 + \frac{64}{3} + \frac{256}{2} - \frac{4096}{48}\right) - \left(-8 - \frac{8}{3} + \frac{16}{2} - \frac{64}{48}\right) = 36$$

Example 1.17 Evaluate the double integral $\iint_R xydA$ over the region R shown in Fig. 1.16.

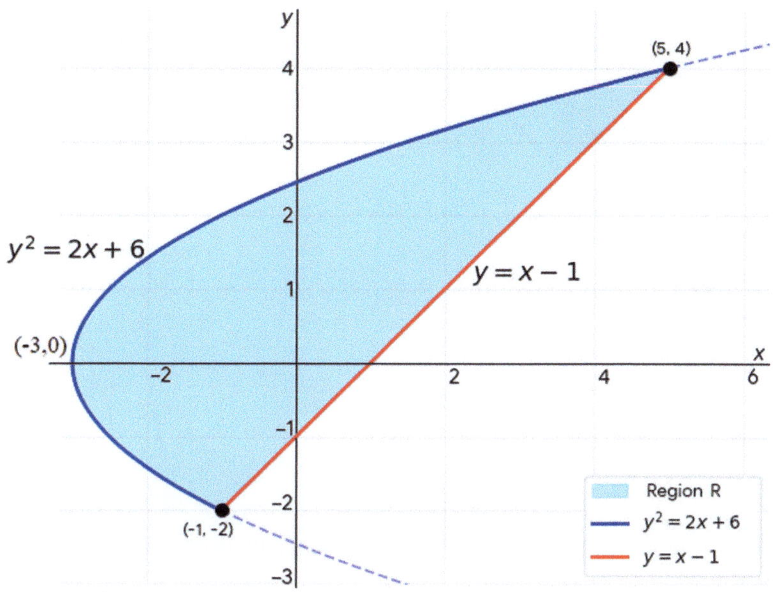

Fig. 1.16 Region R for Example 1.17

Solution This problem was solved in the previous example. Here we will re-evaluate it by reversing the order of integration (integrating with respect to y first, then x).

To facilitate the integration, we decompose the domain R into two non-overlapping regions, R_1 and R_2 (Fig. 1.17). The green line shown in Fig. 1.17 marks the common interface (or boundary) between R_1 and R_2.

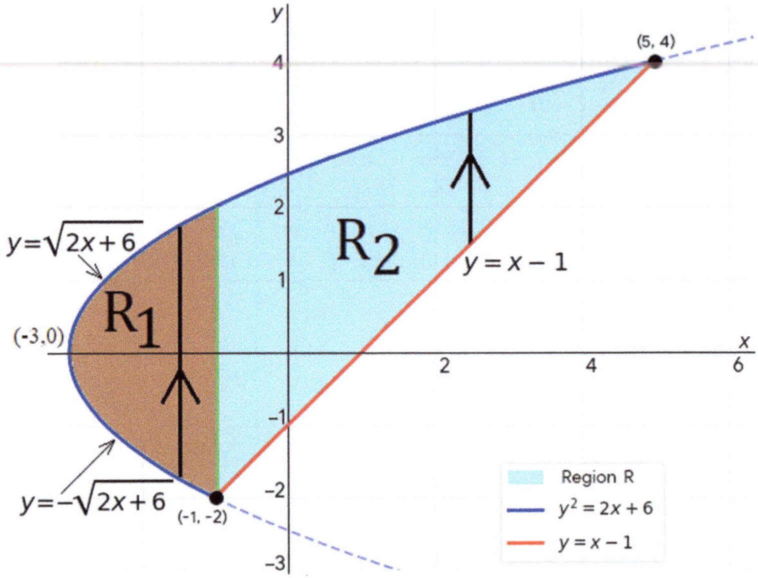

Fig. 1.17 Region R divided into R_1 and R_2

Based on Fig. 1.17,

$$\iint_R xydA = \iint_{R_1} xydA + \iint_{R_2} xydA = \int_{-3}^{-1} \int_{-\sqrt{2x+6}}^{\sqrt{2x+6}} xydydx + \int_{-1}^{5} \int_{x-1}^{\sqrt{2x+6}} xydydx$$

Let's calculate each integral individually. The calculation of $\int_{-3}^{-1} \int_{-\sqrt{2x+6}}^{\sqrt{2x+6}} xydydx$ is shown below:

$$\int_{-\sqrt{2x+6}}^{\sqrt{2x+6}} xydy = x \int_{-\sqrt{2x+6}}^{\sqrt{2x+6}} ydy$$

$$= x\frac{y^2}{2}\Big|_{-\sqrt{2x+6}}^{\sqrt{2x+6}}$$

$$= x\frac{\left(\sqrt{2x+6}\right)^2}{2} - x\frac{\left(-\sqrt{2x+6}\right)^2}{2} = 0$$

$$\int_{-3}^{-1} \int_{-\sqrt{2x+6}}^{\sqrt{2x+6}} xydydx = \int_{-3}^{-1} 0dy = 0$$

The calculation of $\int_{-1}^{5} \int_{x-1}^{\sqrt{2x+6}} xydydx$ is shown below:

$$\int_{x-1}^{\sqrt{2x+6}} xydy = x\int_{x-1}^{\sqrt{2x+6}} ydy = x\frac{y^2}{2}\Big|_{x-1}^{\sqrt{2x+6}}$$
$$= x\frac{\left(\sqrt{2x+6}\right)^2}{2} - x\frac{(x-1)^2}{2}$$
$$= x\frac{(2x+6)}{2} - x\frac{x^2-2x+1}{2}$$
$$= \frac{2x^2+6x-x^3+2x^2-x}{2} = \frac{-x^3+4x^2+5x}{2}$$

$$\int_{-1}^{5}\int_{x-1}^{\sqrt{2x+6}} xydydx = \int_{-1}^{5} \frac{-x^3+4x^2+5x}{2}dx = -\frac{1}{8}x^4 + \frac{2}{3}x^3 + \frac{5}{4}x^2\Big|_{-1}^{5}$$
$$= -\frac{1}{8}5^4 + \frac{2}{3}5^3 + \frac{5}{4}5^2 - \left(-\frac{1}{8}(-1)^4 + \frac{2}{3}(-1)^3 + \frac{5}{4}(-1)^2\right)$$
$$= \frac{875}{24} - \frac{11}{24} = \frac{864}{24}$$
$$= 36$$

Therefore,

$$\iint_R xydA = \iint_{R_1} xydA + \iint_{R_2} xydA$$
$$= \int_{-3}^{-1}\int_{-\sqrt{2x+6}}^{\sqrt{2x+6}} xydydx + \int_{-1}^{5}\int_{x-1}^{\sqrt{2x+6}} xydydx$$
$$= 0 + 36 = 36$$

The value obtained is identical to the result from the preceding problem.

Example 1.18 Calculate $\int_0^4 \int_{\sqrt{x}}^2 \cos\left(y^3\right)dydx$.

Solution The given order is to integrate with respect to y first:

$$\int_{\sqrt{x}}^{2} \cos\left(y^3\right)dy$$

The antiderivative of $\cos\left(y^3\right)$ is not an elementary function (it cannot be expressed using standard functions like polynomials, exponentials, logs, or trigonometric functions).

Because the inner integral is unsolvable using standard techniques, we must change the order of integration.

The current bounds define the region R as $R = \{(x, y) | 0 \le x \le 4, \sqrt{x} \le y \le 2\}$ (Fig. 1.18).

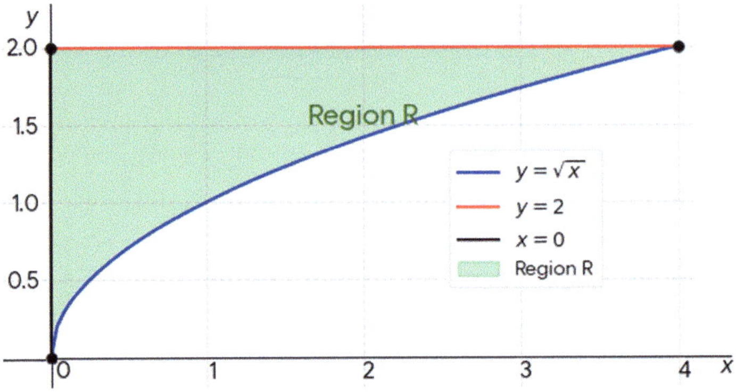

Fig. 1.18 Integration region R

Based on Fig. 1.19, $\int_0^4 \int_{\sqrt{x}}^2 \cos(y^3) dy dx = \int_0^2 \int_0^{y^2} \cos(y^3) dx dy$.

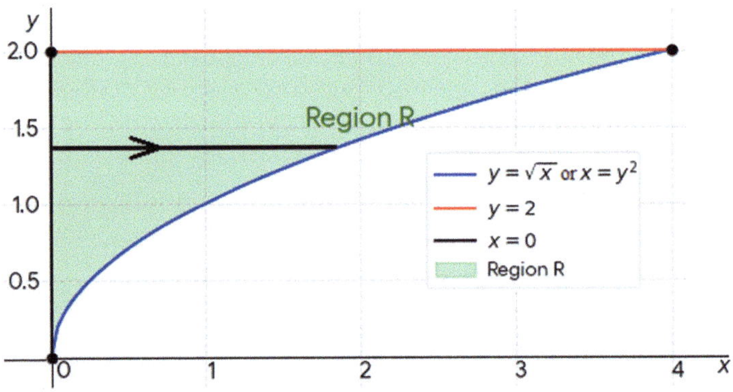

Fig. 1.19 Evaluation of the double integral using the order of integration $dxdy$

The double integral $\int_0^2 \int_0^{y^2} \cos(y^3) dx dy$ is calculated by evaluating the inner and outer integrals sequentially:

$$\int_0^{y^2} \cos(y^3) dx = \cos(y^3) x \Big|_0^{y^2} = \cos(y^3)(y^2 - 0) = y^2 \cos(y^3)$$

$$\int_0^2 \int_0^{y^2} \cos(y^3) dx dy = \int_0^2 y^2 \cos(y^3) dy = \frac{1}{3} \int_0^2 3y^2 \cos(y^3) dy = \frac{1}{3} \sin(y^3) \Big|_0^2$$

$$= \frac{1}{3}\sin\left(2^3\right) - \frac{1}{3}\sin\left(0^3\right) = \frac{1}{3}\sin(8) \approx 0.3298$$

Example 1.19 Calculate $\int_0^4 \int_{\sqrt{x}}^2 \sin\left(y^3\right) dydx$.

Solution The given order is to integrate with respect to y first:

$$\int_{\sqrt{x}}^{2} \sin\left(y^3\right) dy$$

The antiderivative of $\sin\left(y^3\right)$ is not an elementary function (it cannot be expressed using standard functions like polynomials, exponentials, logs, or trigonometric functions).

Because the inner integral is unsolvable using standard techniques, we must change the order of integration.

The current bounds define the region R as $R = \left\{ (x, y) | \sqrt{x} \leq y \leq 2, 0 \leq x \leq 4 \right\}$ (Fig. 1.20).

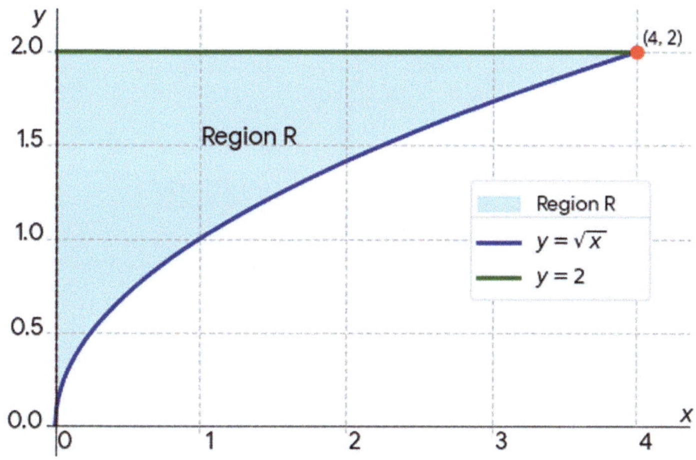

Fig. 1.20 Integration region R

Based on Fig. 1.21, $\int_0^4 \int_{\sqrt{x}}^2 \sin\left(y^3\right) dydx = \int_0^2 \int_0^{y^2} \sin\left(y^3\right) dxdy$.

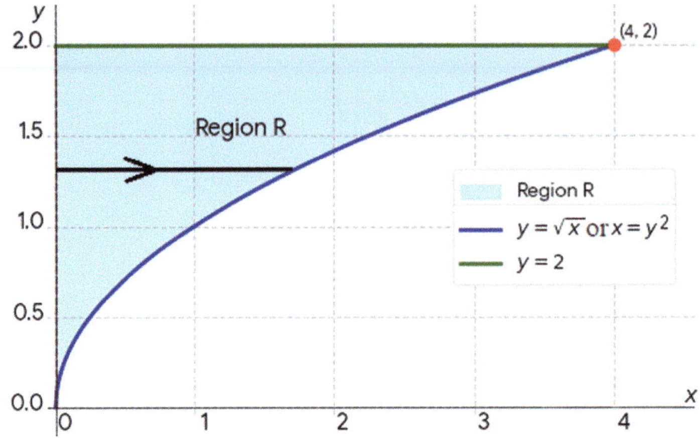

Fig. 1.21 Evaluation of the double integral using the order of integration *dxdy*

The double integral $\int_0^2 \int_0^{y^2} \sin(y^3) dxdy$ is calculated by evaluating the inner and outer integrals sequentially:

$$\begin{aligned}\int_0^{y^2} \sin(y^3) dx &= \sin(y^3) \int_0^{y^2} dx \\ &= \sin(y^3) x \Big|_0^{y^2} \\ &= \sin(y^3) \times y^2 - \sin(y^3) \times 0 \\ &= y^2 \sin(y^3)\end{aligned}$$

$$\begin{aligned}\int_0^2 \int_0^{y^2} \sin(y^3) dxdy &= \int_0^2 y^2 \sin(y^3) dy = \frac{1}{3} \int_0^2 3y^2 \sin(y^3) dy \\ &= -\frac{1}{3} \cos(y^3) \Big|_0^2 = \frac{1}{3} \cos(y^3) \Big|_2^0 \\ &= \frac{1}{3} - \frac{1}{3} \cos(8) = \frac{1 - \cos(8)}{3} \approx 0.3818\end{aligned}$$

Example 1.20 Calculate $\int_0^2 \int_{\frac{y}{2}}^1 ye^{x^3} dxdy$.

Solution The given order is to integrate with respect to x first:

$$\int_{\frac{y}{2}}^1 ye^{x^3} dx$$

Since y is a constant with respect to the inner integration, we have, $y\int_{\frac{y}{2}}^{1} e^{x^3}dx$. The antiderivative of e^{x^3} is not an elementary function (it cannot be expressed using standard functions like polynomials, exponentials, logs, or trigonometric functions).

Because the inner integral is unsolvable using standard techniques, we must change the order of integration.

The current bounds define the region R as $R = \left\{(x, y)|0 \le y \le 2, \frac{y}{2} \le x \le 1\right\}$ (Fig. 1.22).

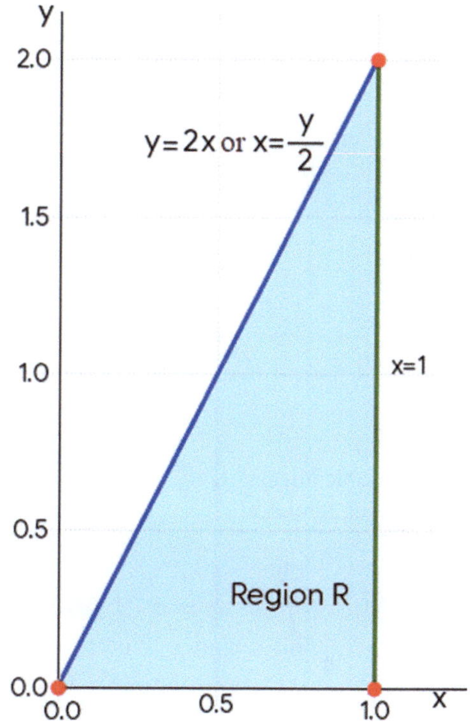

Fig. 1.22 Integration region R

Based on Fig. 1.23, $\int_0^2 \int_{\frac{y}{2}}^{1} ye^{x^3}dxdy = \int_0^1 \int_0^{2x} ye^{x^3}dydx$.

Fig. 1.23 Evaluation of the double integral using the order of integration *dydx*

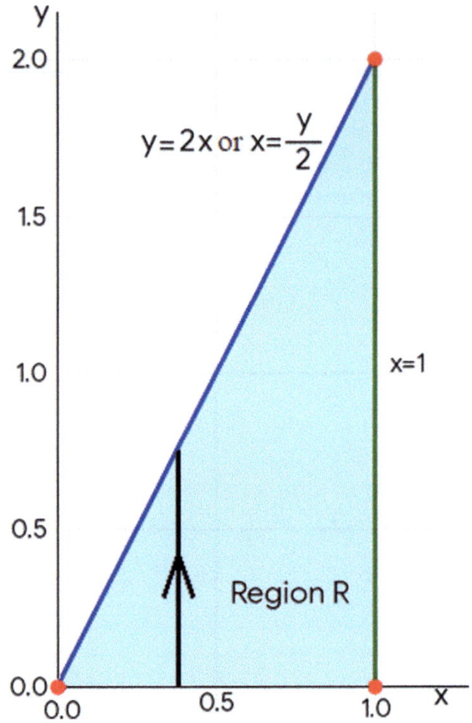

The double integral $\int_0^1 \int_0^{2x} ye^{x^3} dydx$ is calculated by evaluating the inner and outer integrals sequentially:

$$\int_0^{2x} ye^{x^3} dy = e^{x^3} \int_0^{2x} ydy = e^{x^3} \frac{y^2}{2}\bigg|_0^{2x} = e^{x^3} \frac{(2x)^2}{2} - e^{x^3} \frac{0^2}{2}$$
$$= 2x^2 e^{x^3} - 0 = 2x^2 e^{x^3}$$

$$\int_0^1 \int_0^{2x} ye^{x^3} dydx = \int_0^1 2x^2 e^{x^3} dx = 2 \int_0^1 x^2 e^{x^3} dx$$
$$= \frac{2}{3} \int_0^1 3x^2 e^{x^3} dx = \frac{2}{3} e^{x^3} \bigg|_0^1 = \frac{2}{3} e^1 - \frac{2}{3} e^0$$
$$= \frac{2}{3}(e - 1) \approx 1.1455$$

Example 1.21 Calculate $\int_0^1 \int_y^1 \sqrt{1 - x^2} dxdy$.

Solution The given order is to integrate with respect to x first:

$$\int_{y}^{1} \sqrt{1-x^2}dx$$

The antiderivative of $\sqrt{1-x^2}$ is not an elementary function (it cannot be expressed using standard functions like polynomials, exponentials, logs, or trigonometric functions).

Because the inner integral is unsolvable using standard techniques, we must change the order of integration.

The current bounds define the region R as $R = \{(x, y)|y \leq x \leq 1, 0 \leq y \leq 1\}$ (Fig. 1.24).

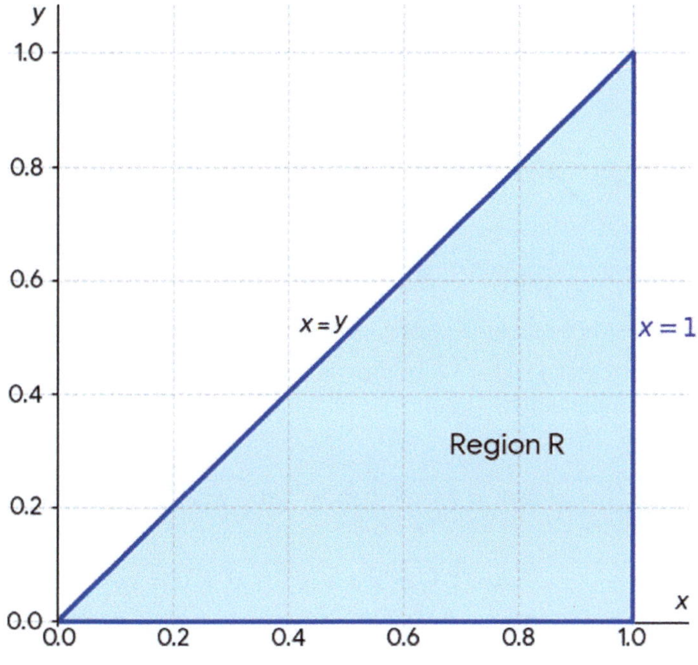

Fig. 1.24 Integration region R

Based on Fig. 1.25, $\int_0^1 \int_y^1 \sqrt{1-x^2}dxdy = \int_0^1 \int_0^x \sqrt{1-x^2}dydx$.

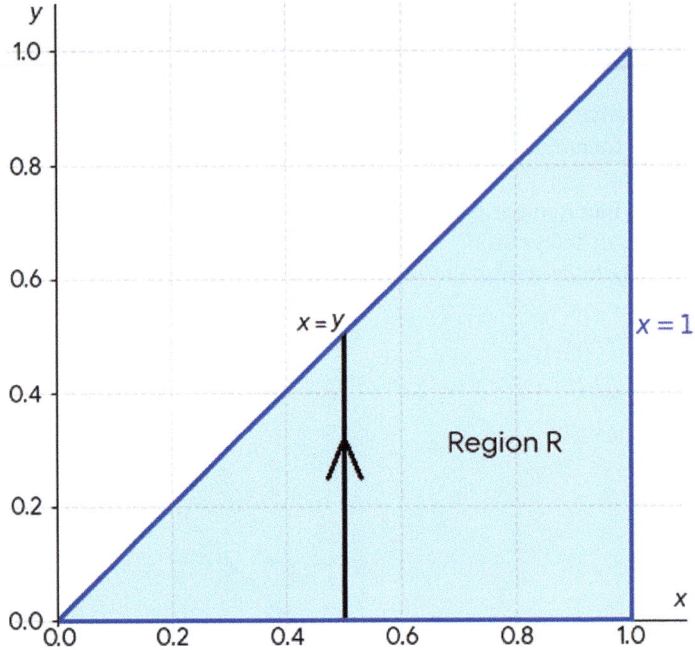

Fig. 1.25 Evaluation of the double integral using the order of integration *dydx*

The double integral $\int_0^1 \int_0^x \sqrt{1-x^2}dydx$ is calculated by evaluating the inner and outer integrals sequentially:

$$\int_0^x \sqrt{1-x^2}dy = \sqrt{1-x^2}\int_0^x dy = \sqrt{1-x^2}y\Big|_0^x$$
$$= \sqrt{1-x^2}\times x - \sqrt{1-x^2}\times 0 = x\sqrt{1-x^2}$$

$$\int_0^1\int_0^x \sqrt{1-x^2}dydx = \int_0^1 x\sqrt{1-x^2} = \frac{1}{-2}\int_0^1 -2x\sqrt{1-x^2} = \frac{1}{-2}\frac{\left(1-x^2\right)^{\frac{1}{2}+1}}{\frac{1}{2}+1}\Bigg|_0^1$$
$$= \frac{1}{-2}\frac{\left(1-x^2\right)^{\frac{3}{2}}}{\frac{3}{2}}\Bigg|_0^1 = \frac{\left(1-x^2\right)^{\frac{3}{2}}}{-3}\Bigg|_0^1 = \frac{\left(1-x^2\right)^{\frac{3}{2}}}{3}\Bigg|_1^0$$
$$= \frac{\left(1-0^2\right)^{\frac{3}{2}}}{3} - \frac{\left(1-1^2\right)^{\frac{3}{2}}}{3}$$
$$= \frac{1}{3} - 0 = \frac{1}{3}$$

Example 1.22 Calculate $\int_0^\pi \int_x^\pi \frac{\sin(y)}{y}dydx$.

Solution The given order is to integrate with respect to y first:

$$\int_{x}^{\pi} \frac{\sin(y)}{y} dx$$

The antiderivative of $\frac{\sin(y)}{y}$ is not an elementary function (it cannot be expressed using standard functions like polynomials, exponentials, logs, or trigonometric functions).

Because the inner integral is unsolvable using standard techniques, we must change the order of integration.

The current bounds define the region R as $R = \{(x, y)|x \le y \le \pi, 0 \le x \le \pi\}$ (Fig. 1.26).

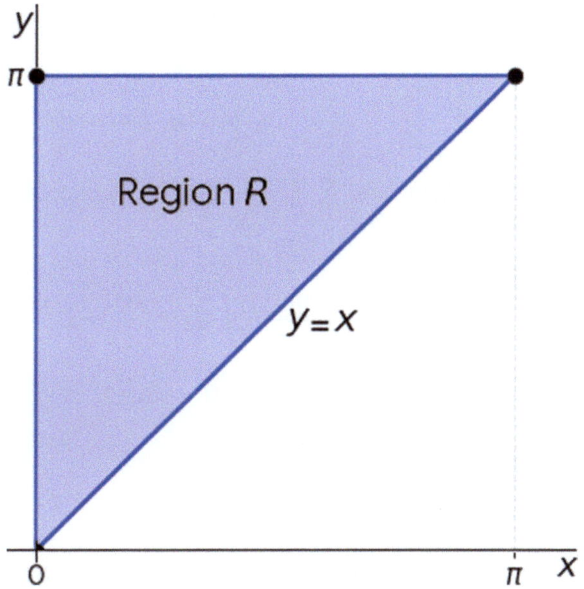

Fig. 1.26 Integration region R

Based on Fig. 1.27, $\int_0^{\pi} \int_x^{\pi} \frac{\sin(y)}{y} dydx = \int_0^{\pi} \int_0^{y} \frac{\sin(y)}{y} dxdy$.

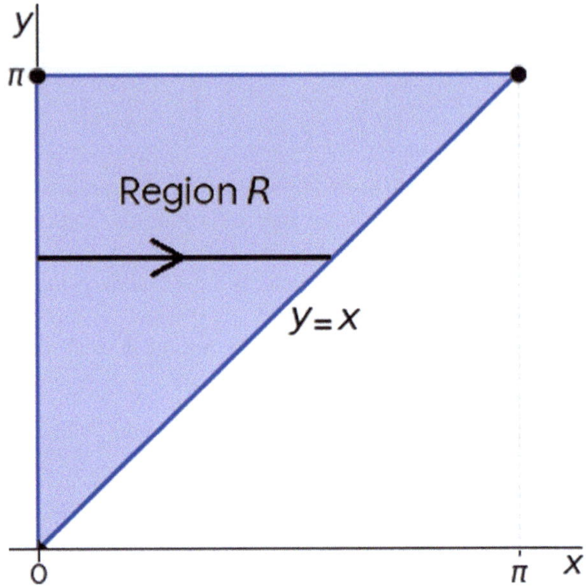

Fig. 1.27 Evaluation of the double integral using the order of integration $dxdy$

The double integral $\int_0^{\pi} \int_0^{y} \frac{\sin(y)}{y} dxdy$ is calculated by evaluating the inner and outer integrals sequentially:

$$\int_0^y \frac{\sin(y)}{y} dx = \frac{\sin(y)}{y} \int_0^y dx = \frac{\sin(y)}{y} x \Big|_0^y = \frac{\sin(y)}{y} y - \frac{\sin(y)}{y} 0$$
$$= \sin(y) - 0 = \sin(y)$$

$$\int_0^{\pi} \int_0^{y} \frac{\sin(y)}{y} dxdy = \int_0^{\pi} \sin(y) dy = -\cos(y)\Big|_0^{\pi} = \cos(y)\Big|_{\pi}^{0}$$
$$= \cos(0) - \cos(\pi) = 1 - (-1) = 2$$

Exercise 1.1 Calculate $\int_0^2 \int_{\frac{y}{2}}^1 ye^{x^3} dxdy$. (Ans. $\frac{2}{3}(e-1) \approx 1.1455$).

Example 1.23 Calculate $\int_0^{\infty} \int_0^x xe^{-\frac{x^2}{y}} dydx$.

Solution The given order is to integrate with respect to y first:

$$\int_0^x xe^{-\frac{x^2}{y}} dy$$

The antiderivative of $xe^{-\frac{x^2}{y}}$ is not an elementary function (it cannot be expressed using standard functions like polynomials, exponentials, logs, or trigonometric functions).

Because the inner integral is unsolvable using standard techniques, we must change the order of integration.

The current bounds define the region R as $R = \{(x, y)|0 \leq x \leq \infty, 0 \leq y \leq x\}$ (Fig. 1.28).

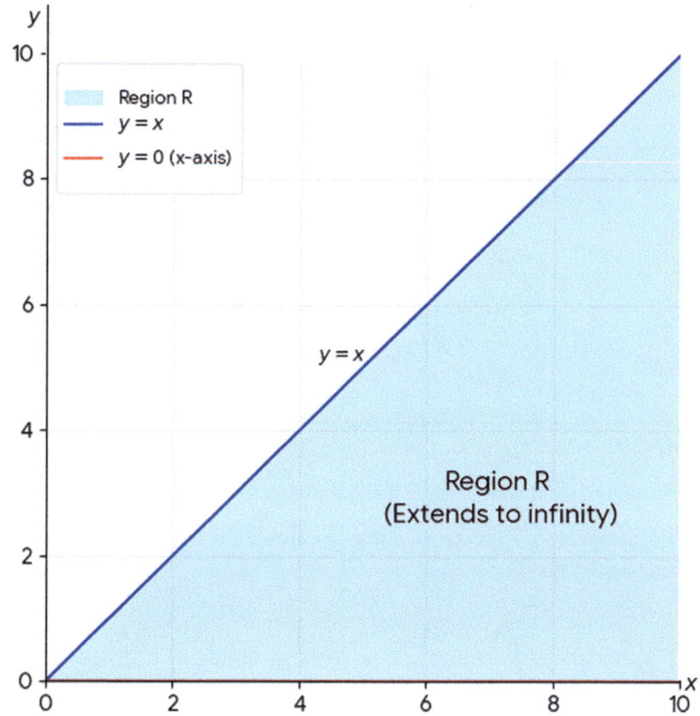

Fig. 1.28 Integration region R

Based on Fig. 1.29, $\int_0^{\infty} \int_0^{x} xe^{-\frac{x^2}{y}} dydx = \int_0^{\infty} \int_y^{\infty} xe^{-\frac{x^2}{y}} dxdy.$

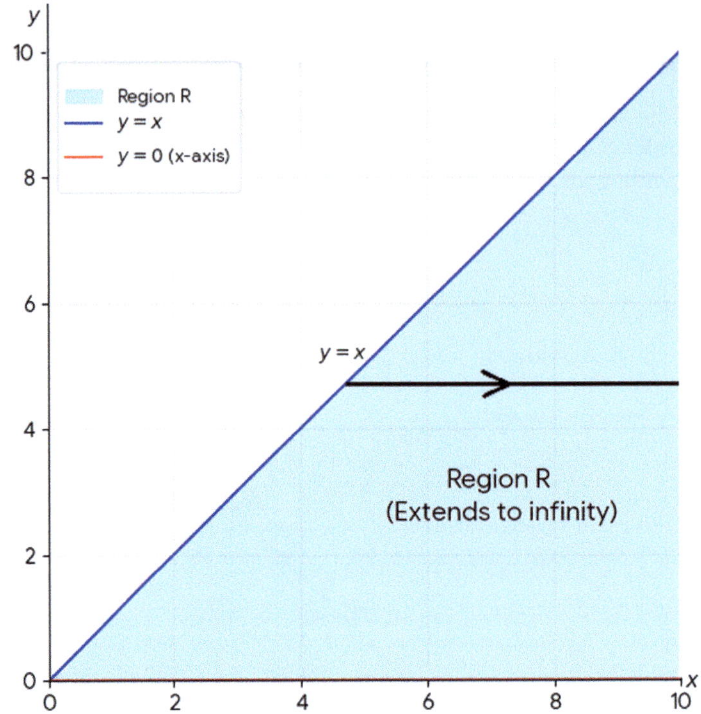

Fig. 1.29 Evaluation of the double integral using the order of integration *dxdy*

The double integral $\int_0^\infty \int_y^\infty xe^{-\frac{x^2}{y}} dxdy$ is calculated by evaluating the inner and outer integrals sequentially:

$$\int_y^\infty xe^{-\frac{x^2}{y}} dx = \int_y^\infty \frac{y}{-2} \times \frac{-2}{y} \times xe^{-\frac{x^2}{y}} dx = \frac{y}{-2} \int_y^\infty \frac{-2x}{y} e^{-\frac{x^2}{y}} dx$$
$$= \frac{y}{-2} e^{-\frac{x^2}{y}} \Big|_y^\infty = \frac{y}{2} e^{-\frac{x^2}{y}} \Big|_\infty^y$$
$$= \frac{y}{2} e^{-\frac{y^2}{y}} - \lim_{x\to\infty} \frac{y}{2} e^{-\frac{x^2}{y}} = \frac{y}{2} e^{-y} - \frac{y}{2} \lim_{x\to\infty} e^{-\frac{x^2}{y}}$$
$$= \frac{y}{2} e^{-y} - \frac{y}{2} \times 0 = \frac{y}{2} e^{-y}$$

$$\int_0^\infty \int_y^\infty xe^{-\frac{x^2}{y}} dxdy = \int_0^\infty \frac{y}{2} e^{-y} dy$$

The Tabular Integration (Integration by Parts) method used to evaluate $\int_0^\infty \frac{y}{2} e^{-y} dy$ is shown in Fig. 1.30.

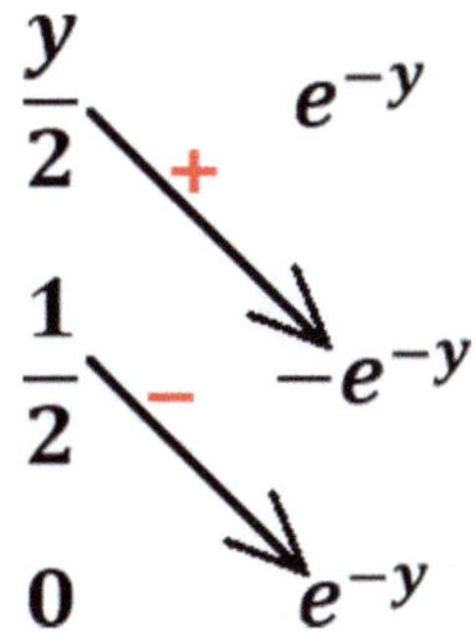

Fig. 1.30 Evaluation of $\int \frac{y}{2}e^{-y}dy$

Based on Fig. 1.30,

$$\int \frac{y}{2}e^{-y}dy = -\frac{y}{2}e^{-y} - \frac{1}{2}e^{-y} + C = -\frac{y+1}{2}e^{-y} + C = -\frac{y+1}{2e^y} + C$$

$$\int_0^\infty \frac{y}{2}e^{-y}dy = -\frac{y+1}{2e^y}\bigg|_0^\infty = \frac{y+1}{2e^y}\bigg|_\infty^0$$
$$= \frac{0+1}{2e^0} - \lim_{y\to\infty}\frac{y+1}{2e^y} = \frac{1}{2} - \lim_{y\to\infty}\frac{1}{2e^y}$$
$$= \frac{1}{2} - 0 = \frac{1}{2}$$

Note that L'Hôpital's Rule is used to evaluate the above limit. The value of $\int_0^\infty \int_y^\infty xe^{-\frac{x^2}{y}}dxdy$ is:

$$\int_0^\infty \int_y^\infty xe^{-\frac{x^2}{y}}dxdy = \int_0^\infty \frac{y}{2}e^{-y}dy = \frac{1}{2} = 0.5$$

Example 1.24 Calculate $\int_0^8 \int_{\sqrt[3]{x}}^2 \frac{1}{y^4+1}dydx$.

Solution The given order is to integrate with respect to y first:

$$\int_0^x \frac{1}{y^4+1}dy$$

The antiderivative of $\frac{1}{y^4+1}$ is not an elementary function (it cannot be expressed using standard functions like polynomials, exponentials, logs, or trigonometric functions).

Because the inner integral is unsolvable using standard techniques, we must change the order of integration.

The current bounds define the region R as $R = \{(x, y)|0 \le x \le 8, \sqrt[3]{x} \le y \le 2\}$ (Fig. 1.31).

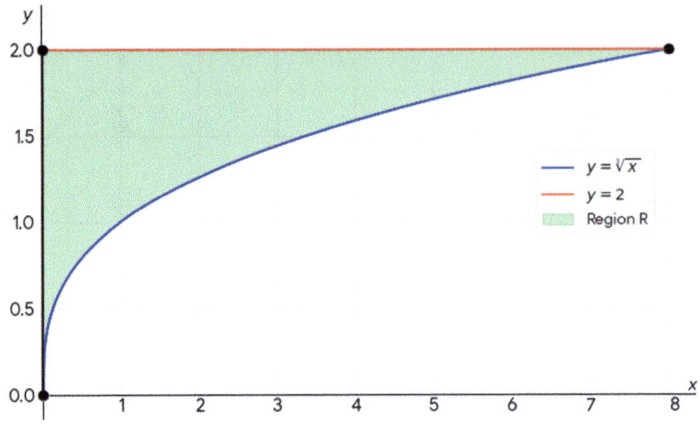

Fig. 1.31 Integration region R

Based on Fig. 1.32, $\int_0^8 \int_{\sqrt[3]{x}}^2 \frac{1}{y^4+1} dydx = \int_0^2 \int_0^{y^3} \frac{1}{y^4+1} dxdy$.

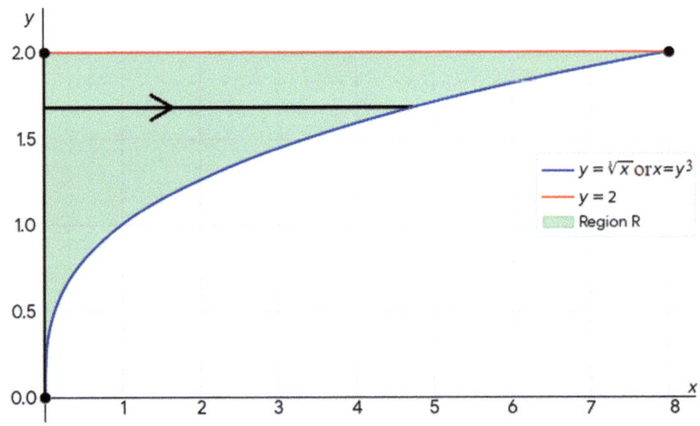

Fig. 1.32 Evaluation of the double integral using the order of integration $dxdy$

The double integral $\int_0^2 \int_0^{y^3} \frac{1}{y^4+1} dxdy$ is calculated by evaluating the inner and outer integrals sequentially:

$$\int_0^{y^3} \frac{1}{y^4+1} dx = \frac{1}{y^4+1} \int_0^{y^3} dx = \frac{1}{y^4+1} x \Big|_0^{y^3}$$

$$= \frac{1}{y^4+1} \times y^3 - \frac{1}{y^4+1} \times 0 = \frac{y^3}{y^4+1}$$

$$
\begin{aligned}
\int_0^2 \int_0^{y^3} \frac{1}{y^4+1} dxdy &= \int_0^2 \frac{y^3}{y^4+1} dy = \frac{1}{4} \int_0^2 \frac{4y^3}{y^4+1} dy \\
&= \frac{1}{4} \ln\left(\left|y^4+1\right|\right) \Big|_0^2 = \frac{1}{4} \ln\left(y^4+1\right) \Big|_0^2 \\
&= \frac{1}{4} \ln\left(2^4+1\right) - \frac{1}{4} \ln\left(0^4+1\right) = \frac{1}{4} \ln(17) - \frac{1}{4} \ln(1) \\
&= \frac{1}{4} \ln(17) - 0 = \frac{1}{4} \ln(17) \\
&\approx 0.7083
\end{aligned}
$$

Example 1.25 Change the order of integration from *dydx* to *dxdy* for the integral $\int_0^{\sqrt{18}} \int_x^{\sqrt{36-x^2}} f(x, y) dydx$.

Solution The region of integration is $R = \left\{ (x, y) | 0 \le x \le \sqrt{18}, x \le y \le \sqrt{36 - x^2} \right\}$ (Fig. 1.33).

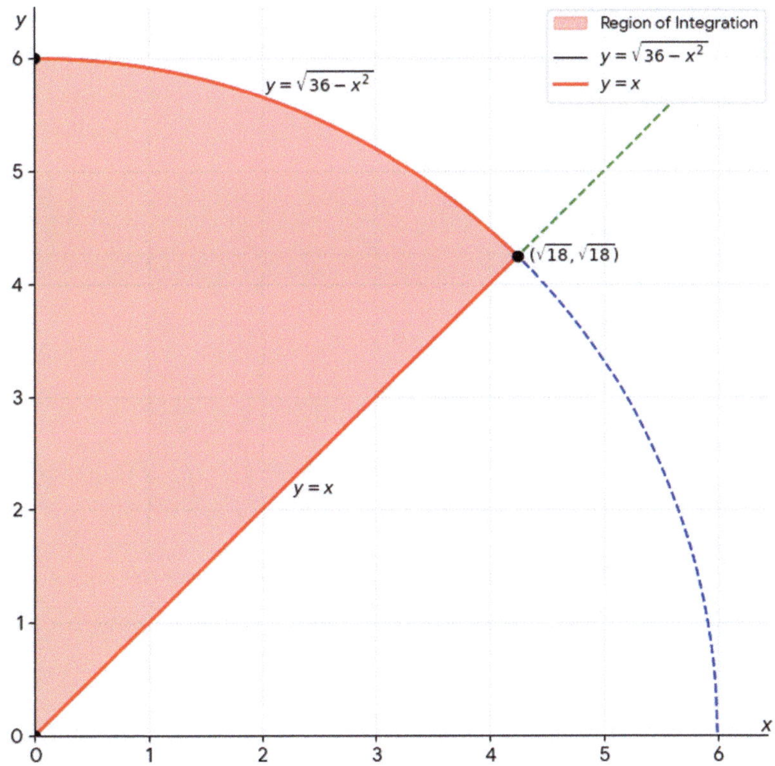

Fig. 1.33 Region R for Example 1.25

To facilitate the integration, we decompose the domain R into two non-overlapping regions, R_1 and R_2. The yellow line shown in Fig. 1.34 marks the common interface (or boundary) between R_1 and R_2.

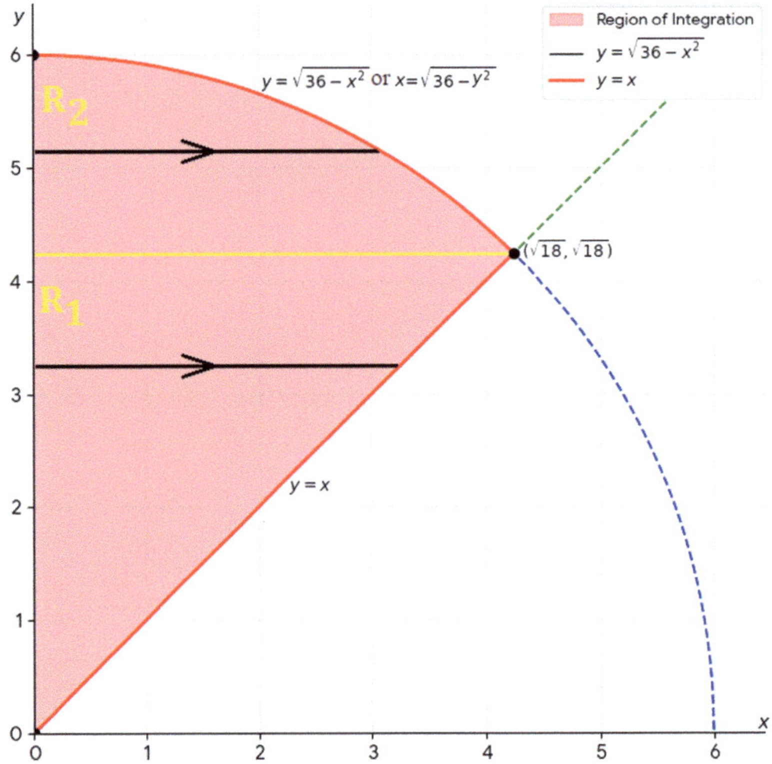

Fig. 1.34 Region R divided into R_1 and R_2

Based on Fig. 1.34,

$$\int_0^{\sqrt{18}}\int_x^{\sqrt{36-x^2}} f(x,y)dydx = \int_0^{\sqrt{18}}\int_0^{y} f(x,y)dxdy + \int_{\sqrt{18}}^{6}\int_0^{\sqrt{36-y^2}} f(x,y)dxdy$$

Example 1.26 Evaluate the double integral $\iint_R f(x,y)dA$ over the region R shown in Fig. 1.35, where the integrand $f(x,y)$ is defined as:

$$f(x,y) = \begin{cases} 1x < y \\ xx > y \end{cases}$$

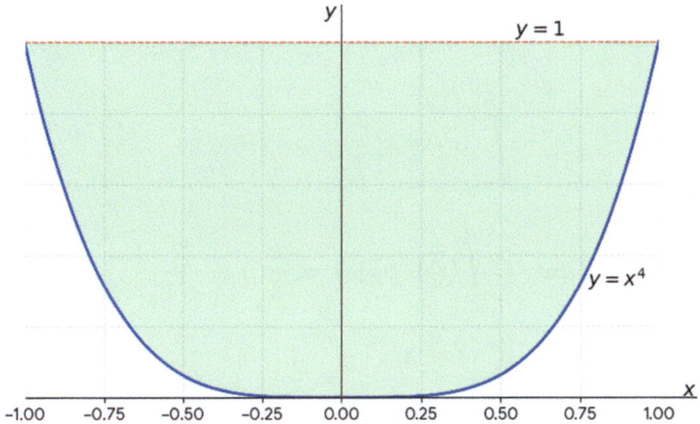

Fig. 1.35 Region R for Example 1.26

Solution To facilitate the integration, we decompose the domain R into two non-overlapping regions, R_1 and R_2 (Fig. 1.36). In region $R_1, f(x, y) = 1$, and in region $R_2, f(x, y) = x$.

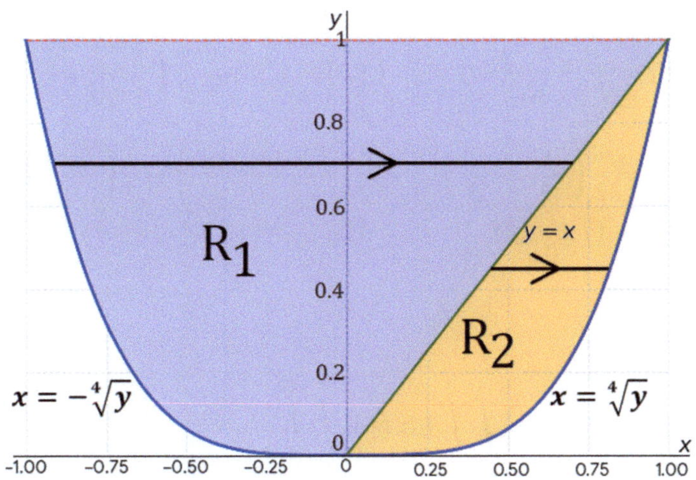

Fig. 1.36 Region R divided into R_1 and R_2

Based on Fig. 1.36,

$$\iint_R f(x, y)dA = \iint_{R_1} f(x, y)dxdy + \iint_{R_2} f(x, y)dxdy \Rightarrow \iint_R f(x, y)dxdy$$

$$= \int_0^1 \int_{-\sqrt[4]{y}}^{y} 1dxdy + \int_0^1 \int_{y}^{\sqrt[4]{y}} xdxdy$$

Let's calculate each integral individually. The calculation of $\int_0^1 \int_{-\sqrt[4]{y}}^{y} 1dxdy$ is shown below:

$$\int_{-\sqrt[4]{y}}^{y} 1dx = x\Big|_{-\sqrt[4]{y}}^{y} = y + \sqrt[4]{y}$$

$$\int_0^1 \int_{-\sqrt[4]{y}}^{y} 1dxdy = \int_0^1 \left(y + \sqrt[4]{y}\right)dy = \frac{y^2}{2} + \frac{y^{\frac{1}{4}+1}}{\frac{1}{4}+1}\Bigg|_0^1$$

$$= \frac{y^2}{2} + \frac{4y^{\frac{5}{4}}}{5}\Bigg|_0^1 = \frac{1}{2} + \frac{4}{5} - 0 = \frac{13}{10} = 1.3$$

The calculation of $\int_0^1 \int_y^{\sqrt[4]{y}} xdxdy$ is shown below:

$$\int_y^{\sqrt[4]{y}} xdx = \frac{x^2}{2}\Bigg|_y^{\sqrt[4]{y}} = \frac{\sqrt[4]{y}^2}{2} - \frac{y^2}{2} = \frac{\sqrt{y}}{2} - \frac{y^2}{2} = \frac{\sqrt{y} - y^2}{2}$$

$$\int_0^1 \int_y^{\sqrt[4]{y}} xdxdy = \int_0^1 \frac{\sqrt{y} - y^2}{2}dy = \frac{1}{2}\int_0^1 \left(\sqrt{y} - y^2\right)dy = \frac{1}{2}\left(\frac{y^{\frac{1}{2}+1}}{\frac{1}{2}+1} - \frac{y^{2+1}}{2+1}\right)\Bigg|_0^1$$

$$= \frac{1}{2}\left(\frac{2y^{\frac{3}{2}}}{3} - \frac{y^3}{3}\right)\Bigg|_0^1 = \frac{1}{2}\left(\frac{2\times 1^{\frac{3}{2}}}{3} - \frac{1^3}{3}\right) - \frac{1}{2}\left(\frac{2\times 0^{\frac{3}{2}}}{3} - \frac{0^3}{3}\right)$$

$$= \frac{1}{6} - 0 = \frac{1}{6} \approx 0.1667$$

Therefore,

$$\iint_R f(x,y)dA = \int_0^1 \int_{\sqrt[4]{y}}^{y} 1dxdy + \int_0^1 \int_y^{\sqrt[4]{y}} xdxdy \Rightarrow \iint_R f(x,y)dA$$

$$= \frac{13}{10} + \frac{1}{6} = \frac{39+5}{30} = \frac{44}{30} = \frac{22}{15}$$

$$\approx 1.4667$$

Example 1.27 Evaluate the double integral $\iint_R f(x,y)dA$ over the region R shown in Fig. 1.37, where the integrand $f(x,y)$ is defined as:

$$f(x,y) = \begin{cases} 1 & x < y \\ x & x > y \end{cases}$$

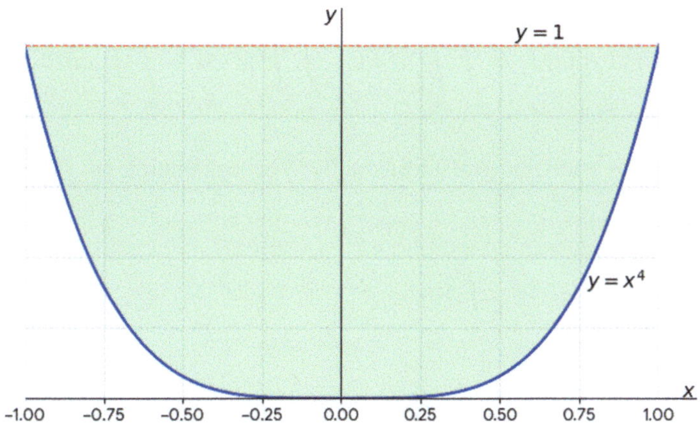

Fig. 1.37 Region R for Example 1.27

Solution This problem was solved in the previous example. Here we will re-evaluate it by reversing the order of integration (integrating with respect to y first, then x).

To facilitate the integration, we decompose the domain R into three non-overlapping regions, R_1, R_2 and R_3 (Fig. 1.38). In regions R_1 and $R_2, f(x, y) = 1$, and in region $R_3, f(x, y) = x$.

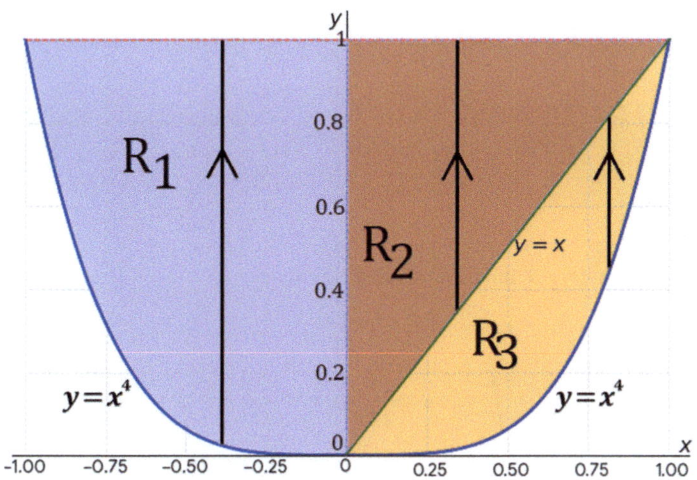

Fig. 1.38 Region R divided into R_1, R_2 and R_3

Based on Fig. 1.38,

$$\iint_R f(x, y)dA = \iint_{R_1} f(x, y)dydx + \iint_{R_2} f(x, y)dydx + \iint_{R_3} f(x, y)dydx$$

$$\Rightarrow \iint_R f(x,y)dydx = \int_{-1}^{0}\int_{x^4}^{1} 1dydx + \int_{0}^{1}\int_{x}^{1} 1dydx + \int_{0}^{1}\int_{x^4}^{x} xdydx$$

Let's calculate each integral individually. The calculation of $\int_{-1}^{0}\int_{x^4}^{1} 1dydx$ is shown below:

$$\int_{x^4}^{1} 1dy = y\Big|_{x^4}^{1} = 1 - x^4$$

$$\int_{-1}^{0}\int_{x^4}^{1} 1dydx = \int_{-1}^{0}(1-x^4)dx = x - \frac{x^5}{5}\Big|_{-1}^{0} = 0 - \left(-1 - \frac{(-1)^5}{5}\right)$$
$$= -\left(-1 + \frac{1}{5}\right) = 1 - \frac{1}{5} = \frac{4}{5}$$
$$= 0.8$$

The calculation of $\int_{0}^{1}\int_{x}^{1} 1dydx$ is shown below:

$$\int_{x}^{1} 1dy = y\Big|_{x}^{1} = 1 - x$$

$$\int_{0}^{1}\int_{x}^{1} 1dydx = \int_{0}^{1}(1-x)dx = \left(x - \frac{x^2}{2}\right)\Big|_{0}^{1} = 1 - \frac{1}{2} - 0 = \frac{1}{2} = 0.5$$

The calculation of $\int_{0}^{1}\int_{x^4}^{x} xdydx$ is shown below:

$$\int_{x}^{1} xdy = x\int_{x}^{1} dy = xy\Big|_{x}^{1} = x \times 1 - x \times x = x - x^2$$

$$\int_{0}^{1}\int_{x^4}^{x} xdydx = \int_{0}^{1}(x - x^2)dx = \left(\frac{x^2}{2} - \frac{x^3}{3}\right)\Big|_{0}^{1} = \frac{1}{2} - \frac{1}{3} - 0 = \frac{1}{6} \approx 0.1667$$

Therefore,

$$\iint_R f(x,y)dA = \int_{-1}^{0}\int_{x^4}^{1} 1dydx + \int_{0}^{1}\int_{x}^{1} 1dydx + \int_{0}^{1}\int_{x^4}^{x} xdydx$$

$$\Rightarrow \iint_R f(x,y)dA = \frac{4}{5} + \frac{1}{2} + \frac{1}{6} = \frac{24+15+5}{30} = \frac{44}{30} = \frac{22}{15} \approx 1.4667$$

The value obtained is identical to the result from the preceding problem.

Example 1.28 Evaluate the double integral $\iint_R f(x,y)dA$ over the region R shown in Fig. 1.39, where the integrand $f(x,y)$ is defined as:

$$f(x,y) = \begin{cases} y & x < y \\ x & x > y \end{cases}$$

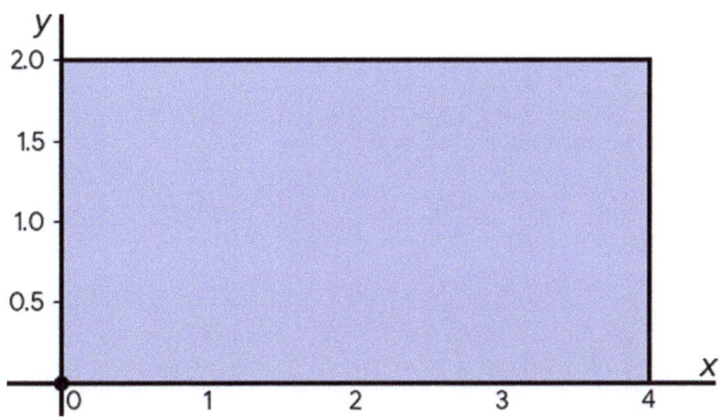

Fig. 1.39 Region R for Example 1.28

Solution To facilitate the integration, we decompose the domain R into two non-overlapping regions, R_1 and R_2. The red line shown in Fig. 1.40 marks the common interface (or boundary) between R_1 and R_2. In region R_1, $f(x,y) = y$, and in region R_2, $f(x,y) = x$.

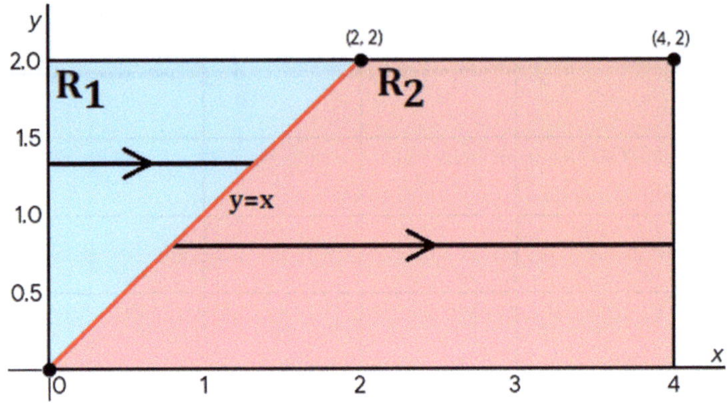

Fig. 1.40 Region R divided into R_1 and R_2

Based on Fig. 1.40,

$$\iint\limits_{R} f(x, y)dA = \iint\limits_{R_1} ydA + \iint\limits_{R_2} xdA = \int_0^2 \int_0^y ydxdy + \int_0^2 \int_y^4 xdxdy$$

We will calculate each integral separately.

$$\int_0^y ydx = y\int_0^y dx = yx\Big|_0^y = y \times y - y \times 0 = y^2 - 0 = y^2$$

$$\int_0^2 \int_0^y ydxdy = \int_0^2 y^2 dy = \frac{y^3}{3}\Big|_0^2 = \frac{2^3}{3} - \frac{0^3}{3} = \frac{8}{3}$$

and

$$\int_y^4 xdx = \frac{x^2}{2}\Big|_y^4 = \frac{4^2}{2} - \frac{y^2}{2} = 8 - \frac{y^2}{2}$$

$$\int_0^2 \int_y^4 xdxdy = \int_0^2 \left(8 - \frac{y^2}{2}\right)dy = 8y - \frac{y^3}{6}\Big|_0^2 = \left(8 \times 2 - \frac{2^3}{6}\right) - 0$$

$$= 16 - \frac{8}{6} = \frac{88}{6} = \frac{44}{3} \approx 14.6667$$

Therefore,

$$\iint\limits_R f(x,y)dA = \iint\limits_{R_1} ydA + \iint\limits_{R_2} xdA = \int\limits_0^2 \int\limits_0^y ydxdy + \int\limits_0^2 \int\limits_y^4 xdxdy$$
$$= \frac{8}{3} + \frac{44}{3} = \frac{52}{3} \approx 17.3333$$

Example 1.29 Evaluate the double integral $\iint\limits_R f(x,y)dA$ over the region R shown in Fig. 1.41, where the integrand $f(x, y)$ is defined as:

$$f(x,y) = \begin{cases} y & x < y \\ x & x > y \end{cases}$$

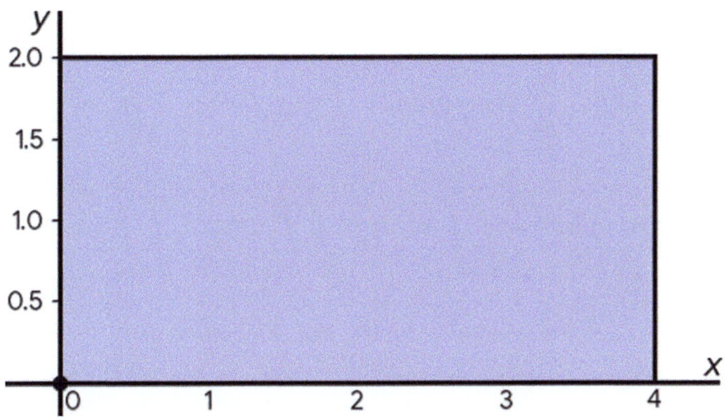

Fig. 1.41 Region R for Example 1.29

Solution This problem was solved in the previous example. Here we will re-evaluate it by reversing the order of integration (integrating with respect to y first, then x).

To facilitate the integration, we decompose the domain R into three non-overlapping regions, R_1, R_2 and R_3 (Fig. 1.42). In region $R_1, f(x, y) = y$, and in regions R_2 and $R_3, f(x, y) = x$.

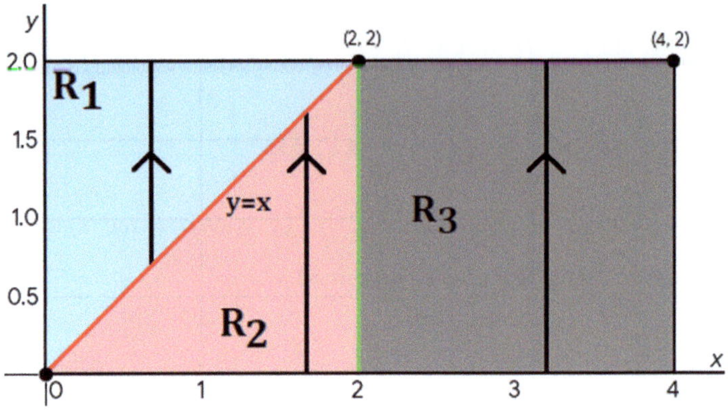

Fig. 1.42 Region R divided into R_1, R_2 and R_3

Based on Fig. 1.42,

$$\iint_R f(x, y)dA = \iint_{R_1} f(x, y)dydx + \iint_{R_2} f(x, y)dydx + \iint_{R_3} f(x, y)dydx$$

$$\Rightarrow \iint_R f(x, y)dydx = \int_0^2 \int_x^2 ydydx + \int_0^2 \int_0^x xdydx + \int_2^4 \int_0^2 xdydx$$

Let's calculate each integral individually. The calculation of $\int_0^2 \int_x^2 ydydx$ is shown below:

$$\int_x^2 ydy = \frac{y^2}{2}\bigg|_x^2 = \frac{2^2}{2} - \frac{x^2}{2} = 2 - \frac{x^2}{2}$$

$$\int_0^2 \int_x^2 ydydx = \int_0^2 \left(2 - \frac{x^2}{2}\right)dx = 2x - \frac{x^3}{6}\bigg|_0^2 = \left(2 \times 2 - \frac{2^3}{6}\right) - 0 = \frac{8}{3} \approx 2.6667$$

The calculation of $\int_0^2 \int_0^x xdydx$ is shown below:

$$\int_0^x xdy = x\int_0^x dy = xy\bigg|_0^x = x \times x - x \times 0 = x^2 - 0 = x^2$$

$$\int_0^2 \int_0^x xdydx = \int_0^2 x^2dx = \frac{x^3}{3}\bigg|_0^2 = \frac{2^3}{3} - 0 = \frac{8}{3} \approx 2.6667$$

The calculation of $\int_2^4 \int_0^2 x dy dx$ is shown below:

$$\int_2^4 \int_0^2 x dy dx = \int_0^2 dy \int_2^4 x dx = \left(y \Big|_0^2 \right) \left(\frac{x^2}{2} \Big|_2^4 \right)$$
$$= (2-0)\left(\frac{4^2}{2} - \frac{2^2}{2} \right) = 2 \times (8-2) = 12$$

Therefore,

$$\iint_R f(x, y) dA = \iint_{R_1} f(x, y) dy dx + \iint_{R_2} f(x, y) dy dx + \iint_{R_3} f(x, y) dy dx$$
$$= \int_0^2 \int_x^2 y dy dx + \int_0^2 \int_0^x y dy dx + \int_2^4 \int_0^2 x dy dx$$
$$= \frac{8}{3} + \frac{8}{3} + 12 = \frac{52}{3} \approx 17.3333$$

The value obtained is identical to the result from the preceding problem.

Example 1.30 Calculate the area of the region R shown in Fig. 1.43.

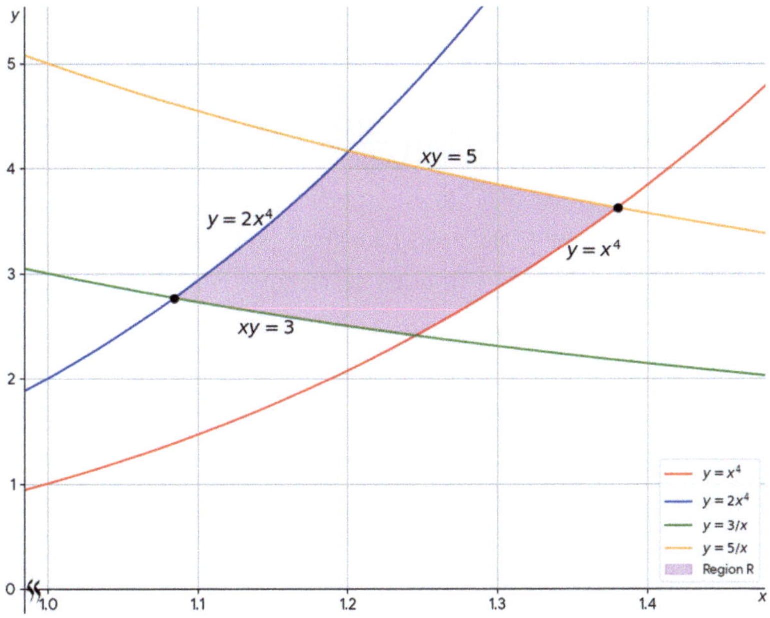

Fig. 1.43 Region R for Example 1.30

Solution The area of the region is $A = \iint_R dA$. Let's apply a change of variables. The region R is bounded by:

$$y = x^4 \Rightarrow \frac{y}{x^4} = 1 \Rightarrow u = 1$$

$$y = 2x^4 \Rightarrow \frac{y}{x^4} = 2 \Rightarrow u = 2$$

$$xy = 3 \Rightarrow v = 3$$

$$xy = 5 \Rightarrow v = 5$$

The change of variables is:

$$u = \frac{y}{x^4}$$

$$v = xy$$

This transforms the complex region R in the xy-plane into a simple rectangular region S in the uv-plane:

$$S = \{(u, v)|1 \leq u \leq 2, 3 \leq v \leq 5\}$$

The area A is given by the integral:

$$A = \iint\limits_S \left|\det\left(\frac{\partial(x, y)}{\partial(u, v)}\right)\right| dudv$$

Since the functions $u = \frac{y}{x^4}$ and $v = xy$ are given explicitly in terms of x and y, it is easier to compute the $\left|\det\left(\frac{\partial(x,y)}{\partial(u,v)}\right)\right|$ using the identity:

$$\left|\det\left(\frac{\partial(x, y)}{\partial(u, v)}\right)\right| = \frac{1}{\left|\det\left(\frac{\partial(u,v)}{\partial(x,y)}\right)\right|} = \frac{1}{\left|\det\left(\begin{bmatrix} \frac{\partial u}{\partial x} & \frac{\partial u}{\partial y} \\ \frac{\partial v}{\partial x} & \frac{\partial v}{\partial y} \end{bmatrix}\right)\right|}$$

The partial derivatives are:

$$\frac{\partial u}{\partial x} = \frac{\partial}{\partial x}\left(\frac{y}{x^4}\right) = -4yx^{-5} = -4 \times \frac{y}{x^4} \times \frac{1}{x} = \frac{-4u}{x}$$

$$\frac{\partial u}{\partial y} = \frac{\partial}{\partial y}\left(\frac{y}{x^4}\right) = \frac{1}{x^4}$$

$$\frac{\partial v}{\partial x} = \frac{\partial}{\partial x}(xy) = y$$

$$\frac{\partial v}{\partial y} = \frac{\partial}{\partial y}(xy) = x$$

Therefore,

$$\left|\det\left(\frac{\partial(x,y)}{\partial(u,v)}\right)\right| = \frac{1}{\left|\det\left(\frac{\partial(u,v)}{\partial(x,y)}\right)\right|} = \frac{1}{\left|\det\left(\begin{bmatrix} \frac{\partial u}{\partial x} & \frac{\partial u}{\partial y} \\ \frac{\partial v}{\partial x} & \frac{\partial v}{\partial y} \end{bmatrix}\right)\right|} = \frac{1}{\left|\det\left(\begin{bmatrix} -\frac{4u}{x} & \frac{1}{x^4} \\ y & x \end{bmatrix}\right)\right|}$$

$$= \frac{1}{\left|-\frac{4u}{x} \times x - \frac{1}{x^4} \times y\right|} = \frac{1}{\left|-4u - \frac{y}{x^4}\right|} = \frac{1}{|-4u - u|}$$

$$= \frac{1}{|-5u|} = \frac{1}{5u}$$

and

$$A = \iint\limits_S \left|\frac{\partial(x,y)}{\partial(u,v)}\right| dudv = \iint\limits_S \left|-\frac{1}{5u}\right| dudv = \iint\limits_S \frac{1}{5u} dudv = \int_3^5 \int_1^2 \frac{1}{5u} dudv$$

The calculation of $\int_3^5 \int_1^2 \frac{1}{5u} dudv$ is shown below:

$$\int_1^2 \frac{1}{5u} du = \frac{1}{5}\int_1^2 \frac{1}{u} du = \frac{1}{5}\ln(u)\Big|_1^2 = \frac{1}{5}\ln(2) - \frac{1}{5}\ln(1) = \frac{1}{5}\ln(2) \approx 0.1386$$

$$A = \int_3^5 \int_1^2 \frac{1}{5u} dudv = \int_3^5 0.1386 dv = 0.1386v\Big|_3^5 = 0.1386(5-3) = 0.2773$$

Another method for calculating $\int_3^5 \int_1^2 \frac{1}{5u} dudv$ is:

$$A = \int_3^5 \int_1^2 \frac{1}{5u} dudv = \frac{1}{5}\int_3^5 \int_1^2 \frac{1}{u} dudv = \frac{1}{5}\int_1^2 \frac{1}{u} du \int_3^5 dv = \frac{1}{5}\ln(u)\Big|_1^2 \times v\Big|_3^5$$

$$= \frac{1}{5}(\ln(2) - \ln(1))(5-3) = \frac{2}{5}\ln(2) \approx 0.2773$$

Therefore, the area of the region R is 0.2773 square units.

Example 1.31 Evaluate the double integral $\iint_R e^{-(x^2+y^2)} dA$ over the region R depicted in Fig. 1.44.

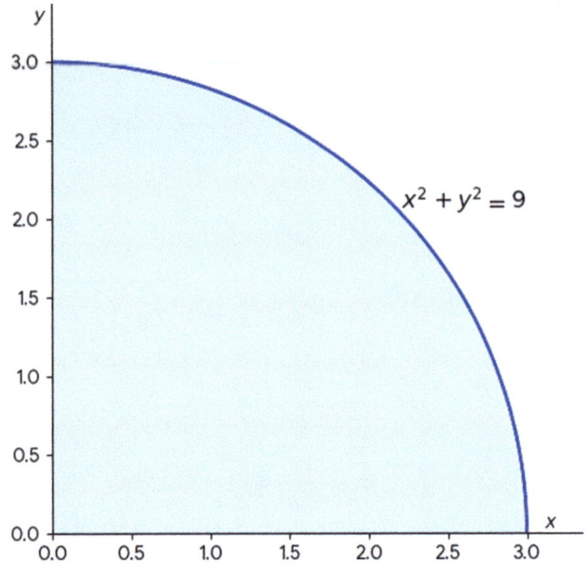

Fig. 1.44 Region R for Example 1.31

Solution The region R in Cartesian coordinates is defined by:

$$R = \left\{ (x, y) | 0 \leq x \leq 3, 0 \leq y \leq \sqrt{9 - x^2} \right\}$$

The integral in the *dxdy* order is:

$$I = \int_0^3 \int_0^{\sqrt{9-y^2}} e^{-(x^2+y^2)} dxdy$$

The integral in the *dydx* order is:

$$I = \int_0^3 \int_0^{\sqrt{9-x^2}} e^{-(x^2+y^2)} dydx$$

The term $e^{-(x^2+y^2)}$ prevents solving the inner integral ($\int_0^{\sqrt{9-y^2}} e^{-(x^2+y^2)} dx$ or $\int_0^{\sqrt{9-x^2}} e^{-(x^2+y^2)} dy$) using elementary functions, making the integral unsolvable in Cartesian form. We use the transformation:

$$x = r\cos(\theta)$$

$$y = r\sin(\theta)$$

The exponent simplifies:

$$x^2 + y^2 = r^2\cos^2(\theta) + r^2\sin^2(\theta) = r^2\left(\cos^2(\theta) + \sin^2(\theta)\right) = r^2 \times 1 = r^2$$

The integrand becomes:

$$e^{-(x^2+y^2)} = e^{-r^2}$$

The absolute value of the Jacobian determinant ($\left|\det\left(\frac{\partial(x,y)}{\partial(r,\theta)}\right)\right|$) is:

$$\begin{aligned}\left|\det\left(\frac{\partial(x,y)}{\partial(r,\theta)}\right)\right| &= \left|\det\left(\begin{bmatrix}\frac{\partial x}{\partial r} & \frac{\partial x}{\partial \theta}\\ \frac{\partial y}{\partial r} & \frac{\partial y}{\partial \theta}\end{bmatrix}\right)\right| = \left|\det\left(\begin{bmatrix}\cos(\theta) & -r\sin(\theta)\\ \sin(\theta) & r\cos(\theta)\end{bmatrix}\right)\right|\\ &= \left|r\cos^2(\theta) + r\sin^2(\theta)\right|\\ &= \left|r\left(\cos^2(\theta) + \sin^2(\theta)\right)\right| = |r \times 1| = |r| = r\end{aligned}$$

and

$$dA = dxdy = dydx = rdrd\theta$$

The region R is the quarter-disk in the first quadrant bounded by $x^2 + y^2 = 9$. In polar coordinates, this region is described as $S = \left\{(r,\theta)|0 \le r \le 3, 0 \le \theta \le \frac{\pi}{2}\right\}$. Thus,

$$\begin{aligned}\iint_R e^{-(x^2+y^2)}dA &= \int_0^{\frac{\pi}{2}}\int_0^3 e^{-r^2} rdrd\theta = \int_0^{\frac{\pi}{2}} d\theta \times \int_0^3 re^{-r^2}dr\\ &= \int_0^{\frac{\pi}{2}} d\theta \times \frac{1}{-2}\int_0^3 -2re^{-r^2}dr\\ &= \int_0^{\frac{\pi}{2}} d\theta \times \frac{1}{2}\int_3^0 -2re^{-r^2}dr = \theta\Big|_0^{\frac{\pi}{2}} \times \frac{1}{2}e^{-r^2}\Big|_3^0\\ &= \frac{\pi}{2} \times \frac{1}{2}\left(1 - e^{-9}\right) = \frac{\pi}{4}\left(1 - e^{-9}\right)\\ &\approx 0.7853\end{aligned}$$

Example 1.32 Evaluate the double integral $\iint_R (x^2+y^2)dA$ over the region R depicted in Fig. 1.45.

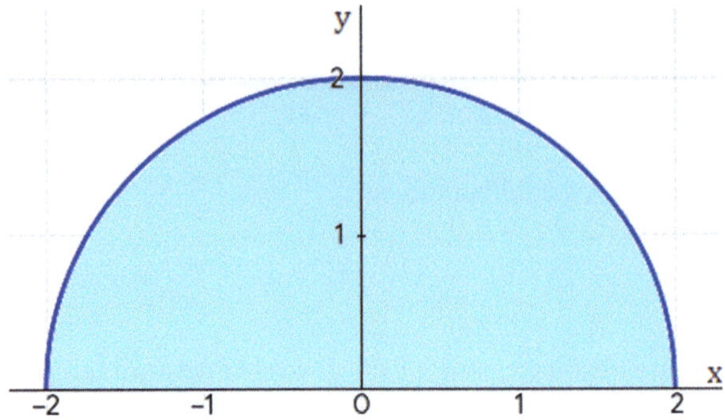

Fig. 1.45 Region R for Example 1.32

Solution Let's convert the double integral from Cartesian coordinates (x, y) and $dA = dxdy = dydx$ to polar coordinates (r, θ) and $dA = rdrd\theta$.

$$x^2 + y^2 \overset{\begin{cases} x = r\cos(\theta) \\ y = r\sin(\theta) \end{cases}}{\Longrightarrow} (r\cos(\theta))^2 + (r\sin(\theta))^2 = r^2\cos^2(\theta) + r^2\sin^2(\theta)$$
$$= r^2\left(\cos^2(\theta) + \sin^2(\theta)\right)$$
$$= r^2 \times 1 = r^2$$

$$\iint_R (x^2 + y^2)dA = \int_0^{\pi}\int_0^{2} r^2 \times rdrd\theta = \int_0^{\pi}\int_0^{2} r^3 drd\theta$$
$$= \int_0^{2} r^3 dr \int_0^{\pi} d\theta = \frac{r^4}{4}\bigg|_0^2 \times \theta\bigg|_0^{\pi}$$
$$= \left(\frac{2^4}{4} - \frac{0^4}{4}\right) \times (\pi - 0) = 4\pi$$

Example 1.33 Evaluate the double integral $\iint_R e^{x^2+y^2} dA$ over the region R depicted in Fig. 1.46.

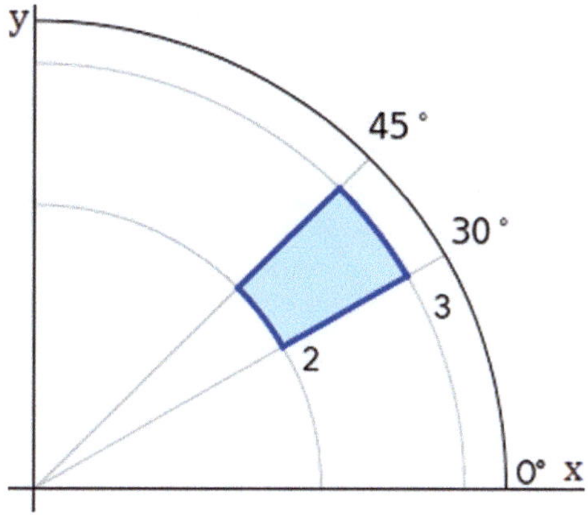

Fig. 1.46 Region R for Example 1.33

Solution Let's convert the double integral from Cartesian coordinates (x, y) and $dA = dxdy = dydx$ to polar coordinates (r, θ) and $dA = rdrd\theta$.

$$\begin{aligned} x^2 + y^2 &\overset{\begin{cases} x = r\cos(\theta) \\ y = r\sin(\theta) \end{cases}}{\Longrightarrow} (r\cos(\theta))^2 + (r\sin(\theta))^2 \\ &= r^2\cos^2(\theta) + r^2\sin^2(\theta) \\ &= r^2\left(\cos^2(\theta) + \sin^2(\theta)\right) \\ &= r^2 \times 1 = r^2 \end{aligned}$$

$$e^{x^2+y^2} = e^{r^2}$$

$$\begin{aligned} \iint\limits_R e^{x^2+y^2} dA &= \int_{\frac{\pi}{6}}^{\frac{\pi}{4}} \int_2^3 e^{r^2} \times rdrd\theta = \int_{\frac{\pi}{6}}^{\frac{\pi}{4}} \int_2^3 re^{r^2} drd\theta \\ &= \int_2^3 re^{r^2} dr \int_{\frac{\pi}{6}}^{\frac{\pi}{4}} d\theta = \frac{1}{2} \int_2^3 2re^{r^2} dr \int_{\frac{\pi}{6}}^{\frac{\pi}{4}} d\theta \\ &= \frac{1}{2} e^{r^2} \Big|_2^3 \times \theta \Big|_{\frac{\pi}{6}}^{\frac{\pi}{4}} = \left(\frac{1}{2} e^{3^2} - \frac{1}{2} e^{2^2}\right) \times \left(\frac{\pi}{4} - \frac{\pi}{6}\right) \\ &= \frac{1}{2}\left(e^9 - e^4\right) \times \frac{2\pi}{24} = \frac{\left(e^9 - e^4\right)\pi}{24} \\ &\approx 1053.5 \end{aligned}$$

Exercise 1.2 Evaluate the double integral $\iint_R e^{x^2+y^2} dA$ over the region R depicted in Fig. 1.47. (Ans. $\left(e^9 - e^4\right)\pi \approx 25285.064$).

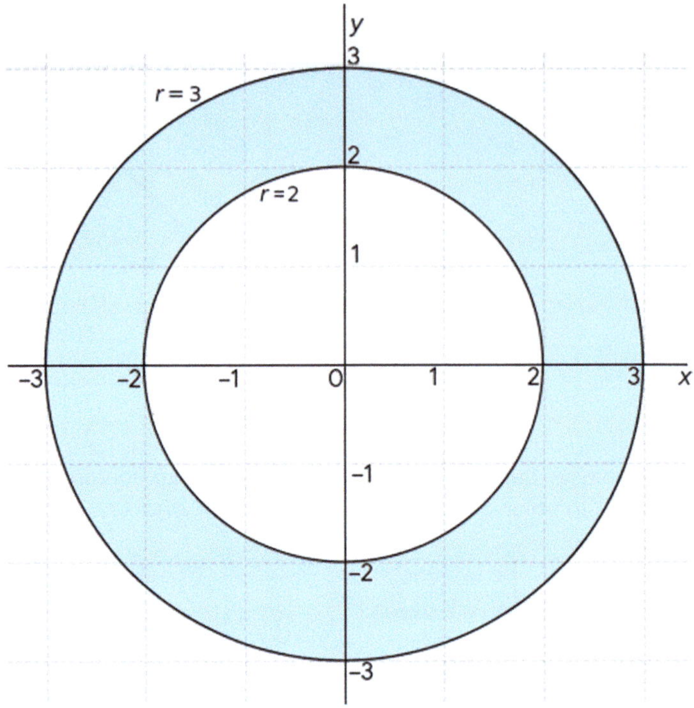

Fig. 1.47 Region R for Exercise 1.2

Example 1.34 Evaluate the double integral $\iint_R (y^2 - x^2) dA$ over the region R depicted in Fig. 1.48.

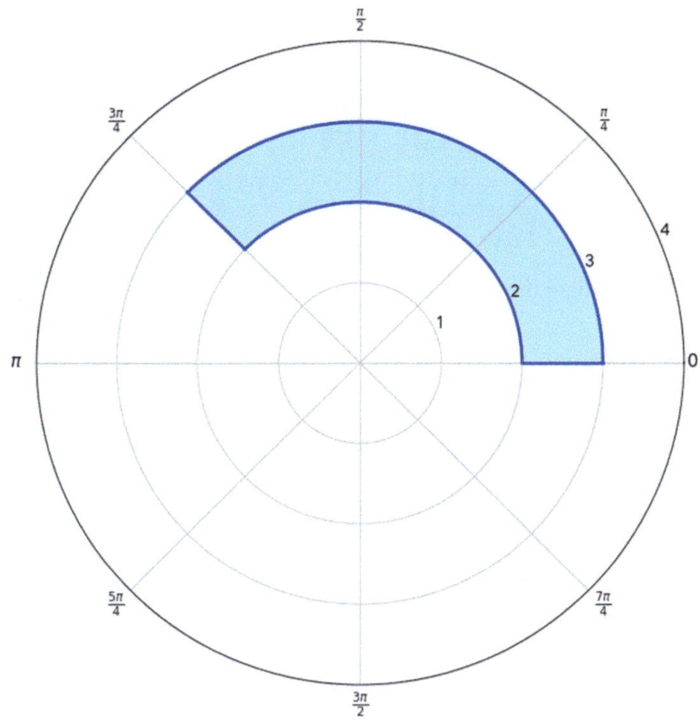

Fig. 1.48 Region R for Example 1.34

Solution Let's convert the double integral from Cartesian coordinates (x, y) and $dA = dxdy = dydx$ to polar coordinates (r, θ) and $dA = rdrd\theta$.

$$y^2 - x^2 \overset{\begin{cases} x = r\cos(\theta) \\ y = r\sin(\theta) \end{cases}}{\Longrightarrow} (r\sin(\theta))^2 - (r\cos(\theta))^2 = r^2\sin^2(\theta) - r^2\cos^2(\theta)$$
$$= r^2\big(\sin^2(\theta) - \cos^2(\theta)\big)$$
$$= r^2 \times -\cos(2\theta)$$
$$= -r^2\cos(2\theta)$$

$$\iint_R (y^2 - x^2)dA = \int_0^{\frac{3\pi}{4}}\int_2^3 -r^2\cos(2\theta) \times rdrd\theta = \int_0^{\frac{3\pi}{4}}\int_2^3 -r^3\cos(2\theta)drd\theta$$
$$= \int_2^3 -r^3dr \int_0^{\frac{3\pi}{4}} \cos(2\theta)d\theta = \int_3^2 r^3dr \int_0^{\frac{3\pi}{4}} \cos(2\theta)d\theta$$

$$= \frac{r^4}{4}\bigg|_3^2 \times \frac{\sin(2\theta)}{2}\bigg|_0^{\frac{3\pi}{4}}$$

$$= \left(\frac{2^4}{4} - \frac{3^4}{4}\right)\left(\frac{\sin\left(2 \times \frac{3\pi}{4}\right)}{2} - \frac{\sin(2 \times 0)}{2}\right)$$

$$= \left(4 - \frac{81}{4}\right)\left(-\frac{1}{2} - 0\right) = \frac{65}{8} \approx 8.1250$$

Example 1.35 Evaluate the double integral $\int_0^{\infty}\int_{-\infty}^{\infty} e^{-(x^2+y^2)}dxdy$.

Solution The integration region is the open upper half-plane in the Cartesian coordinate system (Fig. 1.49).

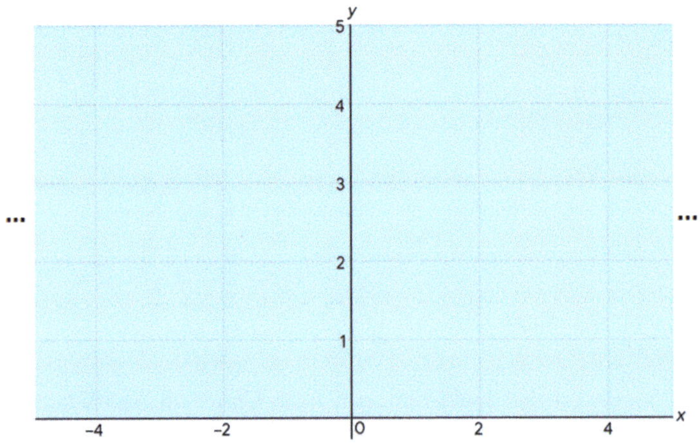

Fig. 1.49 Integration region for Example 1.35

Let's convert the double integral from Cartesian coordinates (x, y) and $dA = dxdy = dydx$ to polar coordinates (r, θ) and $dA = rdrd\theta$.

$$x^2 + y^2 \overset{\begin{cases} x = r\cos(\theta) \\ y = r\sin(\theta) \end{cases}}{\Longrightarrow} (r\cos(\theta))^2 + (r\sin(\theta))^2 = r^2\cos^2(\theta) + r^2\sin^2(\theta)$$

$$= r^2\left(\cos^2(\theta) + \sin^2(\theta)\right)$$

$$= r^2 \times 1 = r^2$$

$$e^{-(x^2+y^2)} = e^{-r^2}$$

$$\int_0^{\infty}\int_{-\infty}^{\infty} e^{-(x^2+y^2)}dxdy = \int_0^{\pi}\int_0^{\infty} e^{-r^2} \times rdrd\theta = \int_0^{\pi}\int_0^{\infty} re^{-r^2}drd\theta = \int_0^{\infty} re^{-r^2}dr\int_0^{\pi} d\theta$$

$$= \frac{1}{-2}\int_0^{\infty} -2re^{-r^2}dr \int_0^{\pi} d\theta = \frac{1}{-2}e^{-r^2}\Big|_0^{\infty} \times \theta\Big|_0^{\pi}$$

$$= \frac{1}{2}e^{-r^2}\Big|_{\infty}^{0} \times \theta\Big|_0^{\pi} = \left(\frac{1}{2}e^{-0^2} - \lim_{r\to\infty}\frac{1}{2}e^{-r^2}\right) \times (\pi - 0)$$

$$= \left(\frac{1}{2} - 0\right) \times \pi = \frac{\pi}{2}$$

Exercise 1.3 Evaluate the double integral $\int_{-\infty}^{\infty}\int_{-\infty}^{\infty} e^{-(x^2+y^2)}dxdy$. (Ans. π).

Example 1.36 Evaluate the double integral $\iint_R (x + 2y)dxdy$ over the region R depicted in Fig. 1.50.

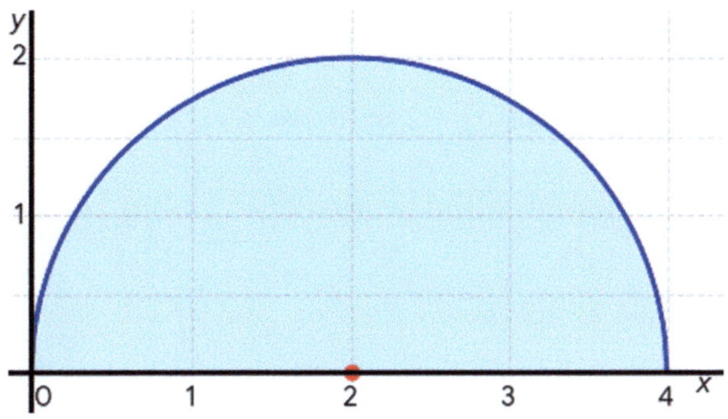

Fig. 1.50 Region defined by $(x - 2)^2 + y^2 \leq 4$

Solution The region R is the upper half of the disk centered at (2, 0) with radius 2, i.e.

$$(x - 2)^2 + y^2 \leq 4, y \geq 0$$

Use polar coordinates about the center (2, 0):

$$\begin{cases} x = 2 + r\cos(\theta) \\ y = r\sin(\theta) \end{cases} 0 \leq r \leq 2, 0 \leq \theta \leq \pi$$

The integrand becomes:

$$x + 2y = 2 + r\cos(\theta) + 2r\sin(\theta) = 2 + r(\cos(\theta) + 2\sin(\theta))$$

Hence,

$$\iint_R (x + 2y)dxdy = \int_0^{\pi}\int_0^{2} (2 + r(\cos(\theta) + 2\sin(\theta))) \times rdrd\theta$$

$$= \int_0^{\pi}\int_0^{2} \left(2r + r^2(\cos(\theta) + 2\sin(\theta))\right) dr d\theta$$

$$= \int_0^{\pi}\int_0^{2} 2r dr d\theta + \int_0^{\pi}\int_0^{2} r^2(\cos(\theta) + 2\sin(\theta)) dr d\theta$$

Let's calculate each integral individually.

$$\int_0^{\pi}\int_0^{2} 2r dr d\theta = \int_0^{2} 2r dr \times \int_0^{\pi} d\theta = r^2\Big|_0^2 \times \theta\Big|_0^{\pi} = (2^2 - 0)(\pi - 0) = 4\pi$$

and

$$\int_0^{\pi}\int_0^{2} r^2(\cos(\theta) + 2\sin(\theta)) dr d\theta = \int_0^{2} r^2 dr \int_0^{\pi} (\cos(\theta) + 2\sin(\theta)) d\theta$$

$$= \frac{r^3}{3}\Big|_0^2 \times (\sin(\theta) - 2\cos(\theta))\Big|_0^{\pi}$$

$$= \left(\frac{2^3}{3} - \frac{0^3}{3}\right)$$

$$(\sin(\pi) - 2\cos(\pi) - \sin(0) + 2\cos(0))$$

$$= \frac{8}{3} \times 4 = \frac{32}{3}$$

Therefore,

$$\iint_R (x + 2y) dx dy = \int_0^{\pi}\int_0^{2} 2r dr d\theta + \int_0^{\pi}\int_0^{2} r^2(\cos(\theta) + 2\sin(\theta)) dr d\theta$$

$$= 4\pi + \frac{32}{3} \approx 23.2330$$

Example 1.37 Evaluate the double integral $\iint_R (4x - 5y) dx dy$ over the region R depicted in Fig. 1.51.

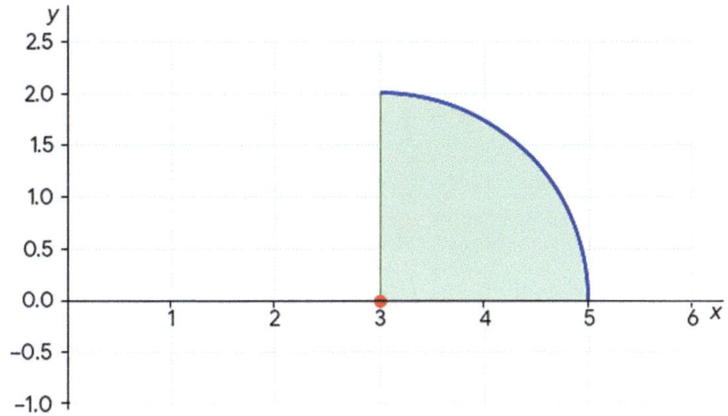

Fig. 1.51 Region R for Example 1.37

Solution We'll use polar coordinates about the center (3, 0):

$$\begin{cases} x = 3 + r\cos(\theta) \\ y = r\sin(\theta) \end{cases} 0 \le r \le 2, 0 \le \theta \le \frac{\pi}{2}$$

The integrand becomes:

$$4x - 5y = 4(3 + r\cos(\theta)) - 5r\sin(\theta) = 12 + r(4\cos(\theta) - 5\sin(\theta))$$

Hence,

$$\begin{aligned}\iint_R (4x - 5y)dxdy &= \int_0^{\frac{\pi}{2}}\int_0^2 (12 + r(4\cos(\theta) - 5\sin(\theta))) \times rdrd\theta \\ &= \int_0^{\frac{\pi}{2}}\int_0^2 \left(12r + r^2(4\cos(\theta) - 5\sin(\theta))\right)drd\theta \\ &= \int_0^{\frac{\pi}{2}}\int_0^2 12rdrd\theta + \int_0^{\frac{\pi}{2}}\int_0^2 r^2(4\cos(\theta) - 5\sin(\theta))drd\theta\end{aligned}$$

Let's calculate each integral individually.

$$\begin{aligned}\int_0^{\frac{\pi}{2}}\int_0^2 12rdrd\theta &= \int_0^2 12rdr\int_0^{\frac{\pi}{2}} d\theta = 6r^2\Big|_0^2 \times \theta\Big|_0^{\frac{\pi}{2}} \\ &= \left(6\times 2^2 - 6\times 0^2\right)\times\left(\frac{\pi}{2} - 0\right) = 12\pi\end{aligned}$$

and

$$\int_0^{\frac{\pi}{2}}\int_0^2 r^2(4\cos(\theta)-5\sin(\theta))drd\theta = \int_0^2 r^2 dr \int_0^{\frac{\pi}{2}}(4\cos(\theta)-5\sin(\theta))d\theta$$
$$= \frac{r^3}{3}\Big|_0^2 \times (4\sin(\theta)+5\cos(\theta))\Big|_0^{\frac{\pi}{2}}$$
$$= \left(\frac{2^3}{3}-\frac{0^3}{3}\right)$$
$$\left(4\sin\left(\frac{\pi}{2}\right)+5\cos\left(\frac{\pi}{2}\right)-4\sin(0)-5\cos(0)\right)$$
$$= \frac{8}{3}\times -1 = -\frac{8}{3}$$

Therefore,

$$\iint_R (4x-5y)dxdy = \int_0^{\frac{\pi}{2}}\int_0^2 12rdrd\theta + \int_0^{\frac{\pi}{2}}\int_0^2 r^2(4\cos(\theta)-5\sin(\theta))drd\theta$$
$$= 12\pi - \frac{8}{3}$$
$$\approx 35.0324$$

Exercise 1.4 Evaluate the double integral $\iint_R xdxdy$ and $\iint_R ydxdy$ over the region R depicted in Fig. 1.52. (Ans. $\iint_R xdxdy = \frac{\pi}{4}+\frac{1}{3}\approx 1.1187$, $\iint_R ydxdy = \frac{1}{3}$).

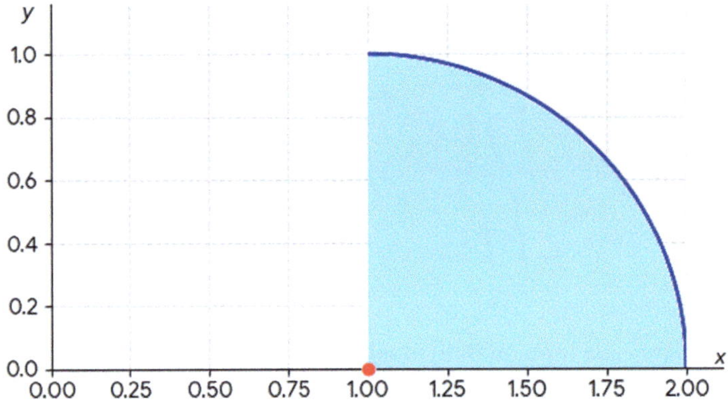

Fig. 1.52 Region R for Exercise 1.4

Exercise 1.5 Evaluate the double integral $\iint_R(3x-4y)dxdy$ over the region R depicted in Fig. 1.53. (Ans. $\pi \approx 3.1415$).

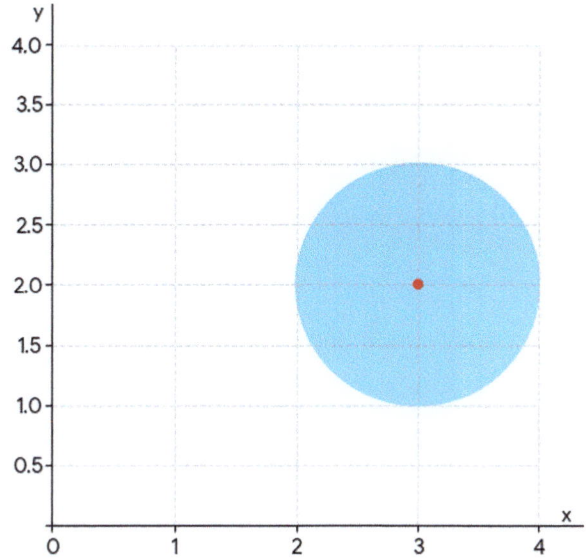

Fig. 1.53 Region R for Exercise 1.5

Example 1.38 Calculate $\int_0^{\sqrt{18}} \int_x^{\sqrt{36-x^2}} \frac{1}{\sqrt{x^2+y^2}} dydx$.

Solution The integration region is the sector inside the circle $r = 6$ between the rays $y = x$ and the positive y-axis (Fig. 1.54). In polar coordinates that is $\theta \in \left[\frac{\pi}{4}, \frac{\pi}{2}\right], \quad r \in [0, 6]$.

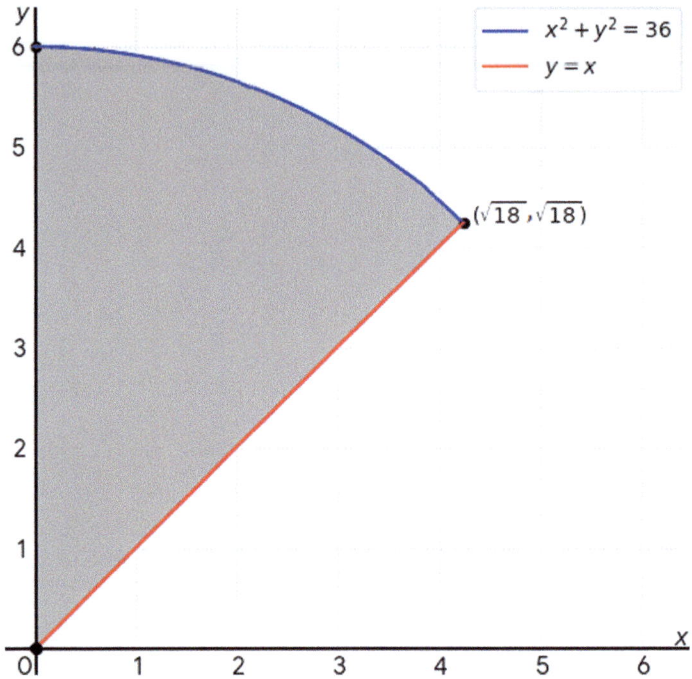

Fig. 1.54 Integration region for Example 1.38

Let's convert the double integral from Cartesian coordinates (x, y) and $dA = dxdy = dydx$ to polar coordinates (r, θ) and $dA = rdrd\theta$.

$$x^2+y^2 \overset{\begin{cases} x = r\cos(\theta) \\ y = r\sin(\theta) \end{cases}}{\Longrightarrow} (r\cos(\theta))^2 + (r\sin(\theta))^2 = r^2\cos^2(\theta) + r^2\sin^2(\theta)$$
$$= r^2\big(\cos^2(\theta) + \sin^2(\theta)\big)$$
$$= r^2 \times 1 = r^2$$

$$\frac{1}{\sqrt{x^2+y^2}} = \frac{1}{\sqrt{r^2}} = \frac{1}{r}$$

Therefore,

$$\int_0^{\sqrt{18}}\int_x^{\sqrt{36-x^2}} \frac{1}{\sqrt{x^2+y^2}}dydx = \int_{\frac{\pi}{4}}^{\frac{\pi}{2}}\int_0^6 \frac{1}{r} \times rdrd\theta = \int_{\frac{\pi}{4}}^{\frac{\pi}{2}}\int_0^6 drd\theta = \int_0^6 dr \int_{\frac{\pi}{4}}^{\frac{\pi}{2}} d\theta$$
$$= r\Big|_0^6 \times \theta\Big|_{\frac{\pi}{4}}^{\frac{\pi}{2}} = (6-0)\times\left(\frac{\pi}{2} - \frac{\pi}{4}\right)$$

$$= 6 \times \frac{\pi}{4} = \frac{3\pi}{2} \approx 4.7124$$

Example 1.39 Evaluate the double integral $\iint_R \ln\left(x^2 + y^2\right)dA$ over the region R depicted in Fig. 1.55.

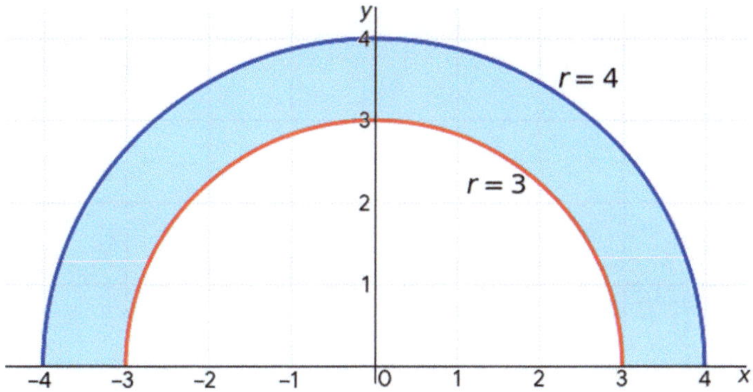

Fig. 1.55 Region R for Example 1.39

Solution Let's convert the double integral from Cartesian coordinates (x, y) and $dA = dxdy = dydx$ to polar coordinates (r, θ) and $dA = rdrd\theta$.

$$x^2 + y^2 \overset{\begin{cases} x = r\cos(\theta) \\ y = r\sin(\theta) \end{cases}}{\Longrightarrow} (r\cos(\theta))^2 + (r\sin(\theta))^2 = r^2\cos^2(\theta) + r^2\sin^2(\theta)$$
$$= r^2\left(\cos^2(\theta) + \sin^2(\theta)\right)$$
$$= r^2 \times 1 = r^2$$

$$\ln\left(x^2 + y^2\right) = \ln\left(r^2\right) = 2\ln(r)$$

Therefore,

$$\iint_R \ln\left(x^2 + y^2\right)dA = \int_0^{\pi}\int_3^4 2\ln(r) \times rdrd\theta$$
$$= \int_0^{\pi}\int_3^4 2r\ln(r)drd\theta$$
$$= \int_0^{\pi} d\theta \int_3^4 2r\ln(r)dr$$

$$= \pi \int_3^4 2r\ln(r)dr$$

The inner integral $\int_3^4 2r\ln(r)dr$ can be calculated using Tabular Integration (Integration by Parts) (Fig. 1.56).

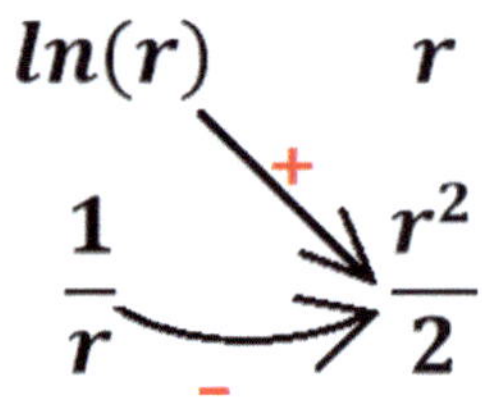

Fig. 1.56 Evaluation of $\int r\ln(r)dr$

Based on Fig. 1.56,

$$\int r\ln(r)dr = \frac{r^2}{2}\ln(r) - \int \frac{1}{r} \times \frac{r^2}{2}dr$$

$$\Rightarrow \int r\ln(r)dr = \frac{r^2}{2}\ln(r) - \int \frac{r}{2}dr$$

$$\Rightarrow \int r\ln(r)dr = \frac{r^2}{2}\ln(r) - \frac{r^2}{4}$$

$$\Rightarrow \int r\ln(r)dr = \frac{2r^2\ln(r) - r^2}{4}$$

Therefore,

$$\int 2r\ln(r)dr = 2\int r\ln(r)dr = \frac{2r^2\ln(r) - r^2}{2}$$

$$\int_3^4 2r\ln(r)dr = \left.\frac{2r^2\ln(r) - r^2}{2}\right|_3^4$$

$$= \frac{2\times 4^2\ln(4) - 4^2}{2} - \frac{2\times 3^2\ln(3) - 3^2}{2}$$

$$= \frac{32\ln(4) - 16}{2} - \frac{18\ln(3) - 9}{2}$$

$$= \frac{32\ln(4) - 18\ln(3) - 7}{2}$$

$$= 16\ln(4) - 9\ln(3) - \frac{7}{2} = 16\ln\left(2^2\right) - 9\ln(3) - \frac{7}{2}$$

$$= 32\ln(2) - 9\ln(3) - \frac{7}{2}$$

$$\approx 8.7932$$

and

$$\iint_R \ln(x^2 + y^2)dA = \pi \int_3^4 2r \ln(r)dr$$
$$= \pi \times \left(32 \ln(4) - 9 \ln(3) - \frac{7}{2}\right)$$
$$\approx 27.6246$$

Exercise 1.6 Evaluate the double integral $\iint_R \frac{1}{x^2+y^2} dA$ over the region R depicted in Fig. 1.57. (Ans. $\frac{3\pi}{4} \ln\left(\frac{4}{3}\right) \approx 0.6778$).

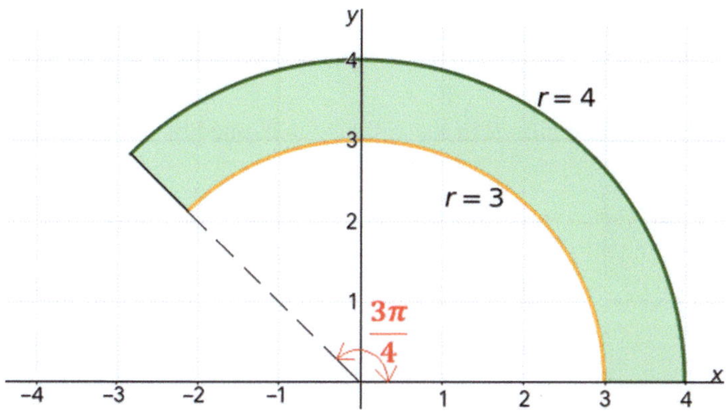

Fig. 1.57 Region R for Exercise 1.6

Example 1.40 Evaluate the double integral $\iint_R e^{\frac{x-y}{x+y}} dA$ over the region R depicted in Fig. 1.58.

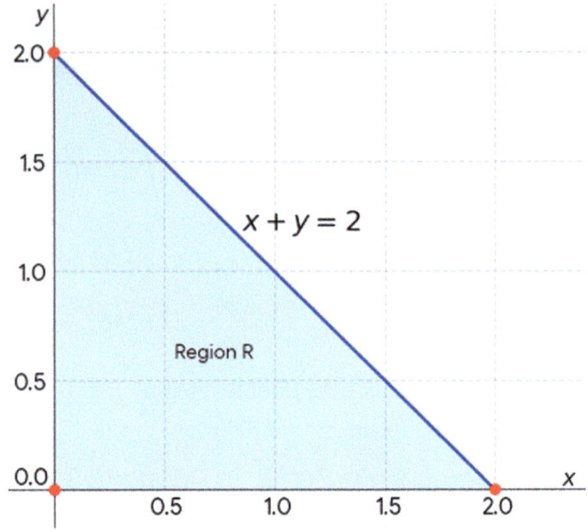

Fig. 1.58 Region R for Example 1.40

Solution The original region R in the xy-plane is defined by the inequalities:

$$x \geq 0$$

$$y \geq 0$$

$$x + y \leq 2$$

We use the transformation $u = x - y$ and $v = x + y$ to find the corresponding bounds for u and v in the new region S.

$$\begin{cases} x - y = u \\ x + y = v \end{cases} \Rightarrow \begin{bmatrix} 1 & -1 \\ 1 & 1 \end{bmatrix} \begin{bmatrix} x \\ y \end{bmatrix} = \begin{bmatrix} u \\ v \end{bmatrix}$$

$$\Rightarrow \begin{bmatrix} x \\ y \end{bmatrix} = \begin{bmatrix} 1 & -1 \\ 1 & 1 \end{bmatrix}^{-1} \begin{bmatrix} u \\ v \end{bmatrix}$$

$$\Rightarrow \begin{bmatrix} x \\ y \end{bmatrix} = \frac{1}{1 \times 1 - (-1 \times 1)} \begin{bmatrix} 1 & 1 \\ -1 & 1 \end{bmatrix} \begin{bmatrix} u \\ v \end{bmatrix}$$

$$\Rightarrow \begin{bmatrix} x \\ y \end{bmatrix} = \frac{1}{2} \begin{bmatrix} 1 & 1 \\ -1 & 1 \end{bmatrix} \begin{bmatrix} u \\ v \end{bmatrix} \Rightarrow \begin{bmatrix} x \\ y \end{bmatrix} = \frac{1}{2} \begin{bmatrix} u + v \\ -u + v \end{bmatrix}$$

$$\Rightarrow \begin{bmatrix} x \\ y \end{bmatrix} = \begin{bmatrix} \frac{u+v}{2} \\ \frac{v-u}{2} \end{bmatrix} \Rightarrow \begin{cases} x = \frac{u+v}{2} \\ y = \frac{v-u}{2} \end{cases}$$

$$x \geq 0 \Rightarrow \frac{u + v}{2} \geq 0 \Rightarrow u + v \geq 0 \Rightarrow u \geq -v$$

$$y \geq 0 \Rightarrow \frac{v-u}{2} \geq 0 \Rightarrow v-u \geq 0 \Rightarrow u \leq v$$

$$x+y \leq 2 \Rightarrow v \leq 2$$

Since x and y must be non-negative, the sum $x+y=v$ must also be non-negative ($v \geq 0$).

The new region S in the uv-plane is bounded by $S = \{(u,v) | -v \leq u \leq v, 0 \leq v \leq 2\}$ (Fig. 1.59).

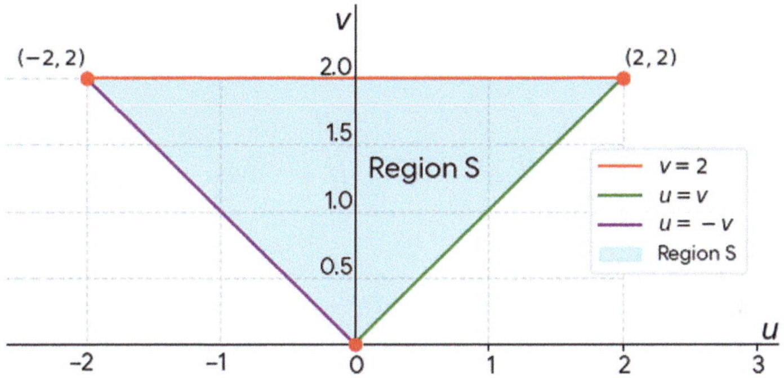

Fig. 1.59 Region S

Now, we compute the $\left|\det\left(\frac{\partial(x,y)}{\partial(u,v)}\right)\right|$:

$$\left|\det\left(\frac{\partial(x,y)}{\partial(u,v)}\right)\right| = \left|\det\left(\begin{bmatrix} \frac{\partial x}{\partial u} & \frac{\partial x}{\partial v} \\ \frac{\partial y}{\partial u} & \frac{\partial y}{\partial v} \end{bmatrix}\right)\right| = \left|\det\left(\begin{bmatrix} \frac{\partial}{\partial u}\left(\frac{u+v}{2}\right) & \frac{\partial}{\partial v}\left(\frac{u+v}{2}\right) \\ \frac{\partial}{\partial u}\left(\frac{v-u}{2}\right) & \frac{\partial}{\partial v}\left(\frac{v-u}{2}\right) \end{bmatrix}\right)\right|$$

$$= \left|\det\left(\begin{bmatrix} \frac{1}{2} & \frac{1}{2} \\ -\frac{1}{2} & \frac{1}{2} \end{bmatrix}\right)\right|$$

$$= \left|\frac{1}{2} \times \frac{1}{2} - \frac{1}{2} \times -\frac{1}{2}\right| = \left|\frac{1}{4} + \frac{1}{4}\right| = \left|\frac{1}{2}\right| = \frac{1}{2} = 0.5$$

The area element is:

$$dA = dxdy = dydx = \left|\det\left(\frac{\partial(x,y)}{\partial(u,v)}\right)\right| dudv = 0.5dudv$$

We now have all the necessary components to calculate the original integral in uv-coordinates. We will now calculate the integral using the order of integration $dvdu$ (Fig. 1.60).

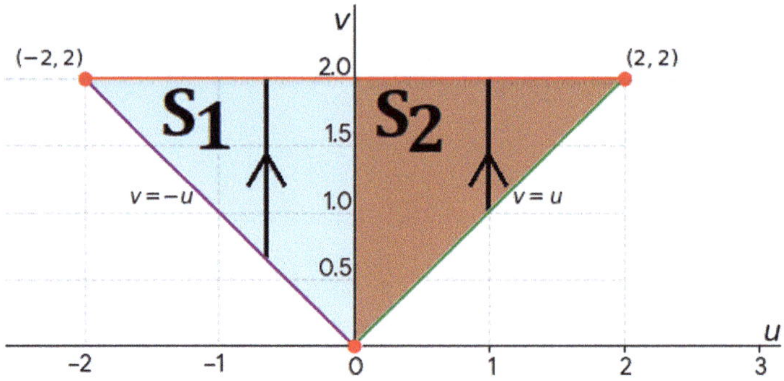

Fig. 1.60 Region S divided into S_1 and S_2

Based on Fig. 1.60,

$$\iint_R e^{\frac{x-y}{x+y}} dA = \iint_{S_1} e^{\frac{x-y}{x+y}} dvdu + \iint_{S_2} e^{\frac{x-y}{x+y}} dvdu$$

$$= \int_{-2}^{0}\int_{-u}^{2} e^{\frac{u}{v}} \times \frac{1}{2} dvdu + \int_{0}^{2}\int_{u}^{2} e^{\frac{u}{v}} \times \frac{1}{2} dvdu$$

$$= \frac{1}{2}\int_{-2}^{0}\int_{-u}^{2} e^{\frac{u}{v}} dvdu + \frac{1}{2}\int_{0}^{2}\int_{u}^{2} e^{\frac{u}{v}} dvdu$$

The inner integral, $\int e^{\frac{u}{v}} dv$, does not possess an elementary antiderivative (a closed-form expression in terms of standard functions). Therefore, evaluating the expression in the order *dvdu* is not feasible. Let's change the order of integration to *dudv* (Fig. 1.61).

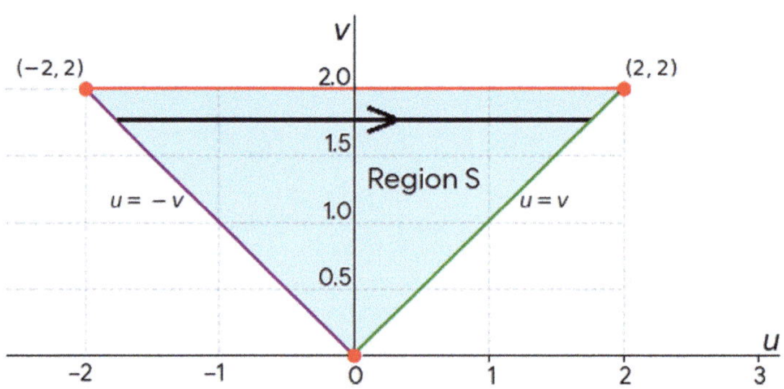

Fig. 1.61 Evaluation of the double integral using the order of integration *dudv*

Based on Fig. 1.61,

$$\iint_R e^{\frac{x-y}{x+y}}dA = \int_0^2 \int_{-v}^{v} e^{\frac{u}{v}} \times \frac{1}{2} dudv = \frac{1}{2}\int_0^2 \int_{-v}^{v} e^{\frac{u}{v}} dudv$$

The calculation of $\frac{1}{2}\int_0^2 \int_{-v}^{v} e^{\frac{u}{v}} dudv$ is shown below:

$$\int_{-v}^{v} e^{\frac{u}{v}} du = v\int_{-v}^{v} \frac{1}{v} e^{\frac{u}{v}} du = ve^{\frac{u}{v}}\Big|_{-v}^{v} = ve^{\frac{v}{v}} - ve^{\frac{-v}{v}} = ve - ve^{-1} = \left(e - e^{-1}\right)v$$

$$\frac{1}{2}\int_0^2 \int_{-v}^{v} e^{\frac{u}{v}} dudv = \frac{1}{2}\int_0^2 \left(e - e^{-1}\right)vdv = \frac{e - e^{-1}}{2}\int_0^2 vdv = \frac{e - e^{-1}}{2}\frac{v^2}{2}\Big|_0^2$$

$$= \frac{e - e^{-1}}{2}\frac{2^2}{2} - \frac{e - e^{-1}}{2}\frac{0^2}{2} = e - e^{-1} \approx 2.3504$$

Example 1.41 Evaluate the double integral $\iint_R (2x - y)dA$ where R denotes the parallelogram enclosed by the lines $y = 2x + 1, y = 2x - 3, y = -3x + 5$ and $y = -3x + 8$ (Fig. 1.62).

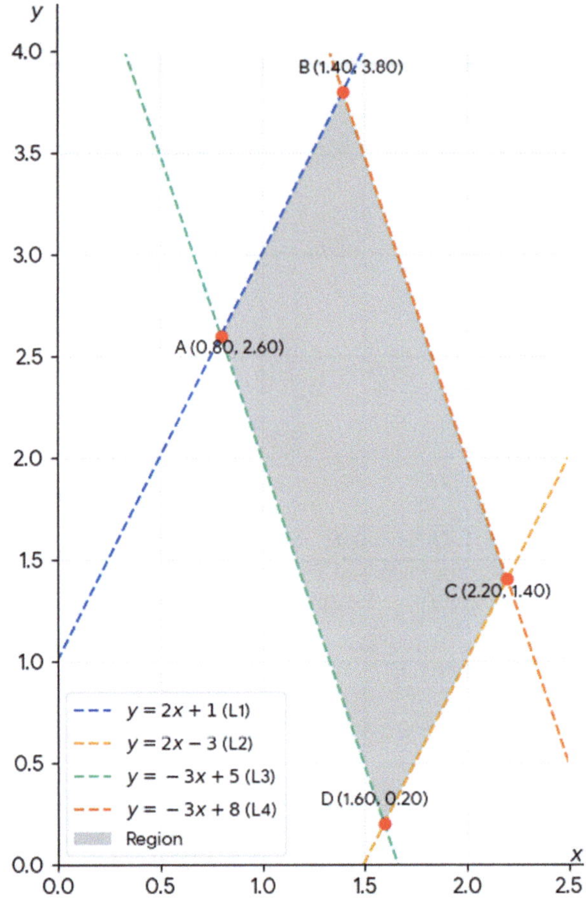

Fig. 1.62 Region R for Example 1.41

Solution Let $u = y - 2x$ and $v = y + 3x$. Then the four bounding lines become the constant-u and constant-v lines.

$$y = 2x + 1 \Rightarrow y - 2x = 1 \Rightarrow u = 1$$

$$y = 2x - 3 \Rightarrow y - 2x = -3 \Rightarrow u = -3$$

$$y = -3x + 5 \Rightarrow y + 3x = 5 \Rightarrow v = 5$$

$$y = -3x + 8 \Rightarrow y + 3x = 8 \Rightarrow v = 8$$

Hence the image of R in the (u, v)-plane is the rectangle:

$$S = \{(u, v)| -3 \leq u \leq 1, 5 \leq v \leq 8\}$$

Let's solve for x and y in terms of u and v:

$$\begin{cases} u = y - 2x \\ v = y + 3x \end{cases} \Rightarrow \begin{bmatrix} -2 & 1 \\ 3 & 1 \end{bmatrix} \begin{bmatrix} x \\ y \end{bmatrix} = \begin{bmatrix} u \\ v \end{bmatrix}$$

$$\Rightarrow \begin{bmatrix} x \\ y \end{bmatrix} = \begin{bmatrix} -2 & 1 \\ 3 & 1 \end{bmatrix}^{-1} \begin{bmatrix} u \\ v \end{bmatrix}$$

$$\Rightarrow \begin{bmatrix} x \\ y \end{bmatrix} = \frac{1}{-2 \times 1 - 1 \times 3} \begin{bmatrix} 1 & -1 \\ -3 & -2 \end{bmatrix} \begin{bmatrix} u \\ v \end{bmatrix}$$

$$\Rightarrow \begin{bmatrix} x \\ y \end{bmatrix} = \frac{1}{-5} \begin{bmatrix} 1 & -1 \\ -3 & -2 \end{bmatrix} \begin{bmatrix} u \\ v \end{bmatrix}$$

$$\Rightarrow \begin{bmatrix} x \\ y \end{bmatrix} = \frac{1}{-5} \begin{bmatrix} u - v \\ -3u - 2v \end{bmatrix}$$

$$\Rightarrow \begin{bmatrix} x \\ y \end{bmatrix} = \begin{bmatrix} \frac{v-u}{5} \\ \frac{3u+2v}{5} \end{bmatrix}$$

$$\Rightarrow \begin{cases} x = \frac{v-u}{5} \\ y = \frac{3u+2v}{5} \end{cases}$$

Hence the integrand becomes:

$$2x - y = 2\left(\frac{v-u}{5}\right) - \left(\frac{3u+2v}{5}\right) = \frac{2v - 2u - 3u - 2v}{5} = -u$$

The absolute value of the Jacobian determinant ($\left|\det\left(\frac{\partial(x,y)}{\partial(u,v)}\right)\right|$) is:

$$\left|\det\left(\frac{\partial(x,y)}{\partial(u,v)}\right)\right| = \left|\det\left(\begin{bmatrix} \frac{\partial x}{\partial u} & \frac{\partial x}{\partial v} \\ \frac{\partial y}{\partial u} & \frac{\partial y}{\partial v} \end{bmatrix}\right)\right| = \left|\det\left(\begin{bmatrix} \frac{\partial}{\partial u}\left(\frac{v-u}{5}\right) & \frac{\partial}{\partial v}\left(\frac{v-u}{5}\right) \\ \frac{\partial}{\partial u}\left(\frac{3u+2v}{5}\right) & \frac{\partial}{\partial v}\left(\frac{3u+2v}{5}\right) \end{bmatrix}\right)\right|$$

$$= \left|\det\left(\begin{bmatrix} -\frac{1}{5} & \frac{1}{5} \\ \frac{3}{5} & \frac{2}{5} \end{bmatrix}\right)\right| = \left|-\frac{1}{5} \times \frac{2}{5} - \frac{1}{5} \times \frac{3}{5}\right|$$

$$= \left|-\frac{2}{25} - \frac{3}{25}\right| = \left|-\frac{5}{25}\right| = \left|-\frac{1}{5}\right| = \frac{1}{5}$$

The area element is:

$$dA = dxdy = dydx = \left|\det\left(\frac{\partial(x,y)}{\partial(u,v)}\right)\right| dudv = \frac{1}{5} dudv$$

Therefore,

$$\iint_R (2x - y)dA = \iint_S -u \times \frac{1}{5}dudv = -\frac{1}{5}\int_5^8 \int_{-3}^1 ududv = -\frac{1}{5}\int_{-3}^1 udu \int_5^8 dv$$

$$= -\frac{1}{5} \times \left(\frac{1^2}{2} - \frac{(-3)^2}{2}\right) \times (8 - 5) = -\frac{1}{5} \times -4 \times 3$$

$$= \frac{12}{5} = 2.4$$

Example 1.42 Evaluate the double integral $\iint_R \left(\sqrt{\frac{y}{x}} + \sqrt{xy}\right)dA$ where R denotes the region enclosed by the curves $xy = 3$, $xy = 6$, $y = x$ and $y = 6x$.

Solution Let $u = xy$ and $v = \frac{y}{x}$. Then the four bounding curves become the constant-u and constant-v lines.

$$xy = 3 \Rightarrow u = 3$$

$$xy = 6 \Rightarrow u = 6$$

$$y = x \Rightarrow \frac{y}{x} = 1 \Rightarrow v = 1$$

$$y = 6x \Rightarrow \frac{y}{x} = 6 \Rightarrow v = 6$$

Hence the image of R in the (u, v)-plane is the rectangle:

$$S = \{(u, v)|3 \leq u \leq 6, 1 \leq v \leq 6\}$$

Hence the integrand becomes:

$$\sqrt{\frac{y}{x}} + \sqrt{xy} = \sqrt{v} + \sqrt{u}$$

Since the functions $u = xy$ and $v = \frac{y}{x}$ are given explicitly in terms of x and y, it is easier to compute the $\left|\det\left(\frac{\partial(x,y)}{\partial(u,v)}\right)\right|$ using the identity:

$$\left|\det\left(\frac{\partial(x, y)}{\partial(u, v)}\right)\right| = \frac{1}{\left|\det\left(\frac{\partial(u,v)}{\partial(x,y)}\right)\right|} = \frac{1}{\left|\det\left(\begin{bmatrix} \frac{\partial u}{\partial x} & \frac{\partial u}{\partial y} \\ \frac{\partial v}{\partial x} & \frac{\partial v}{\partial y} \end{bmatrix}\right)\right|}$$

The partial derivatives are:

$$\frac{\partial u}{\partial x} = \frac{\partial}{\partial x}(xy) = y$$

$$\frac{\partial u}{\partial y} = \frac{\partial}{\partial y}(xy) = x$$

$$\frac{\partial v}{\partial x} = \frac{\partial}{\partial x}\left(\frac{y}{x}\right) = -\frac{y}{x^2}$$

$$\frac{\partial v}{\partial y} = \frac{\partial}{\partial y}\left(\frac{y}{x}\right) = \frac{1}{x}$$

Therefore,

$$\left|\det\left(\frac{\partial(x,y)}{\partial(u,v)}\right)\right| = \frac{1}{\left|\det\left(\frac{\partial(u,v)}{\partial(x,y)}\right)\right|} = \frac{1}{\left|\det\left(\begin{bmatrix}\frac{\partial u}{\partial x} & \frac{\partial u}{\partial y}\\ \frac{\partial v}{\partial x} & \frac{\partial v}{\partial y}\end{bmatrix}\right)\right|} = \frac{1}{\left|\det\left(\begin{bmatrix} y & x \\ -\frac{y}{x^2} & \frac{1}{x}\end{bmatrix}\right)\right|}$$
$$= \frac{1}{\left|y \times \frac{1}{x} - x \times -\frac{y}{x^2}\right|} = \frac{1}{\left|\frac{y}{x} + \frac{y}{x}\right|} = \frac{1}{\left|2\frac{y}{x}\right|} = \frac{1}{|2v|} = \frac{1}{2v}$$

and

$$\iint_R \left(\sqrt{\frac{y}{x}} + \sqrt{xy}\right) dA = \int_1^6 \int_3^6 (\sqrt{v} + \sqrt{u}) \frac{1}{2v} dudv = \frac{1}{2} \int_1^6 \int_3^6 (\sqrt{v} + \sqrt{u}) \frac{1}{v} dudv$$
$$= \frac{1}{2} \int_1^6 \int_3^6 \left(v^{-\frac{1}{2}} + v^{-1} u^{\frac{1}{2}}\right) dudv$$

The calculation of $\frac{1}{2}\int_1^6 \int_3^6 \left(v^{-\frac{1}{2}} + v^{-1}u^{\frac{1}{2}}\right) dudv$ is shown below:

$$\int_3^6 \left(v^{-\frac{1}{2}} + v^{-1} u^{\frac{1}{2}}\right) du = v^{-\frac{1}{2}} u + v^{-1} \frac{u^{\frac{1}{2}+1}}{\frac{1}{2}+1}\Bigg|_3^6 = v^{-\frac{1}{2}} u + v^{-1} \times \frac{2}{3} u^{\frac{3}{2}}\Bigg|_3^6$$
$$= \left(v^{-\frac{1}{2}} \times 6 + v^{-1} \times \frac{2}{3} 6^{\frac{3}{2}}\right) - \left(v^{-\frac{1}{2}} \times 3 + v^{-1} \times \frac{2}{3} 3^{\frac{3}{2}}\right)$$
$$= 3v^{-\frac{1}{2}} + 2\left(2\sqrt{2} - 1\right)\sqrt{3} v^{-1}$$

$$\frac{1}{2} \int_1^6 \int_3^6 \left(v^{-\frac{1}{2}} + v^{-1} u^{\frac{1}{2}}\right) dudv = \frac{1}{2} \int_1^6 \left(3v^{-\frac{1}{2}} + 2\left(2\sqrt{2} - 1\right)\sqrt{3} v^{-1}\right) dv$$

$$= \frac{3}{2}\int_1^6 v^{-\frac{1}{2}}dv + \frac{2\left(2\sqrt{2}-1\right)\sqrt{3}}{2}\int_1^6 v^{-1}dv$$

$$= \frac{3}{2}\int_1^6 v^{-\frac{1}{2}}dv + \left(2\sqrt{2}-1\right)\sqrt{3}\int_1^6 v^{-1}dv$$

$$= \frac{6}{2}v^{\frac{1}{2}} + \left(2\sqrt{2}-1\right)\sqrt{3}\ln(|v|)\Big|_1^6$$

$$= \left(\frac{6}{2}6^{\frac{1}{2}} + \left(2\sqrt{2}-1\right)\sqrt{3}\ln(|6|)\right)$$

$$- \left(\frac{6}{2}1^{\frac{1}{2}} + \left(2\sqrt{2}-1\right)\sqrt{3}\ln(|1|)\right)$$

$$= \left(3\sqrt{6} + \left(2\sqrt{2}-1\right)\sqrt{3}\ln(|6|)\right)$$

$$- (3+0) = 3\left(\sqrt{6}-1\right) + \left(2\sqrt{2}-1\right)\sqrt{3}\ln(6)$$

$$\approx 10.0228$$

Therefore,

$$\iint_R \left(\sqrt{\frac{y}{x}} + \sqrt{xy}\right)dA = 3\left(\sqrt{6}-1\right) + \left(2\sqrt{2}-1\right)\sqrt{3}\ln(6) \approx 10.0228$$

Exercise 1.7 Evaluate the double integral $\iint_R\left(\frac{x^2}{y^6}\right)dA$ where R denotes the region enclosed by the curves $xy = 3$, $xy = 6$, $y^3 = x$ and $y^3 = 6x$. (Ans. $\frac{35}{96} \approx 0.3646$).

Exercise 1.8 Evaluate the double integral $\iint_R\left(\frac{2x-3y}{2x+3y}\right)^2 dA$ where R denotes the region enclosed by the curves $2x - 3y = 3$, $2x - 3y = 6$, $2x + 3y = 11$ and $2x + 3y = 16$. (Ans. $\frac{105}{704} \approx 0.1491$).

Example 1.43 Calculate the area bounded by the cardioid $r = 1 + \sin(\theta)$ and the circle $r = 1$ (Fig. 1.63).

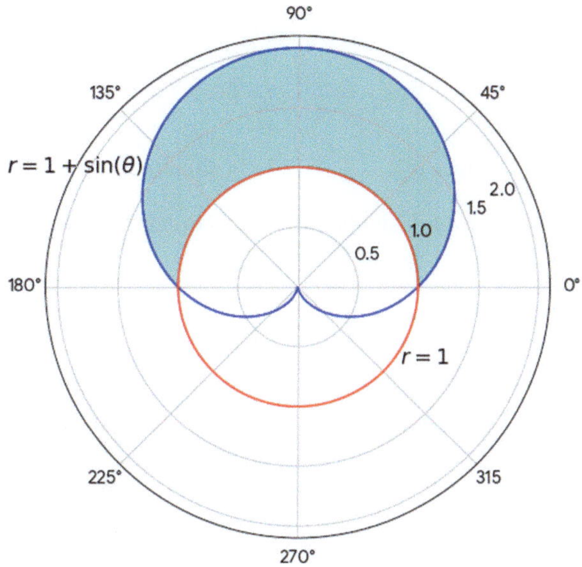

Fig. 1.63 Area bounded by the cardioid $r = 1 + \sin(\theta)$ and the circle $r = 1$

Solution The region between the circle $r = 1$ and the cardioid $r = 1 + \sin(\theta)$ is traced for θ from 0 to π (they meet at $\theta = 0, \pi$) with r running from 1 out to $1 + \sin(\theta)$. Hence the area is:

$$A = \iint\limits_R 1dA = \int_0^{\pi}\int_1^{1+\sin(\theta)} 1 \times rdrd\theta = \int_0^{\pi}\int_1^{1+\sin(\theta)} rdrd\theta$$

The calculation of $\int_0^{\pi}\int_1^{1+\sin(\theta)} rdrd\theta$ is shown below:

$$\int_1^{1+\sin(\theta)} rdr = \frac{r^2}{2}\Bigg|_1^{1+\sin(\theta)} = \frac{(1+\sin(\theta))^2}{2} - \frac{1^2}{2}$$

$$= \frac{1 + 2\sin(\theta) + \sin^2(\theta) - 1}{2} = \sin(\theta) + \frac{\sin^2(\theta)}{2}$$

$$\int_0^{\pi}\int_1^{1+\sin(\theta)} rdrd\theta = \int_0^{\pi}\left(\sin(\theta) + \frac{\sin^2(\theta)}{2}\right)d\theta = \int_0^{\pi}\left(\sin(\theta) + \frac{\frac{1-\cos(2\theta)}{2}}{2}\right)d\theta$$

$$= \int_0^{\pi}\left(\sin(\theta) + \frac{1-\cos(2\theta)}{4}\right)d\theta = -\cos(\theta) + \frac{1}{4}\theta - \frac{\sin(2\theta)}{8}\Bigg|_0^{\pi}$$

$$= \left(-\cos(\pi) + \frac{1}{4} \times \pi - \frac{\sin(2\pi)}{8}\right)$$

$$-\left(-\cos(0)+\frac{1}{4}\times 0-\frac{\sin(2\times 0)}{8}\right)$$
$$=1+\frac{\pi}{4}-(-1)=2+\frac{\pi}{4}\approx 2.7854$$

So the area is $2+\frac{\pi}{4}\approx 2.7854$ square units.

Example 1.44 Calculate the area bounded by the cardioid $r=1+\sin(\theta)$ and the circle $r=1$ (Fig. 1.64).

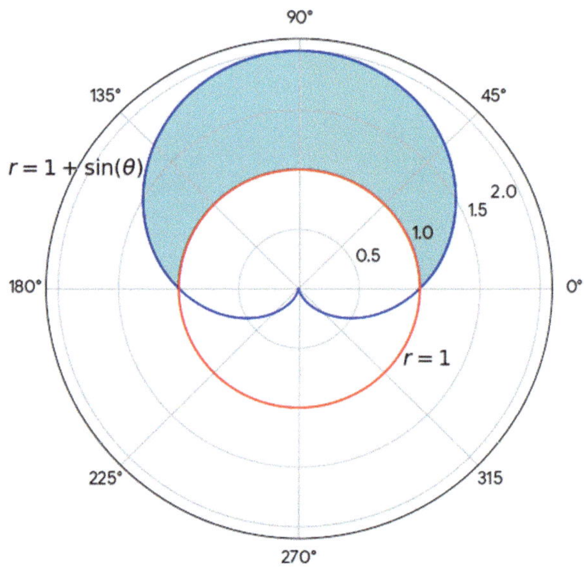

Fig. 1.64 Area bounded by the cardioid $r=1+\sin(\theta)$ and the circle $r=1$

Solution This problem was solved in the previous example. We will now resolve this problem using single-variable integrals.

The area (A) of a region bounded by two polar curves, an outer curve $r_{outer}=f(\theta)$ and an inner curve $r_{inner}=g(\theta)$, between the radial lines $\theta=a$ and $\theta=b$, is given by:

$$A=\int_a^b \frac{1}{2}\left(f(\theta)^2-g(\theta)^2\right)d\theta$$

Therefore,

$$A=\int_0^\pi \frac{1}{2}((1+\sin(\theta))^2-1^2)d\theta=\frac{1}{2}\int_0^\pi \left(1+2\sin(\theta)+\sin^2(\theta)-1\right)d\theta$$

$$= \frac{1}{2}\int_{0}^{\pi} \left(2\sin(\theta) + \sin^2(\theta)\right)d\theta = \int_{0}^{\pi} \sin(\theta)d\theta + \frac{1}{2}\int_{0}^{\pi} \sin^2(\theta)d\theta$$

$$= \int_{0}^{\pi} \sin(\theta)d\theta + \frac{1}{2}\int_{0}^{\pi} \frac{1-\cos(2\theta)}{2}d\theta$$

$$= \int_{0}^{\pi} \sin(\theta)d\theta + \frac{1}{4}\int_{0}^{\pi} d\theta - \frac{1}{4}\int_{0}^{\pi} \cos(2\theta)d\theta$$

$$= -\cos(\theta)|_0^{\pi} + \frac{1}{4}\theta\bigg|_0^{\pi} - \frac{1}{8}\sin(2x)\bigg|_0^{\pi} = \cos(\theta)|_{\pi}^{0} + \frac{1}{4}\theta\bigg|_0^{\pi} + \frac{1}{8}\sin(2x)\bigg|_{\pi}^{0}$$

$$= (\cos(0) - \cos(\pi)) + \left(\frac{\pi}{4} - \frac{0}{4}\right) + \left(\frac{\sin(2\times 0)}{8} - \frac{\sin(2\times\pi)}{8}\right)$$

$$= (1-(-1)) + \frac{\pi}{4} + 0 = 2 + \frac{\pi}{4} \approx 2.7854$$

The value obtained is identical to the result from the preceding problem.

Example 1.45 Calculate the total area enclosed by the complete lemniscate $r^2 = 18\sin(2\theta)$ (Fig. 1.65).

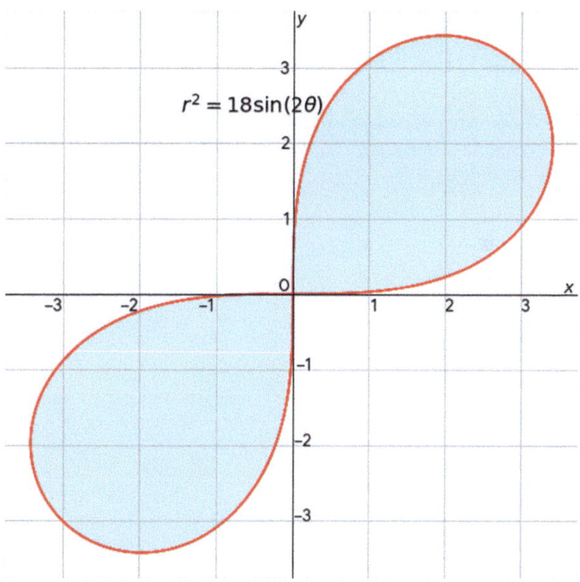

Fig. 1.65 Area enclosed by the complete lemniscate $r^2 = 18\sin(2\theta)$

Solution One complete lobe of the lemniscate is defined by the angular range $0 \leq \theta \leq \frac{\pi}{2}$. The double integral for the area of the first lobe (A_1) is:

$$A_1 = \int_0^{\frac{\pi}{2}} \int_0^{\sqrt{18\sin(2\theta)}} 1 \times r dr d\theta = \int_0^{\frac{\pi}{2}} \int_0^{\sqrt{18\sin(2\theta)}} r dr d\theta$$

The calculation of $\int_0^{\frac{\pi}{2}} \int_0^{\sqrt{18\sin(2\theta)}} r dr d\theta$ is shown below:

$$\int_0^{\sqrt{18\sin(2\theta)}} r dr = \frac{r^2}{2}\Bigg|_0^{\sqrt{18\sin(2\theta)}} = \frac{\left(\sqrt{18\sin(2\theta)}\right)^2}{2} - \frac{0^2}{2}$$

$$= \frac{18\sin(2\theta)}{2} - 0 = 9\sin(2\theta)$$

$$A_1 = \int_0^{\frac{\pi}{2}} \int_0^{\sqrt{18\sin(2\theta)}} r dr d\theta = \int_0^{\frac{\pi}{2}} 9\sin(2\theta) d\theta = -\frac{9}{2}\cos(2\theta)\Bigg|_0^{\frac{\pi}{2}} = \frac{9}{2}\cos(2\theta)\Bigg|_{\frac{\pi}{2}}^{0}$$

$$= \frac{9}{2}\cos(2 \times 0) - \frac{9}{2}\cos\left(2 \times \frac{\pi}{2}\right) = \frac{9}{2} - \frac{9}{2}\cos(\pi) = \frac{9}{2} + \frac{9}{2} = 9$$

The area of one lobe is 9 square units. The complete lemniscate consists of two identical lobes, so the total area is: $A_{total} = 2A_1 = 2 \times 9 = 18$.

Example 1.46 Calculate the total area enclosed by the complete lemniscate $r^2 = 18\sin(2\theta)$ (Fig. 1.66).

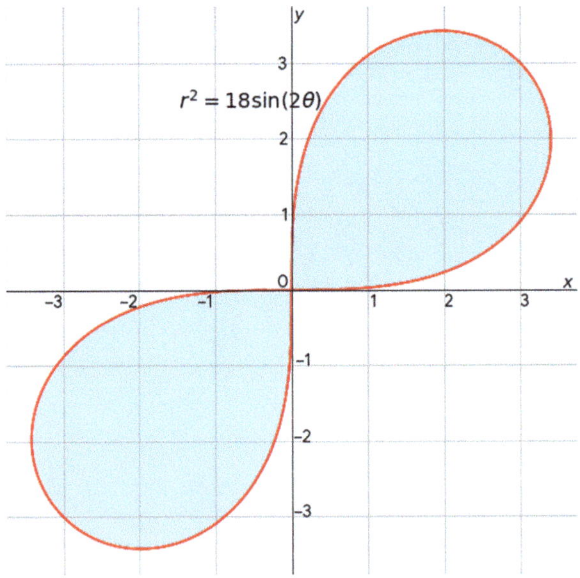

Fig. 1.66 Area enclosed by the complete lemniscate $r^2 = 18\sin(2\theta)$

Solution This problem was solved in the previous example. We will now resolve this problem using single-variable integrals.

The formula for the area enclosed by a polar curve $r = f(\theta)$ is:

$$A = \int_a^b \frac{1}{2} f(\theta)^2 d\theta$$

For one lobe (A_{Lobe}) from $\theta = 0$ to $\theta = \frac{\pi}{2}$:

$$\begin{aligned} A_{Lobe} &= \int_0^{\frac{\pi}{2}} \frac{1}{2} \times 18\sin(2\theta)d\theta = 9\int_0^{\frac{\pi}{2}} \sin(2\theta)d\theta \\ &= -\frac{9}{2}\cos(2\theta)\Big|_0^{\frac{\pi}{2}} = \frac{9}{2}\cos(2\theta)\Big|_{\frac{\pi}{2}}^{0} \\ &= \frac{9}{2}\cos(2\times 0) - \frac{9}{2}\cos\left(2\times\frac{\pi}{2}\right) = \frac{9}{2}\times 1 - \frac{9}{2}\times -1 = 9 \end{aligned}$$

Therefore,

$$A = 2A_{Lobe} = 2\times 9 = 18$$

Example 1.47 Calculate the total area enclosed by the complete lemniscate $r^2 = 18\cos(2\theta)$ (Fig. 1.67).

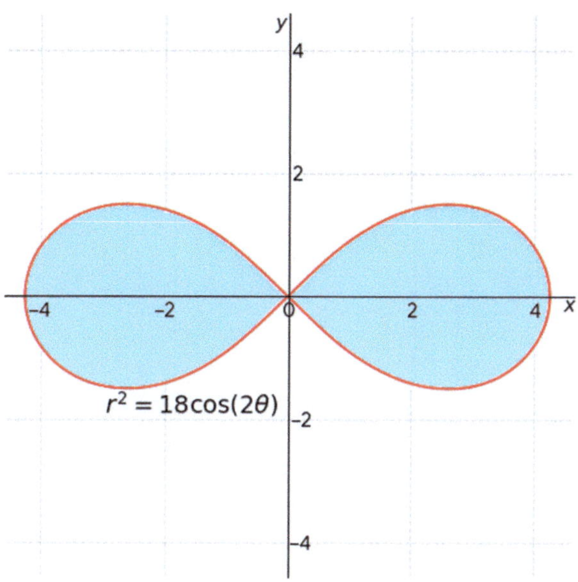

Fig. 1.67 Area enclosed by the complete lemniscate $r^2 = 18\cos(2\theta)$

Solution The region of integration R consists of two lobes. Due to symmetry, we can calculate the area of one lobe and multiply it by 2. For the right lobe, the limits are r from 0 to $\sqrt{18\cos(2\theta)}$ and θ from $-\frac{\pi}{4}$ to $\frac{\pi}{4}$. Therefore,

$$A_{RightLobe} = \int_{-\frac{\pi}{4}}^{\frac{\pi}{4}} \int_{0}^{\sqrt{18\cos(2\theta)}} 1 \times r dr d\theta = \int_{-\frac{\pi}{4}}^{\frac{\pi}{4}} \int_{0}^{\sqrt{18\cos(2\theta)}} r dr d\theta$$

The calculation of $\int_{-\frac{\pi}{4}}^{\frac{\pi}{4}} \int_{0}^{\sqrt{18\cos(2\theta)}} r dr d\theta$ is shown below:

$$\int_{0}^{\sqrt{18\cos(2\theta)}} r dr = \frac{r^2}{2}\Bigg|_{0}^{\sqrt{18\cos(2\theta)}} = \frac{\sqrt{18\cos(2\theta)}^2}{2} - \frac{0^2}{2}$$

$$= \frac{18\cos(2\theta)}{2} - \frac{0}{2} = 9\cos(2\theta)$$

$$\int_{-\frac{\pi}{4}}^{\frac{\pi}{4}} \int_{0}^{\sqrt{18\cos(2\theta)}} r dr d\theta = \int_{-\frac{\pi}{4}}^{\frac{\pi}{4}} 9\cos(2\theta) d\theta = \frac{9}{2}\sin(2\theta)\Bigg|_{-\frac{\pi}{4}}^{\frac{\pi}{4}}$$

$$= \frac{9}{2}\sin\left(2 \times \frac{\pi}{4}\right) - \frac{9}{2}\sin\left(-2 \times \frac{\pi}{4}\right)$$

$$= \frac{9}{2}\sin\left(\frac{\pi}{2}\right) - \frac{9}{2}\sin\left(-\frac{\pi}{2}\right)$$

$$= \frac{9}{2} - \frac{9}{2} \times -1 = \frac{9}{2} + \frac{9}{2} = 18$$

Exercise 1.9 Verify the result of Example 1.47 using single-variable integration.

Example 1.48 Calculate the area between the polar curves $r = 3\cos(\theta)$ and $r = 3(1 + \cos(\theta))$ (Fig. 1.68).

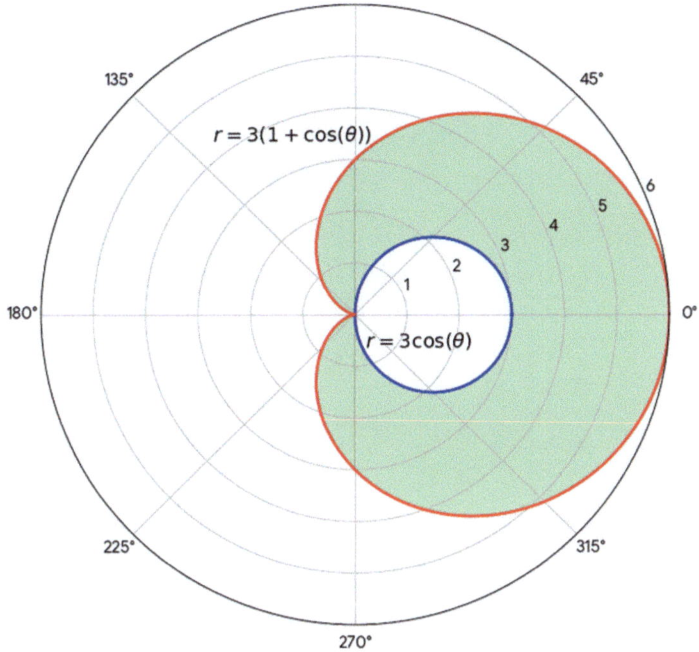

Fig. 1.68 Area between the polar curves $r = 3\cos(\theta)$ and $r = 3(1 + \cos(\theta))$

Solution Due to symmetry, we calculate the area of half the region (Fig. 1.69) and multiply the result by two.

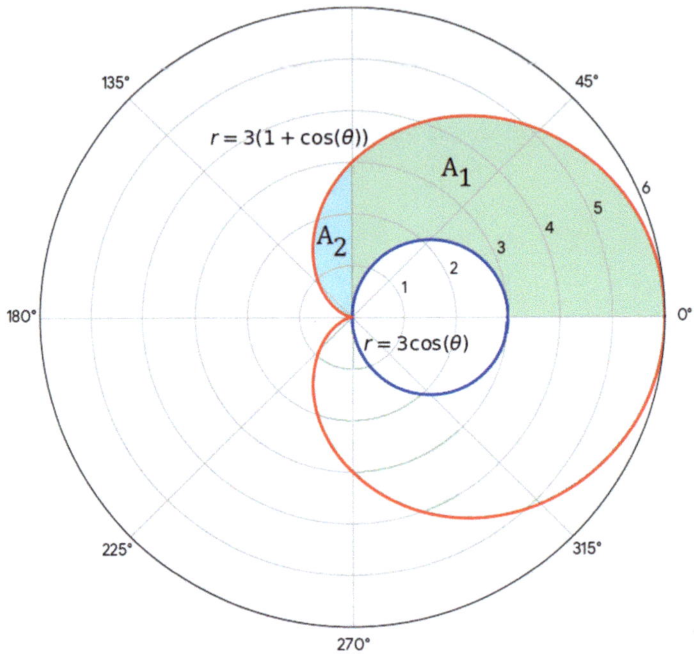

Fig. 1.69 Regions A_1 and A_2

Based on Fig. 1.69,

$$A_1 = \int_0^{\frac{\pi}{2}} \int_{3\cos(\theta)}^{3(1+\cos(\theta))} 1 \times r dr d\theta = \int_0^{\frac{\pi}{2}} \int_{3\cos(\theta)}^{3(1+\cos(\theta))} r dr d\theta$$

$$A_2 = \int_{\frac{\pi}{2}}^{\pi} \int_0^{3(1+\cos(\theta))} 1 \times r dr d\theta = \int_{\frac{\pi}{2}}^{\pi} \int_0^{3(1+\cos(\theta))} r dr d\theta$$

Let's calculate each integral individually. The calculation of $A_1 = \int_0^{\frac{\pi}{2}} \int_{3\cos(\theta)}^{3(1+\cos(\theta))} r dr d\theta$ is shown below:

$$\int_{3\cos(\theta)}^{3(1+\cos(\theta))} r dr = \frac{r^2}{2}\bigg|_{3\cos(\theta)}^{3(1+\cos(\theta))} = \frac{(3(1+\cos(\theta)))^2}{2} - \frac{(3\cos(\theta))^2}{2}$$

$$= \frac{9\left(1+2\cos(\theta)+\cos^2(\theta)\right) - 9\cos^2(\theta)}{2} = \frac{9}{2} + 9cos(\theta)$$

$$A_1 = \int_0^{\frac{\pi}{2}} \int_{3\cos(\theta)}^{3(1+\cos(\theta))} r dr d\theta = \int_0^{\frac{\pi}{2}} \left(\frac{9}{2} + 9\cos(\theta)\right) d\theta = \frac{9}{2}\theta + 9\sin(\theta)\Big|_0^{\frac{\pi}{2}}$$

$$= \left(\frac{9}{2} \times \frac{\pi}{2} + 9\sin\left(\frac{\pi}{2}\right)\right) - \left(\frac{9}{2} \times 0 + 9\sin(0)\right) = \frac{9\pi}{4} + 9 \approx 16.0686$$

The calculation of $A_2 = \int_{\frac{\pi}{2}}^{\pi} \int_0^{3(1+\cos(\theta))} r dr d\theta$ is shown below:

$$\int_0^{3(1+\cos(\theta))} r dr = \frac{r^2}{2}\Big|_0^{3(1+\cos(\theta))} = \frac{(3(1+\cos(\theta)))^2}{2} - \frac{0^2}{2}$$

$$= \frac{9}{2} + 9\cos(\theta) + \frac{9}{2}\cos^2(\theta)$$

$$A_2 = \int_{\frac{\pi}{2}}^{\pi} \int_0^{3(1+\cos(\theta))} r dr d\theta = \int_{\frac{\pi}{2}}^{\pi} \left(\frac{9}{2} + 9\cos(\theta) + \frac{9}{2}\cos^2(\theta)\right) d\theta$$

$$= \int_{\frac{\pi}{2}}^{\pi} \left(\frac{9}{2} + 9\cos(\theta) + \frac{9}{2} \times \frac{1+\cos(2\theta)}{2}\right) d\theta$$

$$= \int_{\frac{\pi}{2}}^{\pi} \left(\frac{27}{4} + 9\cos(\theta) + \frac{9}{4}\cos(2\theta)\right) d\theta = \left(\frac{27}{4}\theta + 9\sin(\theta) + \frac{9}{8}\sin(2\theta)\right)\Big|_{\frac{\pi}{2}}^{\pi}$$

$$= \left(\frac{27}{4} \times \pi + 9\sin(\pi) + \frac{9}{8}\sin(2 \times \pi)\right)$$

$$- \left(\frac{27}{4} \times \frac{\pi}{2} + 9\sin\left(\frac{\pi}{2}\right) + \frac{9}{8}\sin\left(2 \times \frac{\pi}{2}\right)\right)$$

$$= \left(\frac{27\pi}{4} + 0 + 0\right) - \left(\frac{27\pi}{8} + 9 + 0\right) = \frac{27\pi}{8} - 9 \approx 1.6029$$

Therefore,

$$A = 2(A_1 + A_2) = 2\left(\frac{9\pi}{4} + 9 + \frac{27\pi}{8} - 9\right) = \frac{45\pi}{4} \approx 35.3429$$

Example 1.49 Evaluate the double integral $\iint_R \cos(\theta) r dr d\theta$ over the region R depicted in Fig. 1.70.

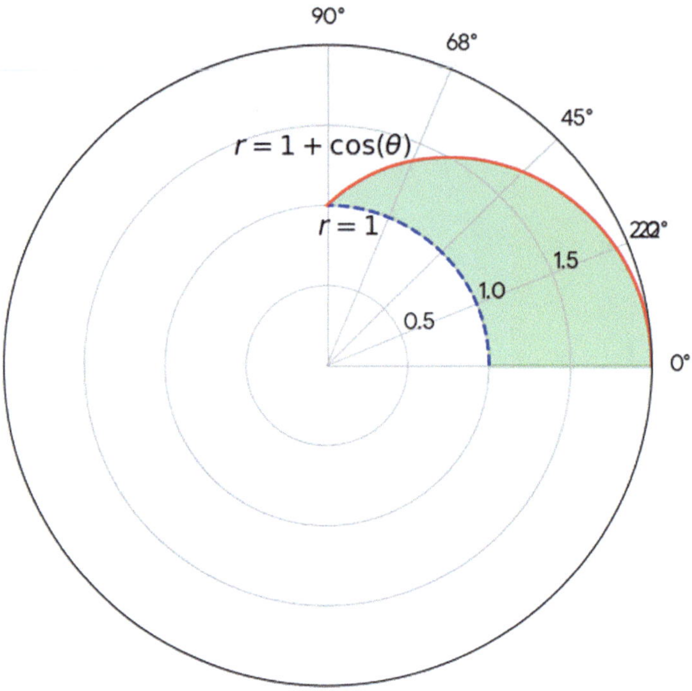

Fig. 1.70 Region R for Example 1.49

Solution Based on Fig. 1.70,

$$\iint_R \cos(\theta) r dr d\theta = \int_0^{\frac{\pi}{2}} \int_1^{1+\cos(\theta)} \cos(\theta) r dr d\theta$$

The calculation of $\int_0^{\frac{\pi}{2}} \int_1^{1+\cos(\theta)} \cos(\theta) r dr d\theta$ is shown below:

$$\int_1^{1+\cos(\theta)} \cos(\theta) r dr = \cos(\theta) \int_1^{1+\cos(\theta)} r dr = \cos(\theta) \frac{r^2}{2} \Bigg|_1^{1+\cos(\theta)}$$

$$= \cos(\theta) \frac{(1+\cos(\theta))^2}{2} - \cos(\theta) \frac{1^2}{2} = \frac{\cos^3(\theta)}{2} + \cos^2(\theta)$$

$$\iint_R \cos(\theta) r dr d\theta = \int_0^{\frac{\pi}{2}} \int_1^{1+\cos(\theta)} \cos(\theta) r dr d\theta = \int_0^{\frac{\pi}{2}} \left(\frac{\cos^3(\theta)}{2} + \cos^2(\theta) \right) d\theta$$

$$= \int_0^{\frac{\pi}{2}} \left(\frac{\cos^3(\theta)}{2} + \frac{1 + \cos(2\theta)}{2} \right) d\theta$$

$$= \frac{1}{2} \int_0^{\frac{\pi}{2}} \cos^3(\theta) d\theta + \frac{1}{2} \int_0^{\frac{\pi}{2}} d\theta + \frac{1}{2} \int_0^{\frac{\pi}{2}} \cos(2\theta) d\theta$$

Let's calculate each integral individually.

$$\frac{1}{2} \int_0^{\frac{\pi}{2}} \cos^3(\theta) d\theta = \frac{1}{2} \int_0^{\frac{\pi}{2}} \cos(\theta) \cos^2(\theta) d\theta = \frac{1}{2} \int_0^{\frac{\pi}{2}} \cos(\theta) \left(1 - \sin^2(\theta)\right) d\theta$$

$$= \frac{1}{2} \int_0^{\frac{\pi}{2}} \cos(\theta) d\theta - \frac{1}{2} \int_0^{\frac{\pi}{2}} \cos(\theta) \sin^2(\theta) d\theta$$

$$= \frac{1}{2} \sin(\theta) \Big|_0^{\frac{\pi}{2}} - \frac{1}{6} \sin^3(\theta) \Big|_0^{\frac{\pi}{2}}$$

$$= \frac{1}{2} \sin(\theta) \Big|_0^{\frac{\pi}{2}} + \frac{1}{6} \sin^3(\theta) \Big|_{\frac{\pi}{2}}^{0}$$

$$= \frac{1}{2} \sin\left(\frac{\pi}{2}\right) - \frac{1}{2} \sin(0) + \frac{1}{6} \sin^3(0) - \frac{1}{6} \sin^3\left(\frac{\pi}{2}\right)$$

$$= \frac{1}{2} - 0 + 0 - \frac{1}{6} = \frac{3-1}{6} = \frac{2}{6} = \frac{1}{3} \approx 0.3333$$

$$\frac{1}{2} \int_0^{\frac{\pi}{2}} d\theta = \frac{1}{2} \left(\frac{\pi}{2} - 0 \right) = \frac{\pi}{4}$$

$$\frac{1}{2} \int_0^{\frac{\pi}{2}} \cos(2\theta) d\theta = \frac{1}{4} \sin(2\theta) \Big|_0^{\frac{\pi}{2}} = \frac{1}{4} \sin\left(2 \times \frac{\pi}{2}\right) - \frac{1}{4} \sin(2 \times 0) = 0 - 0 = 0$$

Therefore,

$$\iint_R \cos(\theta) r dr d\theta = \frac{1}{2} \int_0^{\frac{\pi}{2}} \cos^3(\theta) d\theta + \frac{1}{2} \int_0^{\frac{\pi}{2}} d\theta + \frac{1}{2} \int_0^{\frac{\pi}{2}} \cos(2\theta) d\theta$$

$$= \frac{1}{3} + \frac{\pi}{4} + 0 = \frac{4 + 3\pi}{12} \approx 1.1187$$

Example 1.50 Find the volume (V) of the tetrahedron (a four-sided solid) E bounded by the four planes (Fig. 1.71):

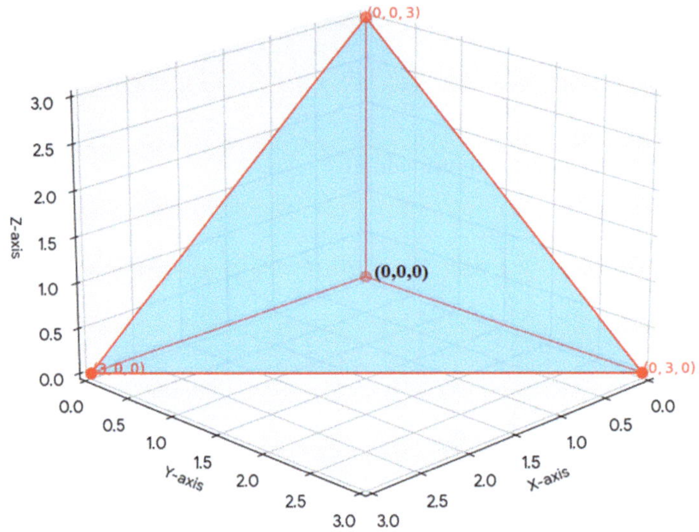

Fig. 1.71 Tetrahedron for Example 1.50

1. $x = 0$ (the yz-plane)
2. $y = 0$ (the xz-plane)
3. $z = 0$ (the xy-plane)
4. $x + y + z = 3$ (a plane intersecting all axes)

Solution The volume of a solid bounded above by a surface $z = f(x, y)$ and below by a region R in the xy-plane can be found using the double integral formula:

$$V = \iint_R f(x, y)dA$$

The tetrahedron E is defined by the inequalities $x \geq 0, y \geq 0, z \geq 0$, and $x + y + z \leq 3$. The top surface of the tetrahedron is the plane $x + y + z = 3$. Solving for z:

$$z = f(x, y) = 3 - x - y$$

The region of integration R is the projection of the tetrahedron onto the xy-plane ($z = 0$). This region is a triangle bounded by $x = 0$, $y = 0$, and the line of intersection of the plane $x + y + z = 3$ with the xy-plane ($z = 0$), which is $x + y = 3$. The region R is defined as:

$$R = \{(x, y)|0 \leq x \leq 3, 0 \leq y \leq 3 - x\}$$

The volume integral is set up as an iterated integral:

$$V = \iint_R f(x, y)dA = \iint_R (3 - x - y)dA = \int_0^3 \int_0^{3-x} (3 - x - y)dydx$$

The calculation of $V = \int_0^3 \int_0^{3-x} (3 - x - y)dydx$ is shown below:

$$\int_0^{3-x} (3 - x - y)dy = \left(3y - xy - \frac{y^2}{2}\right)\Bigg|_0^{3-x}$$

$$= \left(3(3 - x) - x(3 - x) - \frac{(3 - x)^2}{2}\right) - \left(3 \times 0 - x \times 0 - \frac{0^2}{2}\right)$$

$$= \frac{x^2}{2} - 3x + \frac{9}{2}$$

$$V = \int_0^3 \int_0^{3-x} (3 - x - y)dydx = \int_0^3 \left(\frac{x^2}{2} - 3x + \frac{9}{2}\right)dx = \left(\frac{x^3}{6} - \frac{3}{2}x^2 + \frac{9}{2}x\right)\Bigg|_0^3$$

$$= \left(\frac{3^3}{6} - \frac{3}{2} \times 3^2 + \frac{9}{2} \times 3\right) - \left(\frac{0^3}{6} - \frac{3}{2} \times 0^2 + \frac{9}{2} \times 0\right) = \frac{27}{6} = \frac{9}{2} = 4.5$$

The volume of the tetrahedron is $\frac{9}{2} = 4.5$ cubic units.

Let's verify this result. The region E is a tetrahedron in the first octant, and its vertices are $(0, 0, 0)$, $(3, 0, 0)$, $(0, 3, 0)$, and $(0, 0, 3)$.

For a tetrahedron defined by the intercepts a, b, c on the positive axes and the origin, the volume V is given by the formula $V = \frac{1}{6}abc$. In this case $a = b = c = 3$.

$$V = \frac{1}{6} \times 3 \times 3 \times 3 = \frac{27}{6} = \frac{9}{2} = 4.5$$

This confirms the result obtained via the double integral method.

Example 1.51 Calculate the volume (V) of the solid shown in Fig. 1.72.

Fig. 1.72 A quarter-cylinder

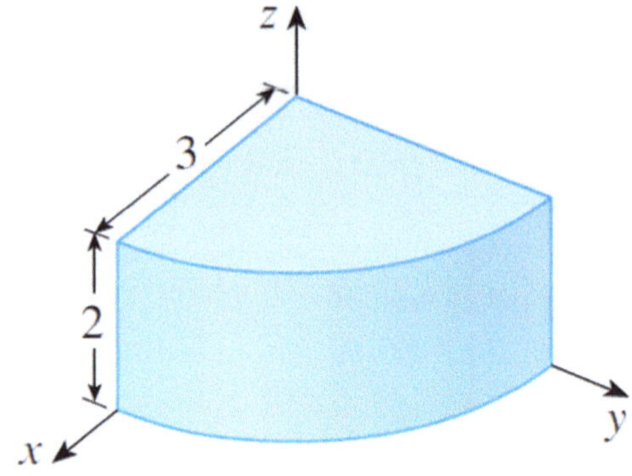

Solution The volume of a solid bounded above by a surface $z = f(x, y)$ and below by a region R in the xy-plane can be found using the double integral formula:

$$V = \iint\limits_R f(x, y)dA$$

In this problem we need to calculate the double integral $V = \iint_R 2dA$, where the region R is shown in Fig. 1.73.

Fig. 1.73 Projection onto the xy-plane

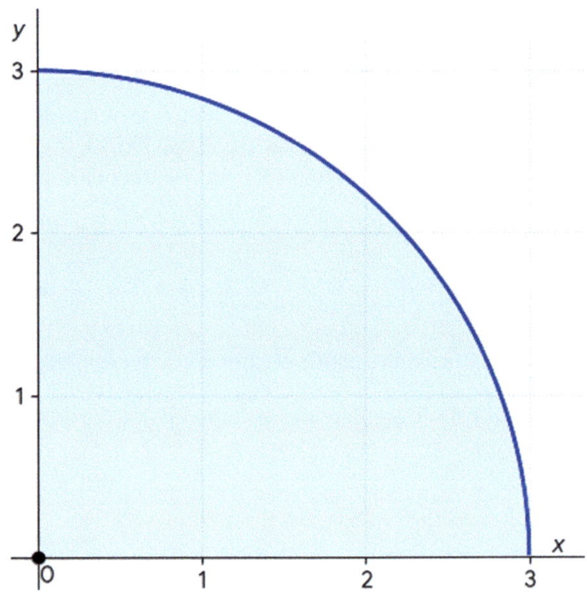

The problem is best solved using polar coordinates.

$$V = \int_{0}^{\frac{\pi}{2}} \int_{0}^{3} 2 \times r dr d\theta = \int_{0}^{3} 2 r dr \int_{0}^{\frac{\pi}{2}} d\theta = r^2 \Big|_{0}^{3} \times \theta \Big|_{0}^{\frac{\pi}{2}}$$
$$= \left(3^2 - 0^2\right) \times \left(\frac{\pi}{2} - 0\right) = \frac{9\pi}{2} \approx 14.1372$$

Let's verify this result. A complete cylinder with a base radius r and height h has a volume $V_{cylinder}$ given by the formula:

$$V_{cylinder} = \pi r^2 h$$

The solid in Fig. 1.73 is a quarter of a cylinder and its volume is $V = \frac{V_{cylinder}}{4} = \frac{\pi r^2 h}{4}$. In this example, $r = 3$ and $h = 2$. Therefore,

$$V = \frac{\pi \times 3^2 \times 2}{4} = \frac{9\pi}{2} \approx 14.1372$$

This confirms the result obtained via the double integral method.

Example 1.52 Calculate the volume (V) of the solid shown in Fig. 1.74.

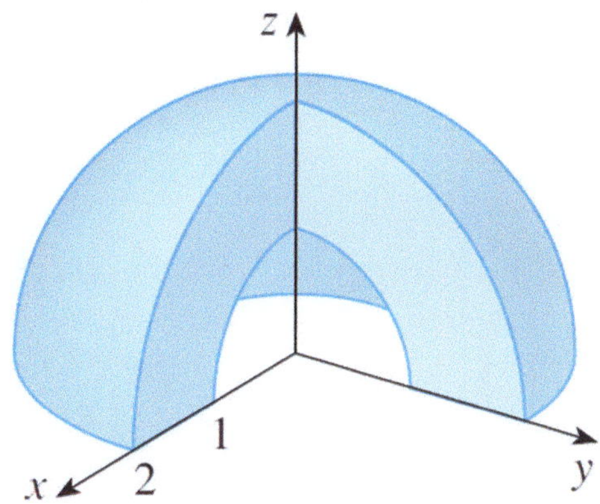

Fig. 1.74 Solid region for Example 1.52

Solution The solid region is bounded by the concentric spheres $x^2 + y^2 + z^2 = 2^2$ and $x^2 + y^2 + z^2 = 1^2$.

The volume V can be computed by integrating $f_1(x, y) = \sqrt{2^2 - \left(x^2 + y^2\right)} - \sqrt{1^2 - \left(x^2 + y^2\right)}$ over the region R_1 shown in Fig. 1.75, and integrating $f_2(x, y) = \sqrt{2^2 - \left(x^2 + y^2\right)}$ over the region R_2 shown in Fig. 1.76.

Fig. 1.75 The region R_1

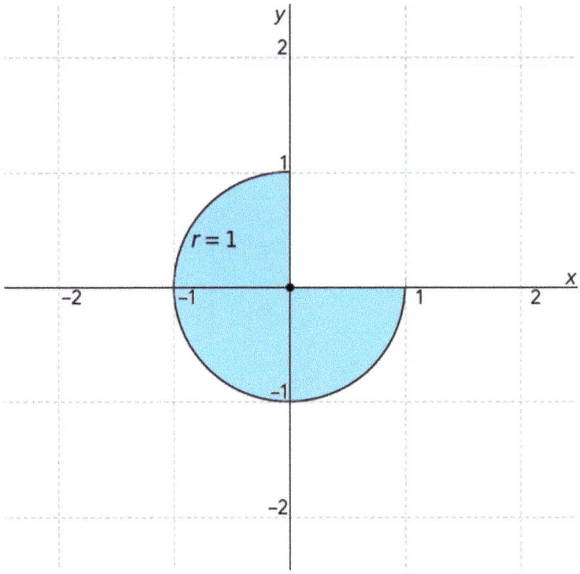

Fig. 1.76 The region R_2

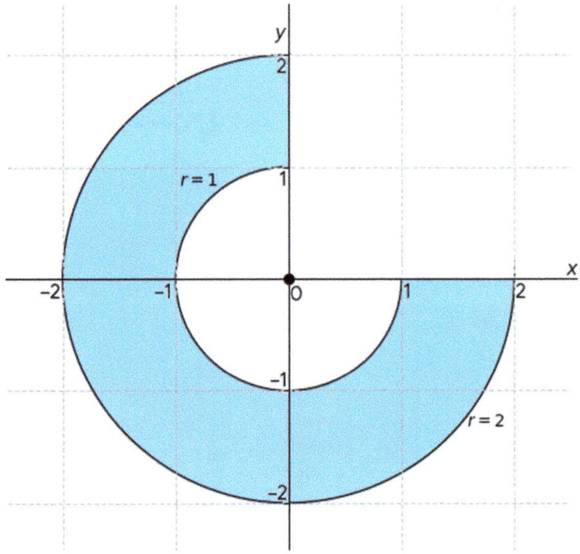

The problem is best solved using polar coordinates.

$$f_1(x, y) = \sqrt{2^2 - (x^2 + y^2)} - \sqrt{1^2 - (x^2 + y^2)}$$

$$\underset{\Longrightarrow}{\begin{cases} x = r\cos(\theta) \\ y = r\sin(\theta) \end{cases}} \quad f_1(r, \theta) = \sqrt{4 - \left(r^2\cos^2(\theta) + r^2\sin^2(\theta)\right)} - \sqrt{1 - \left(r^2\cos^2(\theta) + r^2\sin^2(\theta)\right)}$$

$$= \sqrt{4 - r^2\left(\cos^2(\theta) + \sin^2(\theta)\right)} - \sqrt{1 - r^2\left(\cos^2(\theta) + \sin^2(\theta)\right)}$$
$$= \sqrt{4 - r^2} - \sqrt{1 - r^2}$$

$$f_2(x, y) = \sqrt{2^2 - \left(x^2 + y^2\right)}$$

$$\begin{cases} x = r\cos(\theta) \\ y = r\sin(\theta) \end{cases} \Longrightarrow \quad f_2(r, \theta) = \sqrt{4 - \left(r^2\cos^2(\theta) + r^2\sin^2(\theta)\right)}$$
$$= \sqrt{4 - r^2\left(\cos^2(\theta) + \sin^2(\theta)\right)} = \sqrt{4 - r^2}$$

$$V = \iint\limits_{R_1} f_1(r, \theta)dA + \iint\limits_{R_2} f_2(r, \theta)dA$$
$$= \int_{\frac{\pi}{2}}^{2\pi}\int_0^1 \left(\sqrt{4 - r^2} - \sqrt{1 - r^2}\right) \times rdrd\theta + \int_{\frac{\pi}{2}}^{2\pi}\int_1^2 \sqrt{4 - r^2} \times rdrd\theta$$
$$= \int_{\frac{\pi}{2}}^{2\pi}\int_0^1 r\sqrt{4 - r^2}drd\theta - \int_{\frac{\pi}{2}}^{2\pi}\int_0^1 r\sqrt{1 - r^2}drd\theta + \int_{\frac{\pi}{2}}^{2\pi}\int_1^2 r\sqrt{4 - r^2}drd\theta$$

Let's calculate each integral individually.

$$\int_{\frac{\pi}{2}}^{2\pi}\int_0^1 r\sqrt{4 - r^2}drd\theta = \int_0^1 r\sqrt{4 - r^2}dr \int_{\frac{\pi}{2}}^{2\pi} d\theta$$
$$= \frac{1}{-2}\int_0^1 -2r\sqrt{4 - r^2}dr \int_{\frac{\pi}{2}}^{2\pi} d\theta$$
$$= \frac{1}{2}\int_1^0 -2r\sqrt{4 - r^2}dr \int_{\frac{\pi}{2}}^{2\pi} d\theta$$
$$= \left(\frac{1}{2}\frac{(4 - r^2)^{\frac{1}{2}+1}}{\frac{1}{2}+1}\Bigg|_1^0\right) \times \left(2\pi - \frac{\pi}{2}\right)$$
$$= \left(\frac{(4 - r^2)^{\frac{3}{2}}}{3}\Bigg|_1^0\right) \times \frac{3\pi}{2}$$

$$= \left(\frac{\left(4-0^2\right)^{\frac{3}{2}}}{3} - \frac{\left(4-1^2\right)^{\frac{3}{2}}}{3}\right) \times \frac{3\pi}{2}$$

$$= \frac{8-3\sqrt{3}}{3} \times \frac{3\pi}{2}$$

$$= \frac{\left(8-3\sqrt{3}\right)\pi}{2}$$

$$\int_{\frac{\pi}{2}}^{2\pi}\int_{0}^{1} r\sqrt{1-r^2}drd\theta = \int_{0}^{1} r\sqrt{1-r^2}dr \int_{\frac{\pi}{2}}^{2\pi} d\theta$$

$$= \frac{1}{-2}\int_{0}^{1} -2r\sqrt{1-r^2}dr \int_{\frac{\pi}{2}}^{2\pi} d\theta$$

$$= \frac{1}{2}\int_{1}^{0} -2r\sqrt{1-r^2}dr \int_{\frac{\pi}{2}}^{2\pi} d\theta$$

$$= \left(\frac{1}{2}\frac{\left(1-r^2\right)^{\frac{1}{2}+1}}{\frac{1}{2}+1}\Bigg|_{1}^{0}\right) \times \left(2\pi - \frac{\pi}{2}\right)$$

$$= \left(\frac{\left(1-r^2\right)^{\frac{3}{2}}}{3}\Bigg|_{1}^{0}\right) \times \frac{3\pi}{2}$$

$$= \left(\frac{\left(1-0^2\right)^{\frac{3}{2}}}{3} - \frac{\left(1-1^2\right)^{\frac{3}{2}}}{3}\right) \times \frac{3\pi}{2}$$

$$= \frac{1}{3} \times \frac{3\pi}{2} = \frac{\pi}{2}$$

$$\int_{\frac{\pi}{2}}^{2\pi}\int_{1}^{2} r\sqrt{4-r^2}drd\theta = \int_{1}^{2} r\sqrt{4-r^2}dr \int_{\frac{\pi}{2}}^{2\pi} d\theta$$

$$= \frac{1}{-2}\int_{1}^{2} -2r\sqrt{4-r^2}dr \int_{\frac{\pi}{2}}^{2\pi} d\theta$$

$$= \frac{1}{2}\int_{2}^{1} -2r\sqrt{4-r^2}dr \int_{\frac{\pi}{2}}^{2\pi} d\theta$$

$$= \left(\frac{1}{2} \frac{(4 - r^2)^{\frac{1}{2}+1}}{\frac{1}{2}+1} \Bigg|_1^2 \right) \times \left(2\pi - \frac{\pi}{2} \right)$$

$$= \left(\frac{(4 - r^2)^{\frac{3}{2}}}{3} \Bigg|_1^2 \right) \times \frac{3\pi}{2}$$

$$= \left(\frac{(4 - 1^2)^{\frac{3}{2}}}{3} - \frac{(4 - 2^2)^{\frac{3}{2}}}{3} \right) \times \frac{3\pi}{2}$$

$$= \frac{3\sqrt{3}}{3} \times \frac{3\pi}{2}$$

$$= \frac{3\sqrt{3}\pi}{2}$$

Therefore,

$$V = \iint\limits_{R_1} f_1(r, \theta) dA + \iint\limits_{R_2} f_2(r, \theta) dA$$

$$= \int_{\frac{\pi}{2}}^{2\pi} \int_0^1 r\sqrt{4 - r^2} dr d\theta - \int_{\frac{\pi}{2}}^{2\pi} \int_0^1 r\sqrt{1 - r^2} dr d\theta + \int_{\frac{\pi}{2}}^{2\pi} \int_1^2 r\sqrt{4 - r^2} dr d\theta$$

$$= \frac{\left(8 - 3\sqrt{3}\right)\pi}{2} - \frac{\pi}{2} + \frac{3\sqrt{3}\pi}{2} = 4\pi - \frac{\pi}{2} = \frac{7\pi}{2} \approx 10.9956$$

Example 1.53 Find the volume of the solid that lies above the cone $z = \sqrt{x^2 + y^2}$ and below the sphere $x^2 + y^2 + z^2 = z$ (Fig. 1.77).

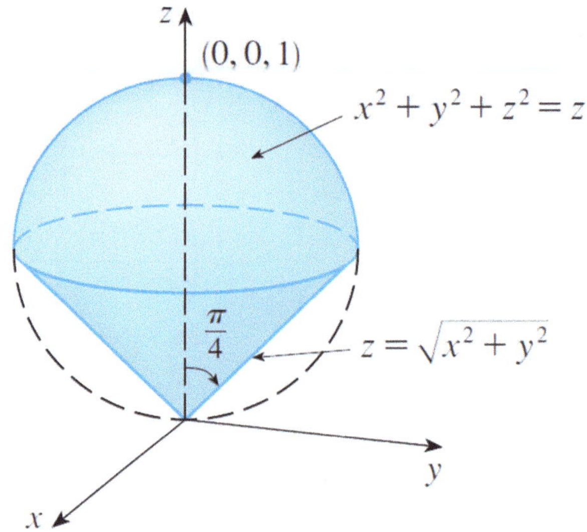

Fig. 1.77 Solid region for Example 1.53

Solution To find the volume (V) using a double integral, we typically use the formula:

$$\iint_R \big(z_{top}(x, y) - z_{bottom}(x, y)\big)dA$$

where R is the projection of the solid onto the xy-plane, z_{top} is the upper surface, and z_{bottom} is the lower surface. In this example:

Lower Surface (Cone): $z_{bottom} = \sqrt{x^2 + y^2}$
Upper Surface (Sphere):

$$\begin{aligned}
x^2 + y^2 + z^2 = z &\Rightarrow x^2 + y^2 + z^2 - z = 0 \\
&\Rightarrow x^2 + y^2 + \left(z - \frac{1}{2}\right)^2 - \frac{1}{4} = 0 \\
&\Rightarrow \left(z - \frac{1}{2}\right)^2 = \frac{1}{4} - \left(x^2 + y^2\right) \\
&\Rightarrow z - \frac{1}{2} = \pm\sqrt{\frac{1}{4} - \left(x^2 + y^2\right)} \\
&\Rightarrow z = \frac{1}{2} \pm \sqrt{\frac{1}{4} - \left(x^2 + y^2\right)} \\
&\Rightarrow z = \frac{1 \pm \sqrt{1 - 4\left(x^2 + y^2\right)}}{2} \\
&\Rightarrow z_{top} = \frac{1 + \sqrt{1 - 4\left(x^2 + y^2\right)}}{2}
\end{aligned}$$

The surfaces $z_{top} = \frac{1+\sqrt{1-4(x^2+y^2)}}{2}$ and $z_{bottom} = \sqrt{x^2+y^2}$ are shown in Fig. 1.78.

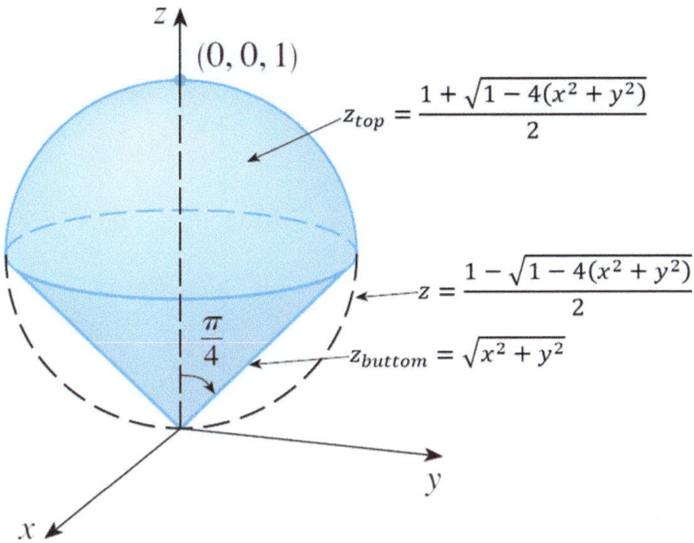

Fig. 1.78 Equations of the top and bottom surfaces

The region R in the xy-plane is defined by the intersection of the cone and the sphere. Set the two z equations equal to find the intersection:

$$\begin{cases} z = \sqrt{x^2+y^2} \\ x^2+y^2+z^2 = z \end{cases} \Rightarrow x^2+y^2+\left(\sqrt{x^2+y^2}\right)^2 = \sqrt{x^2+y^2}$$

$$\Rightarrow x^2+y^2+x^2+y^2 = \sqrt{x^2+y^2}$$

$$\Rightarrow 2(x^2+y^2) = \sqrt{x^2+y^2}$$

$$\Rightarrow 4(x^2+y^2)^2 = x^2+y^2$$

$$\Rightarrow 4(x^2+y^2)^2 - (x^2+y^2) = 0$$

$$\Rightarrow (x^2+y^2)(4(x^2+y^2)-1) = 0$$

$$\Rightarrow \begin{cases} x^2+y^2 = 0 \\ 4(x^2+y^2)-1 = 0 \end{cases}$$

$$\Rightarrow \begin{cases} x^2+y^2 = 0 \\ x^2+y^2 = \frac{1}{4} \end{cases}$$

The equation $x^2+y^2 = 0$ is satisfied only at the origin (0, 0), whereas $x^2+y^2 = \frac{1}{4}$ defines a circle centered at the origin (0, 0) with a radius of $\frac{1}{2}$. The region R is bounded by this circle (Fig. 1.79).

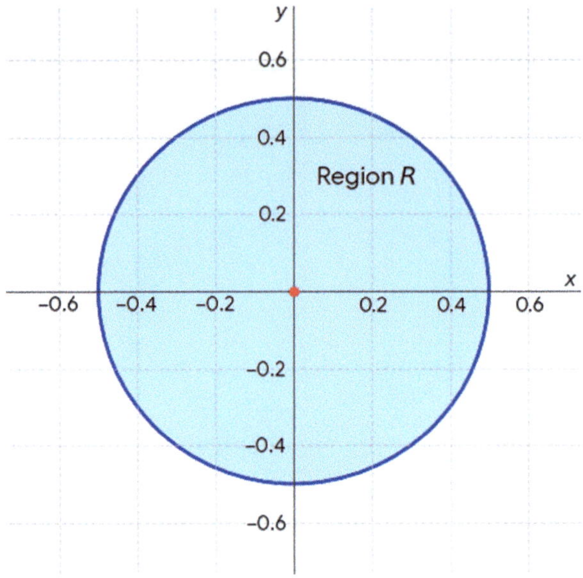

Fig. 1.79 Projection of the solid onto the xy-plane

The volume integral is:

$$V = \iint_R \big(z_{top}(x, y) - z_{bottom}(x, y)\big) dA$$

$$\Rightarrow V = \iint_R \left(\frac{1 + \sqrt{1 - 4\left(x^2 + y^2\right)}}{2} - \sqrt{x^2 + y^2} \right) dA$$

Since the region R is a disk, we switch to polar coordinates, where $x^2 + y^2 = r^2$ and $dA = r dr d\theta$. The region R in polar coordinates is: $R = \left\{ (r, \theta) | 0 \le r \le \frac{1}{2}, 0 \le \theta \le 2\pi \right\}$. Therefore,

$$\begin{aligned} V &= \iint_R \left(\frac{1 + \sqrt{1 - 4\left(x^2 + y^2\right)}}{2} - \sqrt{x^2 + y^2} \right) dA \\ &= \int_0^{2\pi} \int_0^{\frac{1}{2}} \left(\frac{1 + \sqrt{1 - 4r^2}}{2} - \sqrt{r^2} \right) r dr d\theta \\ &= \int_0^{2\pi} \int_0^{\frac{1}{2}} \left(\frac{1 + \sqrt{1 - 4r^2}}{2} - r \right) r dr d\theta \end{aligned}$$

$$= \int_0^{2\pi}\int_0^{\frac{1}{2}} \left(\frac{r + r\sqrt{1-4r^2}}{2} - r^2 \right) dr d\theta$$

First, evaluate the inner r-integral:

$$\int_0^{\frac{1}{2}} \left(\frac{r + r\sqrt{1-4r^2}}{2} - r^2 \right) dr = \int_0^{\frac{1}{2}} \frac{r}{2} dr + \int_0^{\frac{1}{2}} \frac{r\sqrt{1-4r^2}}{2} dr - \int_0^{\frac{1}{2}} r^2 dr$$

$$= \frac{1}{2}\int_0^{\frac{1}{2}} r dr + \frac{1}{2\times -8}\int_0^{\frac{1}{2}} -8r\sqrt{1-4r^2} dr - \int_0^{\frac{1}{2}} r^2 dr$$

$$= \frac{1}{2}\int_0^{\frac{1}{2}} r dr + \frac{1}{16}\int_{\frac{1}{2}}^{0} -8r\sqrt{1-4r^2} dr + \int_{\frac{1}{2}}^{0} r^2 dr$$

$$= \left.\frac{r^2}{4}\right|_0^{\frac{1}{2}} + \frac{1}{16}\left.\frac{\left(1-4r^2\right)^{\frac{1}{2}+1}}{\frac{1}{2}+1}\right|_{\frac{1}{2}}^{0} + \left.\frac{r^3}{3}\right|_{\frac{1}{2}}^{0}$$

$$= \left.\frac{r^2}{4}\right|_0^{\frac{1}{2}} + \left.\frac{\left(1-4r^2\right)^{\frac{3}{2}}}{24}\right|_{\frac{1}{2}}^{0} + \left.\frac{r^3}{3}\right|_{\frac{1}{2}}^{0}$$

$$= \left(\frac{\left(\frac{1}{2}\right)^2}{4} - \frac{0^2}{4} \right)$$

$$+ \left(\frac{\left(1-4\times 0^2\right)^{\frac{3}{2}}}{24} - \frac{\left(1-4\times\left(\frac{1}{2}\right)^2\right)^{\frac{3}{2}}}{24} \right)$$

$$+ \left(\frac{0^3}{3} - \frac{\left(\frac{1}{2}\right)^3}{3} \right)$$

$$= \frac{1}{16} + \frac{1}{24} - \frac{1}{24} = \frac{1}{16}$$

Finally, evaluate the outer θ-integral:

$$V = \int_0^{2\pi}\int_0^{\frac{1}{2}} \left(\frac{r + r\sqrt{1-4r^2}}{2} - r^2 \right) dr d\theta$$

$$= \int_{0}^{2\pi} \frac{1}{16} d\theta = \frac{1}{16} \int_{0}^{2\pi} d\theta = \frac{1}{16} \times (2\pi - 0) = \frac{\pi}{8}$$

$$\approx 0.3927$$

Example 1.54 Calculate the volume of the solid bounded by the surfaces $z = 0, y = 0, x = 2y^2$, and $x + 2y + z = 4$ in the region $y \geq 0$.

Solution The volume (V) of the solid is given by:

$$V = \iint_R \big(z_{top}(x, y) - z_{bottom}(x, y)\big) dA$$

The solid is bounded by:

$$z_{top}(x, y) = 4 - x - 2y$$

$$z_{bottom}(x, y) = 0$$

$$V = \iint_R \big(z_{top}(x, y) - z_{bottom}(x, y)\big) dA = \iint_R (4 - x - 2y) dA$$

The projection region R is on the xy-plane ($z = 0$), bounded by:

$$y = 0$$

$$x = 2y^2$$

$x + 2y = 4$ (The intersection of the top plane with $z = 0$)

$$y \geq 2$$

The region R is shown in Fig. 1.80.

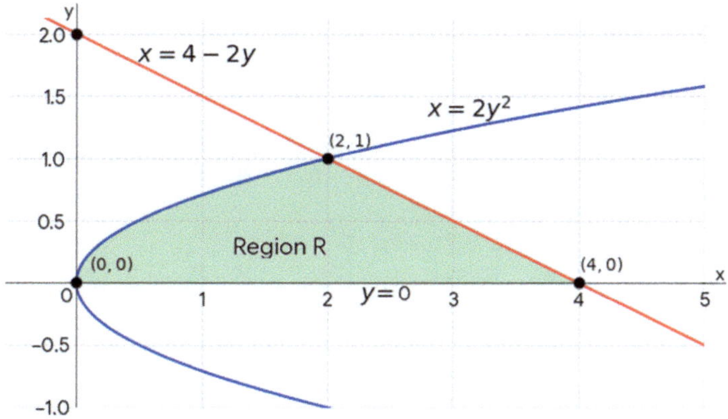

Fig. 1.80 Integration region R

Based on Fig. 1.81, $V = \iint_R (4 - x - 2y)dA = \int_0^1 \int_{2y^2}^{4-2y} (4 - x - 2y)dxdy$.

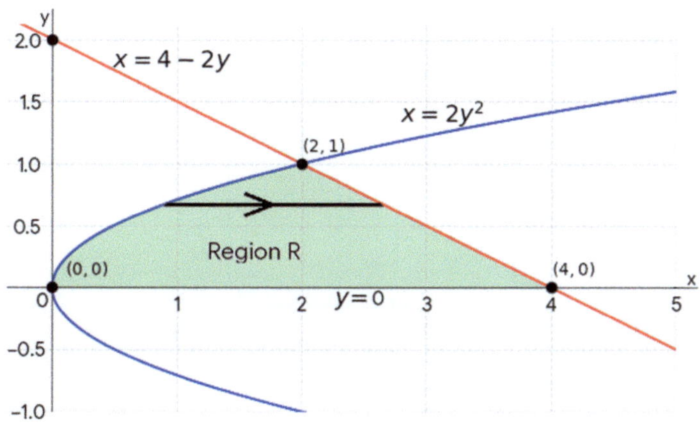

Fig. 1.81 Evaluation of the double integral using the order of integration $dxdy$

The double integral $\int_0^1 \int_{2y^2}^{4-2y} \cos(y^3)dxdy$ is calculated by evaluating the inner and outer integrals sequentially:

$$\int_{2y^2}^{4-2y} (4 - x - 2y)dx = 4x - \frac{x^2}{2} - 2yx \bigg|_{2y^2}^{4-2y}$$

$$= \left(4(4-2y) - \frac{(4-2y)^2}{2} - 2y(4-2y)\right)$$

$$- \left(4(2y^2) - \frac{(2y^2)^2}{2} - 2y(2y^2)\right)$$

$$= \left(16 - 8y - 8 + 8y - 2y^2 - 8y + 4y^2\right) - \left(8y^2 - 2y^4 - 4y^3\right)$$

$$= 8 - 8y + 2y^2 - 8y^2 + 2y^4 + 4y^3$$
$$= 8 - 8y - 6y^2 + 4y^3 + 2y^4$$

$$\int_0^1 \int_{2y^2}^{4-2y} (4 - x - 2y)dxdy = \int_0^1 (8 - 8y - 6y^2 + 4y^3 + 2y^4)dy$$
$$= 8y - 4y^2 - 2y^3 + y^4 + \frac{2}{5}y^5 \Big|_0^1$$
$$= 8 - 4 - 2 + 1 + \frac{2}{5} - 0 = 3 + \frac{2}{5} = \frac{17}{5} = 2.4$$

The volume of the solid is $\frac{17}{5}$ or 3.4 cubic units.

Example 1.55 Find the volume of the solid in the first octant bounded by the plane $x + 2y + 3z = 4$.

Solution The volume (V) of the solid is given by:

$$V = \iint_R \left(z_{top}(x, y) - z_{bottom}(x, y)\right)dA$$

The solid is bounded by:

$$z_{top}(x, y) = \frac{4 - x - 2y}{3}$$

$$z_{bottom}(x, y) = 0$$

$$V = \iint_R \left(z_{top}(x, y) - z_{bottom}(x, y)\right)dA = \iint_R \left(\frac{4 - x - 2y}{3}\right)dA$$

The projection region R is on the xy-plane ($z = 0$), bounded by:

$$x + 2y + 3z = 4 \overset{z=0}{\Longrightarrow} x + 2y = 4$$

$$x = 0$$

$$y = 0$$

The region R is shown in Fig. 1.82.

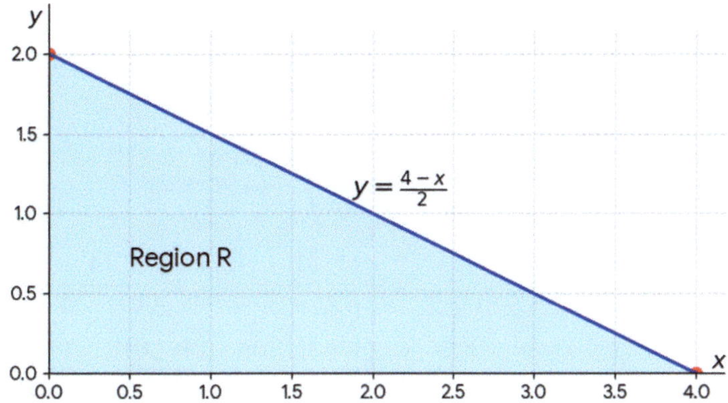

Fig. 1.82 Integration region R

Based on Fig. 1.83, $V = \iint_R \left(\frac{4-x-2y}{3}\right) dA = \int_0^4 \int_0^{\frac{4-x}{2}} \left(\frac{4-x-2y}{3}\right) dydx.$

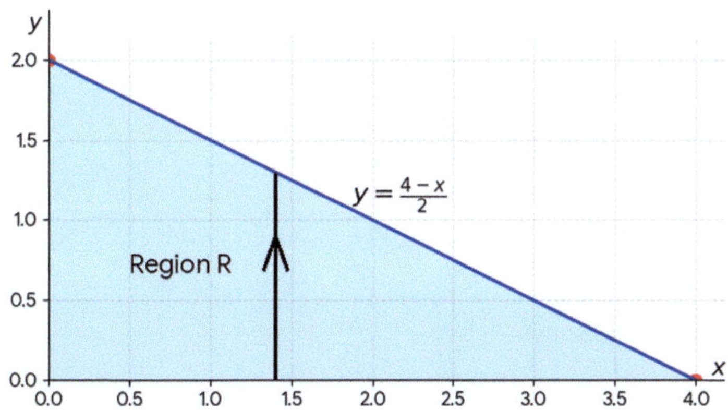

Fig. 1.83 Evaluation of the double integral using the order of integration $dydx$

The double integral $\int_0^4 \int_0^{\frac{4-x}{2}} \left(\frac{4-x-2y}{3}\right) dydx$ is calculated by evaluating the inner and outer integrals sequentially:

$$\int_0^{\frac{4-x}{2}} \left(\frac{4-x-2y}{3}\right) dy = \frac{1}{3} \int_0^{\frac{4-x}{2}} (4-x-2y) dy = \frac{1}{3}\left((4-x)y - y^2\right)\Big|_0^{\frac{4-x}{2}}$$

$$= \frac{1}{3}\left((4-x)\left(\frac{4-x}{2}\right) - \left(\frac{4-x}{2}\right)^2\right) - 0$$

$$= \frac{1}{3}\frac{(4-x)^2}{4} = \frac{(4-x)^2}{12}$$

$$V = \int_0^4 \int_0^{\frac{4-x}{2}} \left(\frac{4 - x - 2y}{3} \right) dy dx = \int_0^4 \frac{(4-x)^2}{12} dx = \frac{1}{12} \int_0^4 (4-x)^2 dx$$

$$= -\frac{1}{12} \int_0^4 -(4-x)^2 dx$$

$$= -\frac{1}{12} \frac{(4-x)^3}{3} \Big|_0^4 = \frac{1}{12} \frac{(4-x)^3}{3} \Big|_4^0$$

$$= \frac{(4-x)^3}{36} \Big|_4^0 = \frac{(4-0)^3}{36} - \frac{(4-4)^3}{36}$$

$$= \frac{64}{36} - 0 = \frac{64}{36} = \frac{16}{9} \approx 1.7778$$

The volume of the solid is $\frac{16}{9} \approx 1.7778$ cubic units.

Example 1.56 Calculate the volume of the solid E that is bounded above by the paraboloid $z = 3 - x^2 - y^2$ and below by the paraboloid $z = x^2 + y^2$.

Solution The volume (V) of the solid is given by:

$$V = \iint_R \left(z_{top}(x, y) - z_{bottom}(x, y) \right) dA$$

The solid is bounded by:

$$z_{top}(x, y) = 3 - x^2 - y^2$$

$$z_{bottom}(x, y) = x^2 + y^2$$

$$V = \iint_R \left(z_{top}(x, y) - z_{bottom}(x, y) \right) dA = \iint_R \left(3 - 2\left(x^2 + y^2\right) \right) dA$$

The projection region R is on the xy-plane ($z = 0$), bounded by:

$$\begin{cases} z = 3 - x^2 - y^2 \\ z = x^2 + y^2 \end{cases} \Rightarrow \begin{cases} 0 = 3 - x^2 - y^2 \\ 0 = x^2 + y^2 \end{cases}$$

$$\Rightarrow 3 - x^2 - y^2 = x^2 + y^2 \Rightarrow 3 = 2\left(x^2 + y^2\right)$$

$$\Rightarrow x^2 + y^2 = \frac{3}{2}$$

The region R is shown in Fig. 1.84.

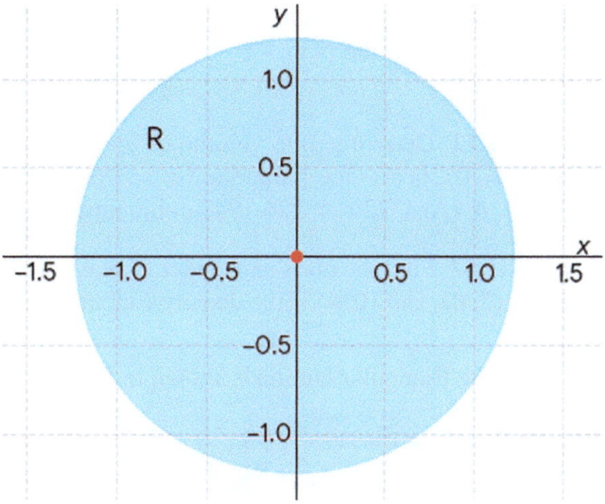

Fig. 1.84 Integration region *RR*

Due to the circular symmetry of the region R and the integrand, we use Polar Coordinates ($x^2 + y^2 = r^2$ and $dA = rdrd\theta$).

$$V = \iint_R (3 - 2(x^2 + y^2))dA = \int_0^{2\pi} \int_0^{\sqrt{\frac{3}{2}}} (3 - 2r^2) \times rdrd\theta = \int_0^{2\pi} \int_0^{\sqrt{\frac{3}{2}}} (3 - 2r^2)rdrd\theta$$

The calculation of $V = \int_0^{2\pi} \int_0^{\sqrt{\frac{3}{2}}} (3 - 2r^2)rdrd\theta$ is shown below:

$$\int_0^{\sqrt{\frac{3}{2}}} (3 - 2r^2)rdr = \int_0^{\sqrt{\frac{3}{2}}} (3r - 2r^3)dr = \left(\frac{3}{2}r^2 - \frac{1}{2}r^4\right)\Bigg|_0^{\frac{\sqrt{3}}{2}}$$

$$= \frac{3}{2}\left(\sqrt{\frac{3}{2}}\right)^2$$

$$-\frac{1}{2}\left(\sqrt{\frac{3}{2}}\right)^4 - 0 = \frac{9}{4} - \frac{9}{8}$$

$$= \frac{18 - 9}{8} = \frac{9}{8} = 1.125$$

$$V = \int_0^{2\pi} \int_0^{\sqrt{\frac{3}{2}}} (3 - 2r^2)rdrd\theta = \int_0^{2\pi} \frac{9}{8}d\theta = \frac{9}{8} \times 2\pi = \frac{9\pi}{4} \approx 7.0686$$

The volume of the solid E is $\frac{9\pi}{4} \approx 7.0686$ cubic units.

Exercise 1.10 Determine the volume of the solid region E bounded by the surface $x^2 + 9y^2 = 16$ and the planes $z - x = 4$ and $z = 0$. (Ans. $\frac{64\pi}{3} \approx 67.0206$ cubic units).

Exercise 1.11 Determine the volume of the solid region E that lies above the xy-plane ($z = 0$), below the paraboloid $z = 16 - x^2 - y^2$, and outside the cylinder $x^2 + y^2 = 4$. (Ans. $72\pi \approx 226.1947$ cubic units).

Example 1.57 Find the mass and center of mass of a triangular lamina with vertices at $(0, 0)$, $(2, 0)$, and $(0, 4)$ if the density function is $\rho(x, y, z) = 2 + x + 2y$.

Solution The triangular lamina is shown in Fig. 1.85.

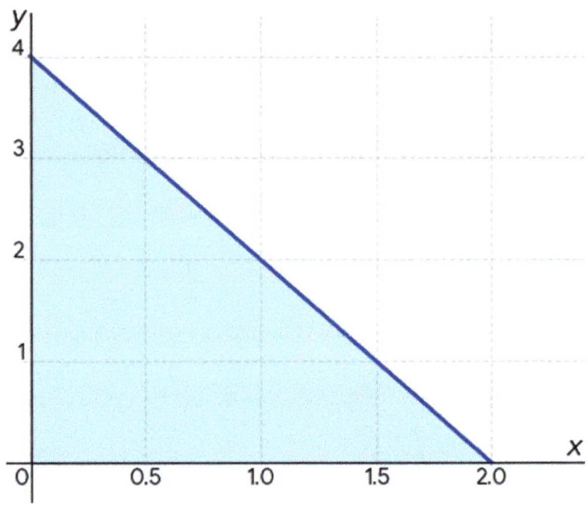

Fig. 1.85 Triangular lamina for Example 1.57

First, we find the equation of the line connecting $(2, 0)$ and $(0, 4)$ to set the integration limits.

$$m = \frac{4 - 0}{0 - 2} = -2$$

$$y - y_0 = m(x - x_0) \Rightarrow y - 4 = -2(x - 0) \Rightarrow y = -2x + 4$$

The region R is shown in Fig. 1.86.

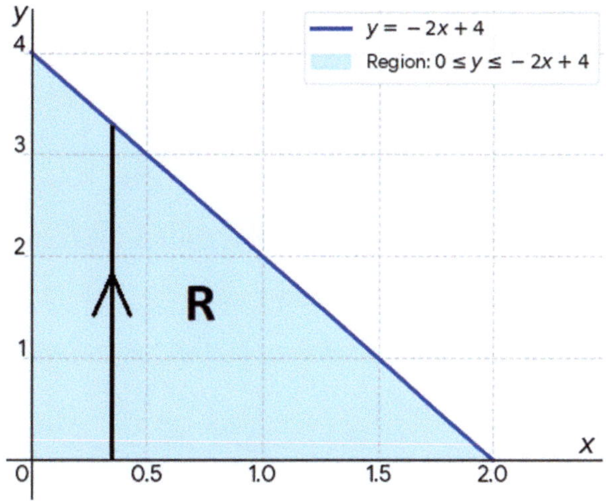

Fig. 1.86 Integration region R

The total mass m is:

$$m = \iint_R \rho(x, y)dA = \int_0^2 \int_0^{-2x+4} (2 + x + 2y)dydx$$

The calculation of $\int_0^2 \int_0^{-2x+4} (2 + x + 2y)dydx$ is shown below:

$$\int_0^{-2x+4} (2 + x + 2y)dy = (2 + x)y + y^2 \Big|_0^{-2x+4}$$
$$= (2 + x)(-2x + 4) + (-2x + 4)^2$$
$$= -4x + 8 - 2x^2 + 4x + 4x^2 - 16x + 16$$
$$= 2x^2 - 16x + 24$$

$$m = \int_0^2 \int_0^{-2x+4} (2 + x + 2y)dydx = \int_0^2 \left(2x^2 - 16x + 24\right)dx$$
$$= \frac{2x^3}{3} - 8x^2 + 24x \Big|_0^2$$
$$= \frac{2 \times 2^3}{3} - 8 \times 2^2 + 24 \times 2 - \left(\frac{2 \times 0^3}{3} - 8 \times 0^2 + 24 \times 0\right)$$
$$= \frac{16}{3} - 32 + 48 - 0$$
$$= \frac{16}{3} + 16 = \frac{64}{3} \approx 21.3333$$

The moment about the y-axis is:

$$M_y = \iint\limits_R x \cdot \rho(x, y)dA = \int\limits_0^2 \int\limits_0^{-2x+4} x(2 + x + 2y)dydx$$

The calculation of $\int_0^2 \int_0^{-2x+4} x(2 + x + 2y)dydx$ is shown below:

$$\begin{aligned}
\int\limits_0^{-2x+4} x(2 + x + 2y)dy &= \int\limits_0^{-2x+4} \left(2x + x^2 + 2xy\right)dy \\
&= (2x + x^2)y + xy^2 \Big|_0^{-2x+4} \\
&= (2x + x^2)(-2x + 4) + x(-2x + 4)^2 \\
&\quad - \left(\left(2x + x^2\right) \times 0 + x \times 0^2\right) \\
&= -4x^2 + 8x - 2x^3 + 4x^2 + 4x^3 - 16x^2 + 16x - 0 \\
&= 2x^3 - 16x^2 + 24x
\end{aligned}$$

$$\begin{aligned}
M_y &= \int\limits_0^2 \int\limits_0^{-2x+4} x(2 + x + 2y)dydx = \int\limits_0^2 \left(2x^3 - 16x^2 + 24x\right)dx \\
&= \frac{x^4}{2} - \frac{16x^3}{3} + 12x^2 \Big|_0^2 \\
&= \left(\frac{2^4}{2} - \frac{16 \times 2^3}{3} + 12 \times 2^2\right) - \left(\frac{0^4}{2} - \frac{16 \times 0^3}{3} + 12 \times 0^2\right) \\
&= 8 - \frac{128}{3} + 48 - 0 \\
&= 56 - \frac{128}{3} = \frac{40}{3} \approx 13.3333
\end{aligned}$$

The moment about the x-axis is:

$$M_x = \iint\limits_R y \cdot \rho(x, y)dA = \int\limits_0^2 \int\limits_0^{-2x+4} y(2 + x + 2y)dydx$$

The calculation of $\int_0^2 \int_0^{-2x+4} y(2 + x + 2y)dydx$ is shown below:

$$\int_0^{-2x+4} y(2+x+2y)dy = \int_0^{-2x+4} (2y+xy+2y^2)dy = y^2 + \frac{xy^2}{2} + \frac{2y^3}{3}\bigg|_0^{-2x+4}$$

$$= (-2x+4)^2 + \frac{x(-2x+4)^2}{2} + \frac{2(-2x+4)^3}{3}$$
$$- \left(0^2 + \frac{x \times 0^2}{2} + \frac{2 \times 0^3}{3}\right)$$
$$= (-2x+4)^2 + \frac{x(-2x+4)^2}{2} + \frac{2(-2x+4)^3}{3}$$
$$= -\frac{10}{3}x^3 + 28x^2 - 72x + \frac{176}{3}$$

$$M_x = \int_0^2 \int_0^{-2x+4} y(2+x+2y)dydx = \int_0^2 \left(-\frac{10}{3}x^3 + 28x^2 - 72x + \frac{176}{3}\right)dx$$
$$= -\frac{10}{12}x^4 + \frac{28x^3}{3} - 36x^2 + \frac{176}{3}x\bigg|_0^2$$
$$= -\frac{10}{12} \times 2^4 + \frac{28 \times 2^3}{3} - 36 \times 2^2 + \frac{176}{3} \times 2$$
$$- \left(-\frac{10}{12} \times 0^4 + \frac{28 \times 0^3}{3} - 36 \times 0^2 + \frac{176}{3} \times 0\right)$$
$$= -\frac{160}{12} + \frac{224}{3} - 144 + \frac{352}{3} - 0$$
$$= \frac{104}{3} \approx 34.6667$$

Using the mass $m = \frac{64}{3}$ and moments $M_x = \frac{104}{3}$ and $M_y = \frac{40}{3}$,

$$\overline{x} = \frac{M_y}{m} = \frac{\frac{40}{3}}{\frac{64}{3}} = \frac{40}{64} = \frac{5}{8} \approx 0.6250$$

$$\overline{y} = \frac{M_x}{m} = \frac{\frac{104}{3}}{\frac{64}{3}} = \frac{104}{64} = \frac{13}{8} \approx 1.6250$$

Therefore, the mass is $\frac{64}{3}$, and the center of mass is located at $\left(\frac{5}{8}, \frac{13}{8}\right)$.

Exercise 1.12 The density at any point on a semicircular lamina (Fig. 1.87) is given by $\rho = 5\sqrt{x^2+y^2}$. Find the total mass and the center of mass of the lamina. (Ans. $m = 45\pi$, $(\overline{x}, \overline{y}) = \left(0, \frac{9}{2\pi}\right)$).

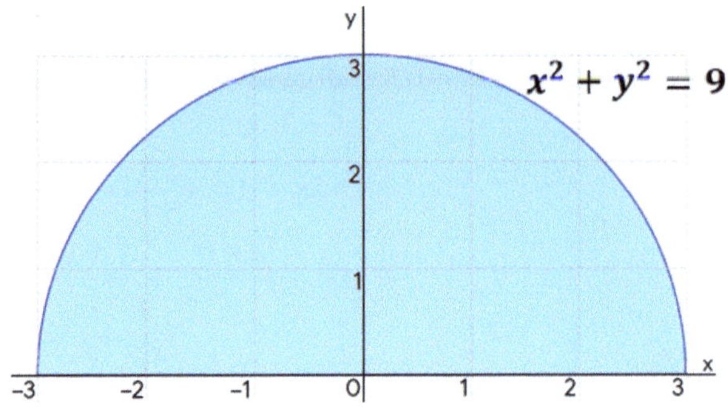

Fig. 1.87 Semicircular lamina for Exercise 1.12

Example 1.58 Find the moments of inertia I_x, I_y, and I_0 (with respect to the origin) for a homogeneous semicircular lamina with a constant density of $\rho(x, y) = 5$ (Fig. 1.88).

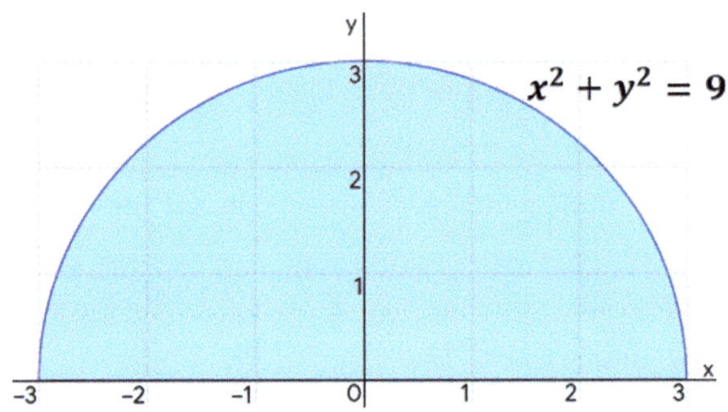

Fig. 1.88 Semicircular lamina for Example 1.58

Solution To solve for the moments of inertia of a semicircular lamina with a constant density $\rho(x, y) = 5$, we will use polar coordinates.

For a semicircular lamina with radius 3 centered at the origin, the region R in polar coordinates is:

$$0 \le r \le 3$$

$$0 \le \theta \le \pi$$

The moment of inertia about the x-axis is defined as $I_x = \iint_R y^2 \cdot \rho(x, y)dA$. In polar coordinates, $y = r\sin(\theta)$ and $dA = rdrd\theta$.

$$I_x = \iint\limits_R y^2 \cdot \rho(x, y)dA \Rightarrow I_x = \int_0^{\pi}\int_0^{3} (r\sin(\theta))^2 \times 5 \times rdrd\theta$$

$$\Rightarrow I_x = 5\int_0^{\pi}\int_0^{3} r^3\sin^2(\theta)drd\theta$$

$$\Rightarrow I_x = 5\int_0^{\pi}\sin^2(\theta)d\theta\int_0^{3} r^3dr$$

Let's calculate each integral individually

$$\int_0^{\pi}\sin^2(\theta)d\theta = 2\int_0^{\frac{\pi}{2}}\sin^2(\theta)d\theta = 2\times\left(\frac{1}{2}\times\frac{\pi}{2}\right) = \frac{\pi}{2}$$

$$\int_0^{3} r^3dr = \frac{r^4}{4}\Big|_0^3 = \frac{3^4}{4} - \frac{0^4}{4} = \frac{81}{4}$$

Thus,

$$I_x = 5\int_0^{\pi}\sin^2(\theta)d\theta\int_0^{3} r^3dr = 5\times\frac{\pi}{2}\times\frac{81}{4} = \frac{405\pi}{8} \approx 159.0431$$

The moment of inertia about the y-axis is defined as $I_y = \iint_R x^2\cdot\rho(x,y)dA$. In polar coordinates, $x = r\cos(\theta)$ and $dA = rdrd\theta$.

$$I_y = \iint_R x^2\cdot\rho(x,y)dA$$

$$\Rightarrow I_y = \int_0^{\pi}\int_0^{3}(r\cos(\theta))^2\times 5\times rdrd\theta$$

$$\Rightarrow I_y = 5\int_0^{\pi}\int_0^{3} r^3\cos^2(\theta)drd\theta$$

$$\Rightarrow I_y = 5\int_0^{\pi}\cos^2(\theta)d\theta\int_0^{3} r^3dr$$

Let's calculate each integral individually.

$$\int_0^{\pi} \cos^2(\theta)d\theta = 2\int_0^{\frac{\pi}{2}} \cos^2(\theta)d\theta = 2 \times \left(\frac{1}{2} \times \frac{\pi}{2}\right) = \frac{\pi}{2}$$

$$\int_0^3 r^3 dr = \frac{r^4}{4}\bigg|_0^3 = \frac{3^4}{4} - \frac{0^4}{4} = \frac{81}{4}$$

Thus,

$$I_y = 5\int_0^{\pi} \cos^2(\theta)d\theta \int_0^3 r^3 dr = 5 \times \frac{\pi}{2} \times \frac{81}{4} = \frac{405\pi}{8} \approx 159.0431$$

The polar moment of inertia about the origin is defined as $I_0 = \iint_R (x^2 + y^2) \cdot \rho(x, y)dA$. In polar coordinates, $x^2 + y^2 = (r\cos(\theta))^2 + (r\sin(\theta))^2 = r^2$ and $dA = rdrd\theta$.

$$I_0 = \iint_R (x^2 + y^2) \cdot \rho(x, y)dA$$

$$\Rightarrow I_0 = \int_0^{\pi}\int_0^3 r^2 \times 5 \times rdrd\theta$$

$$\Rightarrow I_0 = 5\int_0^{\pi}\int_0^3 r^3 drd\theta \Rightarrow I_0 = 5\int_0^{\pi} d\theta \int_0^3 r^3 dr$$

$$\Rightarrow I_0 = 5 \times (\pi - 0) \times \frac{r^4}{4}\bigg|_0^3 \Rightarrow I_0 = 5 \times \pi \times \frac{81}{4}$$

$$\Rightarrow I_0 = \frac{405\pi}{4} \approx 318.0863$$

The calculated moments of inertia for the semicircular lamina are summarized as follows:

$$I_x = \frac{405\pi}{8} \approx 159.0431$$

$$I_y = \frac{405\pi}{8} \approx 159.0431$$

$$I_0 = \frac{405\pi}{4} \approx 318.0863$$

Example 1.59 The boundary of a lamina consists of the semicircles $y = \sqrt{4 - x^2}$ and $y = \sqrt{9 - x^2}$ together with the portions of the x-axis that join them (Fig. 1.89). Find the center of mass of the lamina if the density at any point is proportional to its distance from the origin.

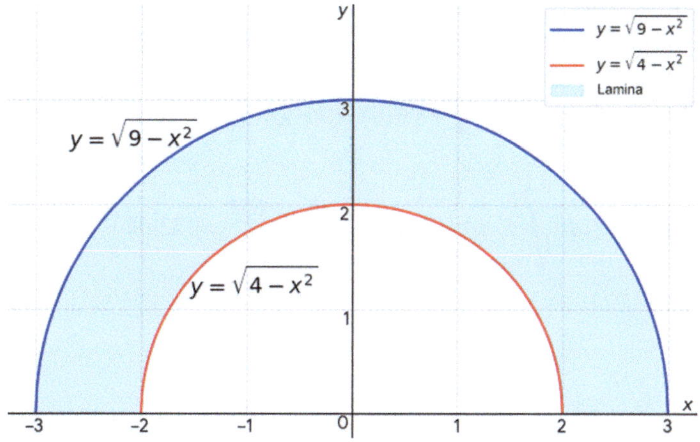

Fig. 1.89 Lamina for Example 1.59

Solution To find the center of mass of this lamina, we will use polar coordinates, as the boundary is defined by semicircles. The region R(Fig. 1.89) is:

$$2 \le r \le 3$$

$$0 \le \theta \le \pi$$

The density is proportional to the distance from the origin (r):

$$\rho(r, \theta) = kr$$

where k is a constant of proportionality.

Using the polar area element $dr = rdrd\theta$:

$$m = \iint\limits_R \rho(x, y)dA = \int\limits_0^\pi \int\limits_2^3 kr \times rdrd\theta = k \int\limits_0^\pi \int\limits_2^3 r^2 drd\theta = k \int\limits_0^\pi d\theta \int\limits_2^3 r^2 dr$$

$$= k \times (\pi - 0) \times \left. \frac{r^3}{3} \right|_2^3 = \frac{19k\pi}{3} \approx 19.8968k$$

Let's calculate the moments. Due to the symmetry of the region and the density function across the y-axis, the center of mass must lie on the y-axis. Mathematically:

$$M_y = \iint_R x\rho(x, y)dA = \int_0^{\pi}\int_2^{3} r\cos(\theta) \times kr \times rdrd\theta = k\int_0^{\pi}\int_2^{3} r^3\cos(\theta)drd\theta$$

$$= k\int_0^{\pi}\cos(\theta)d\theta\int_2^{3} r^3dr = k \times 0 \times \frac{65}{4} = 0$$

$$M_x = \iint_R y\rho(x, y)dA = \int_0^{\pi}\int_2^{3} r\sin(\theta) \times kr \times rdrd\theta = k\int_0^{\pi}\int_2^{3} r^3 sin(\theta)drd\theta$$

$$= k\int_0^{\pi}\sin(\theta)d\theta\int_2^{3} r^3dr = k \times 2 \times \frac{65}{4} = \frac{65k}{2} = 32.5k$$

Therefore,

$$\overline{x} = \frac{M_y}{m} = \frac{0}{\frac{19k\pi}{3}} = 0$$

$$\overline{y} = \frac{M_x}{m} = \frac{\frac{65k}{2}}{\frac{19k\pi}{3}} = \frac{195}{38\pi} \approx 1.6334$$

The calculated mass and center of mass for the given semicircular lamina are as follows:

$$m = \frac{19k\pi}{3} \approx 19.8968k$$

$$(\overline{x}, \overline{y}) = \left(0, \frac{195}{38\pi}\right) \approx (0, 1.6334)$$

Example 1.60 Use MATLAB to evaluate the double integral $\int_1^3\int_0^2 \left(x^2 + 4y\right)dydx$.

Solution The required code is shown in Fig. 1.90.

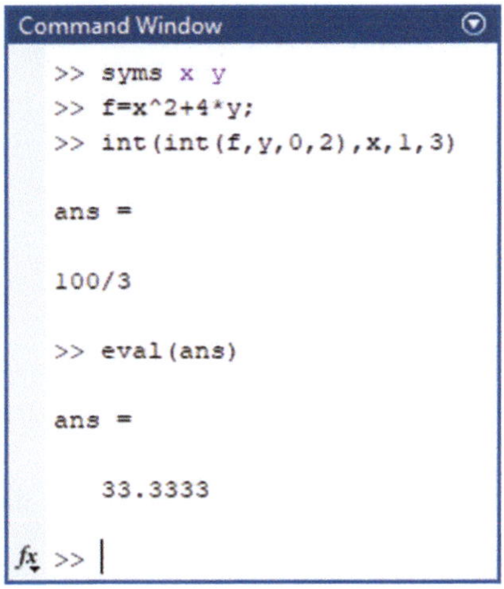

```
>> syms x y
>> f=x^2+4*y;
>> int(int(f,y,0,2),x,1,3)

ans =

100/3

>> eval(ans)

ans =

   33.3333

>>
```

Fig. 1.90 MATLAB code

Example 1.61 Use MATLAB to evaluate the double integral $\int_{-2}^{4}\int_{\frac{y^2-6}{2}}^{y+1} xydxdy$.

Solution The required code is shown in Fig. 1.91.

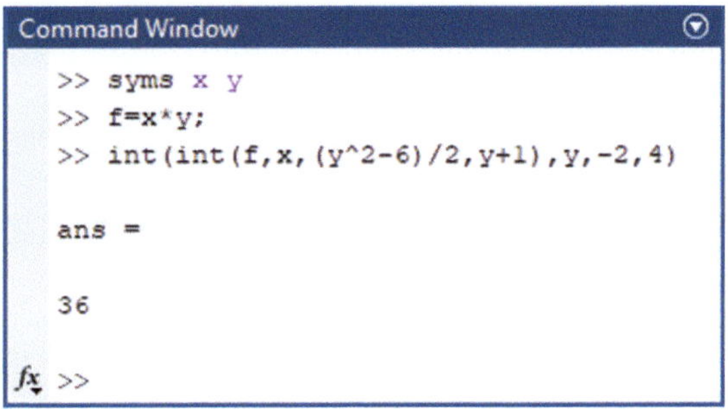

```
>> syms x y
>> f=x*y;
>> int(int(f,x,(y^2-6)/2,y+1),y,-2,4)

ans =

36

>>
```

Fig. 1.91 MATLAB code

Example 1.62 Use MATLAB to evaluate the double integral $\int_{0}^{4}\int_{0}^{\frac{4-x}{2}}\left(\frac{4-x-2y}{3}\right)dydx$.

Solution The required code is shown in Fig. 1.92.

```
Command Window
>> syms x y
>> f=(4-x-2*y)/3;
>> int(int(f,y,0,(4-x)/2),x,0,4)

ans =

16/9

fx >>
```

Fig. 1.92 MATLAB code

Example 1.63 Use MATLAB to evaluate the double integral $\int_{-3}^{3}\int_{-\sqrt{9-x^2}}^{+\sqrt{9-x^2}}\frac{3}{\sqrt{9-(x^2+y^2)}}dydx$.

Solution The required code is shown in Fig. 1.93.

```
Command Window
>> syms x y
>> f=3/sqrt(9-(x^2+y^2));
>> int(int(f,y,-sqrt(9-x^2),sqrt(9-x^2)),x,-3,3)

ans =

18*pi

fx >>
```

Fig. 1.93 MATLAB code

Example 1.64 Use MATLAB to evaluate the double integral $\int_{1}^{3}\int_{x+1}^{2x+3} ye^{-x^2+y}dydx$.

Solution The required code is shown in Fig. 1.94.

Fig. 1.94 MATLAB code

```
Command Window
>> syms x y
>> f=y*exp(-x^2+y);
>> eval(int(int(f,y,x+1,2*x+3),x,1,3))

ans =

  244.1415

fx >>
```

Example 1.65 Use MATLAB to evaluate the double integral $\int_{-\frac{\pi}{6}}^{+\frac{\pi}{6}}\int_{1}^{\cos(3\theta)} rdrd\theta$.

Solution The required code is shown in Fig. 1.95.

```
Command Window
>> syms r theta
>> f=r;
>> int(int(f,r,0,cos(3*theta)),theta,-pi/6,pi/6)

ans =

pi/12

>>
```

Fig. 1.95 MATLAB code

Example 1.66 Use MATLAB to evaluate the double integral $\int_{-3}^{3}\int_{-\sqrt{9-x^2}}^{+\sqrt{9-x^2}}\sqrt{4(x^2+y^2)+1}dydx$.

Solution While most double integrals can be evaluated analytically using the int command, some cases result in complex symbolic expressions rather than numerical values (Fig. 1.96). For this specific problem, applying nested intcommands produces an unwieldy result; therefore, using the integral2 command is more efficient for obtaining a direct numerical solution.

```
Command Window
>> syms x y
>> f=sqrt(4*(x^2+y^2)+1);
>> int(int(f,y,-sqrt(9-x^2),sqrt(9-x^2)),x,-3,3)

ans =

int(37^(1/2)*(9 - x^2)^(1/2) + 2*asinh((9 - x^2)^(1/

>> eval(ans)

ans =

int(37^(1/2)*(9 - x^2)^(1/2) + 2*asinh((9 - x^2)^(1/

>>
```

Fig. 1.96 Nested int commands can't evaluate the integral $\int_{-3}^{3}\int_{-\sqrt{9-x^2}}^{+\sqrt{9-x^2}}\sqrt{4(x^2+y^2)+1}dydx$

The script in Fig. 1.97 utilizes the integral2 command to efficiently compute the numerical value of the given integral.

Fig. 1.97 Evaluation of $\int_{-3}^{3}\int_{-\sqrt{9-x^2}}^{+\sqrt{9-x^2}}\sqrt{4(x^2+y^2)+1}dydx$ using the integral2 command

```
Command Window
>> f=@(x,y) sqrt(4*(x.^2+y.^2)+1);
>> ymin = @(x) -sqrt(9 - x.^2);
>> ymax = @(x) sqrt(9 - x.^2);
>> integral2(f,-3,3,ymin,ymax)

ans =

  117.3187

fx >>
```

In MATLAB's integral2 command, the variable ordering in the function handle determines the integration sequence. Table 1.1 summarizes the required function handle syntax for each pair of integration variables.

Example 1.67 Use MATLAB to evaluate the double integral $\int_{-2}^{2}\int_{-\sqrt{4-y^2}}^{+\sqrt{4-y^2}}\sqrt{x^2+y^2+1}dxdy$.

Solution The required code is shown in Fig. 1.98. Since the integration order is *dxdy*, the function handle is defined as @(y, x).

Fig. 1.98 Evaluation of $\int_{-2}^{2}\int_{-\sqrt{4-y^2}}^{+\sqrt{4-y^2}}\sqrt{x^2+y^2+1}dxdy$ using the integral2 command

```
Command Window
>> fun = @(y, x) sqrt(x.^2 + y.^2 + 1);
>> xmin = @(y) -sqrt(4 - y.^2);
>> xmax = @(y) sqrt(4 - y.^2);
>> integral2(fun, -2, 2, xmin, xmax)

ans =

  21.3217

fx >>
```

Example 1.68 Use MATLAB to evaluate the double integral $\int_{0}^{2}\int_{x^2}^{2x}e^x\cos(y)dydx$.

Solution The required code is shown in Fig. 1.99. Since the integration order is *dydx*, the function handle is defined as @(x, y).

Table 1.1 Function handle definitions for triple integration orders

Integral	Integration order	Function handle
$\int_{y_0}^{y_1}\int_{g(y)}^{h(y)}f(x,y)dxdy$	*dxdy*	@(y, x)
$\int_{x_0}^{x_1}\int_{g(x)}^{h(x)}f(x,y)dydx$	*dydx*	@(x, y)

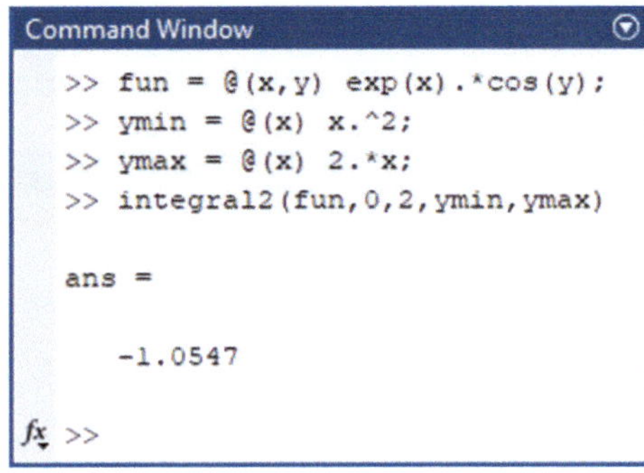

Fig. 1.99 Evaluation of $\int_0^2 \int_{x^2}^{2x} e^x \cos(y)dydx$ using the integral2 command

Note that reversing the function handle to @(y,x) while maintaining the original limit definitions leads to an incorrect result (Fig. 1.100).

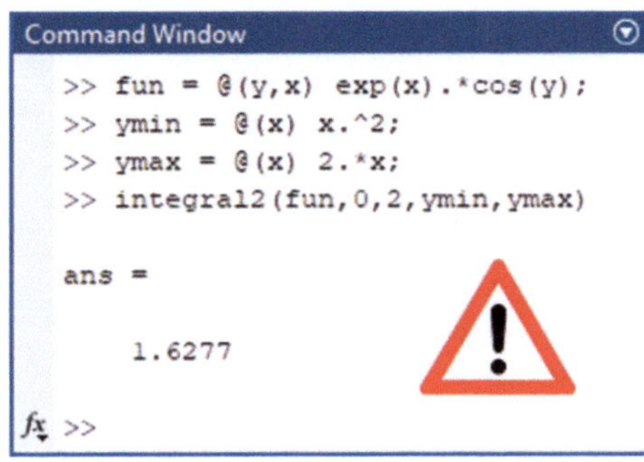

Fig. 1.100 Inverting the function handle to @(y,x) results in an incorrect result

Exercise 1.13 Change the order of integration from *dydx* to *dxdy* for $\int_0^2 \int_{x^2}^{2x} e^x \cos(y)dydx$. (Ans. $\int_0^4 \int_{\frac{y}{2}}^{\sqrt{y}} e^x \cos(y)dxdy$).

Example 1.69 Use MATLAB to evaluate the double integral $\int_0^4 \int_{\frac{y}{2}}^{\sqrt{y}} e^x \cos(y)dxdy$.

Solution The required code is shown in Fig. 1.101. Since the integration order is *dxdy*, the function handle is defined as @(y, x).

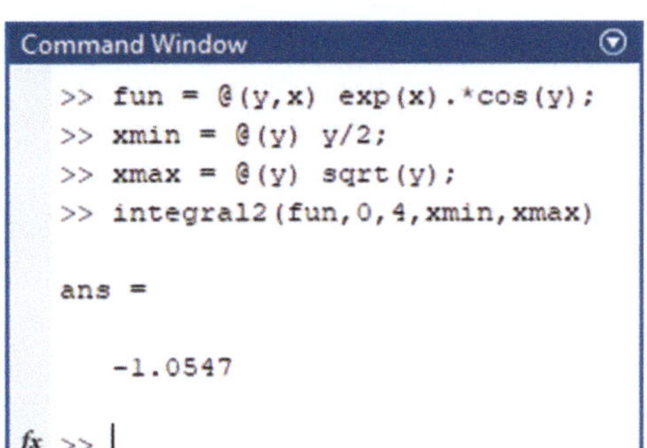

Fig. 1.101 Evaluation of $\int_0^4 \int_{\frac{y}{2}}^{\sqrt{y}} e^x \cos(y)dxdy$ using the integral2 command

You can also utilize the MATLAB script illustrated in Fig. 1.102.

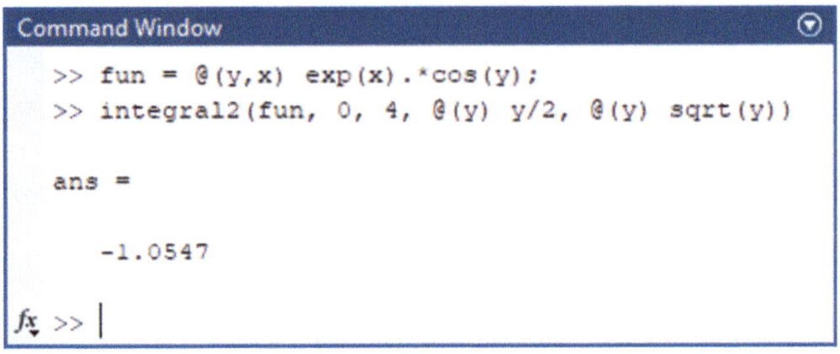

Fig. 1.102 Evalu of $\int_0^4 \int_{\frac{y}{2}}^{\sqrt{y}} e^x \cos(y)dxdy$ using the integral2 command

Example 1.70 Use MATLAB to evaluate the double integral $\int_0^4 \int_{\frac{y}{2}}^{\sqrt{y}} 2dxdy$.

Solution The required code is shown in Fig. 1.103. Since the integration order is *dxdy*, the function handle is defined as @(y, x).

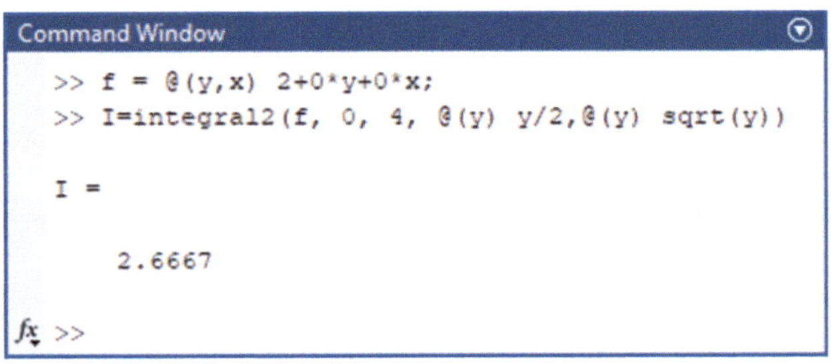

Fig. 1.103 Evaluation of $\int_0^4 \int_{\frac{y}{2}}^{\sqrt{y}} 2dxdy$ using the integral2 command

Example 1.71 Use MATLAB to evaluate the double integral $\int\limits_0^2 \int_{2-\sqrt{1-(x-1)^2}}^{2+\sqrt{1-(x-1)^2}} 1dydx$.

Solution The required code is shown in Fig. 1.104. Since the integration order is *dydx*, the function handle is defined as @(x, y).

```
Command Window
>> f=@(x,y) 1+0*x+0*y;
>> I=integral2(f, 0, 2, @(x) 2-sqrt(1-(x-1).^2), @(x) 2+sqrt(1-(x-1).^2))

I =

    3.1416

fx >>
```

Fig. 1.104 Evaluation of $\int_0^2 \int_{2-\sqrt{1-(x-1)^2}}^{2+\sqrt{1-(x-1)^2}} 1dydx$ using the integral2 command

Example 1.72 Use MATLAB to draw the plot of $r = 3(1 + \cos(\theta))$ for $0 \leq \theta \leq 2\pi$.

Solution The required code is shown in Fig. 1.105. The output of this code is shown in Fig. 1.106.

```
Command Window
>> theta = 0:0.01:2*pi;
>> r = 3 * (1 + cos(theta));
>> polarplot(theta, r);
>> title('r = 3(1 + cos(\theta))');
fx >>
```

Fig. 1.105 MATLAB code

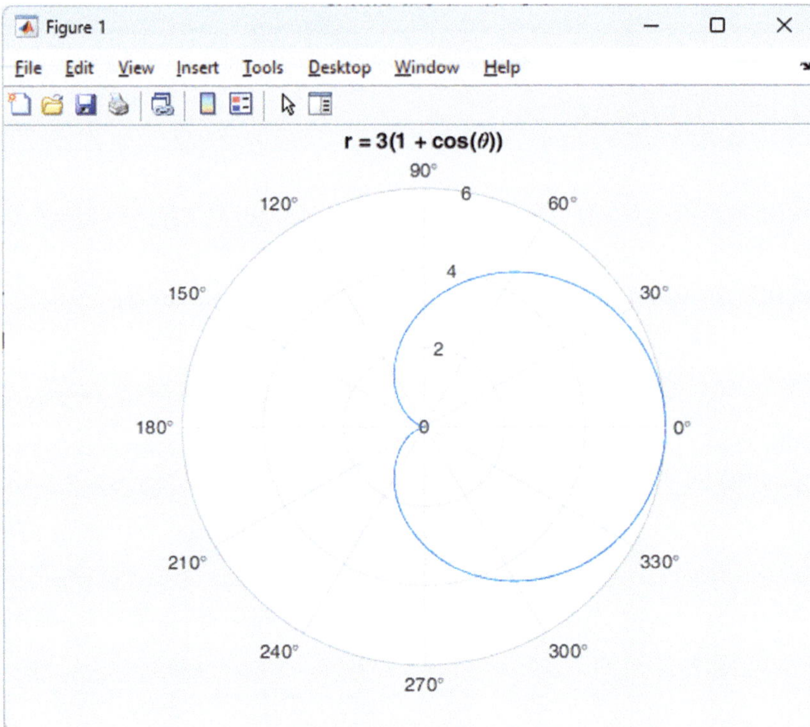

Fig. 1.106 Output generated by the code shown in Fig. 1.105

Example 1.73 Utilize MATLAB to generate the polar plots of $r_1 = 3(1 + \cos(\theta))$ and $r_2 = 5(1 + \sin(\theta))$ for the interval $0 \le \theta \le 2\pi$ on a single set of axes.

Solution The required code is shown in Fig. 1.107. The output of this code is shown in Fig. 1.108.

```
Command Window
>> theta = 0:0.01:2*pi;
>> r1 = 3 * (1 + cos(theta));
>> r2 = 5 * (1 + sin(theta));
>> polarplot(theta, r1);
>> hold on
>> polarplot(theta, r2);
>> legend('r_1 = 3(1 + cos(\theta))','r_2 = 5(1 + sin(\theta))');
fx >>
```

Fig. 1.107 MATLAB code

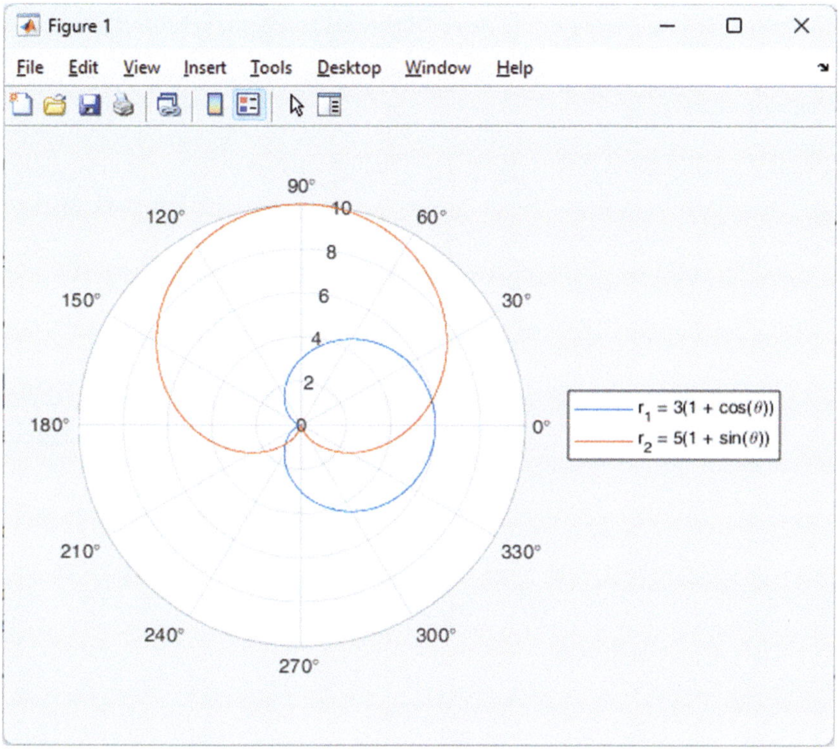

Fig. 1.108 Output generated by the code shown in Fig. 1.107

References

1. Calculus, George Thomas, Maurice Weir, Joel Hass, Pearson, 2014.
2. Calculus, James Stewart, Cengage Learning, 2015.
3. Engineering Mathematics with MATLAB® and Simulink®, Farzin Asadi, Springer, 2025.
4. Applied Numerical Analysis with MATLAB®/Simulink®: For Engineers and Scientists, Farzin Asadi, Springer, 2023.
5. A Problem-Solving Approach to Single-Variable Calculus (Volume I), Farzin Asadi, Springer, 2026.
6. A Problem-Solving Approach to Single-Variable Calculus (Volume II), Farzin Asadi, Springer, 2026.
7. A Problem-Solving Approach to Multi-Variable Calculus (Volume I), Farzin Asadi, Springer, 2026.
8. A Problem-Solving Approach to Multi-Variable Calculus (Volume III), Farzin Asadi, Springer, 2026.
9. A Problem-Solving Approach to Ordinary Differential Equations (Volume I), Farzin Asadi, Springer, 2026.
10. A Problem-Solving Approach to Ordinary Differential Equations (Volume II), Farzin Asadi, Springer, 2026.

11. A Problem-Solving Approach to Ordinary Differential Equations (Volume III), Farzin Asadi, Springer, 2026.
12. A Problem-Solving Approach to Ordinary Differential Equations (Volume IV), Farzin Asadi, Springer, 2026.

Chapter 2
Area of Curved Surfaces

2.1 Introduction

This chapter introduces the mathematical framework for calculating the surface area of a curved surface S. The core intuition relies on projection: We "shadow" a complex 3D surface onto a flat 2D coordinate plane where integration is manageable, then adjust for the "stretch" caused by the surface's tilt.

If your surface is given by $z = f(x, y)$, you can define an implicit function $g(x, y, z)$ by rearranging it:

$$z = f(x, y) \Rightarrow g(x, y, z) = f(x, y) - z = 0$$

The general formula for the surface area $A(S)$ when projecting the surface S onto a region D in a coordinate plane is:

$$A(S) = \iint_D \frac{|\nabla g|}{|\nabla g \cdot \boldsymbol{p}|} dA$$

where:

(A) D is the projection of the surface S onto the coordinate plane.
(B) $|\nabla g|$ is the magnitude of the gradient vector. Recall that the gradient of $g(x, y, z)$ is a vector that is normal (perpendicular) to the surface at every point: $\nabla g = \frac{\partial g}{\partial x}\hat{\boldsymbol{i}} + \frac{\partial g}{\partial y}\hat{\boldsymbol{j}} + \frac{\partial g}{\partial z}\hat{\boldsymbol{k}}$. For the case $g(x, y, z) = f(x, y) - z$, the gradient is: $\nabla g = \frac{\partial f}{\partial x}\hat{\boldsymbol{i}} + \frac{\partial f}{\partial y}\hat{\boldsymbol{j}} - \hat{\boldsymbol{k}}$.
(C) $\boldsymbol{p}$ is the unit normal vector to the projection plane. The choice of $\boldsymbol{p}$ determines the projection plane:

F. Asadi, *A Problem-Solving Approach to Multi-Variable Calculus*,
https://doi.org/10.1007/978-3-032-21115-6_2

C1. When projecting the surface S onto the yz-plane, the projection vector $\boldsymbol{p}$ must be $\hat{\boldsymbol{i}} = 1, 0, 0$. In this case $dA = dydz$. In this case the surface S must be expressed as a function of y and z, i.e., $x = f(y, z)$.

C2. When projecting the surface S onto the zx-plane, the projection vector $\boldsymbol{p}$ must be $\hat{\boldsymbol{j}} = 0, 1, 0$. In this case $dA = dzdx$. In this case the surface S must be expressed as a function of z and x, i.e., $y = f(z, x)$.

C3. When projecting the surface S onto the xy-plane, the projection vector $\boldsymbol{p}$ must be $\hat{\boldsymbol{k}} = 0, 0, 1$. In this case $dA = dxdy$. In this case the surface S must be expressed as a function of x and y, i.e., $z = f(x, y)$.

The surface area is independent of the chosen projection plane.

The most common case is projection onto the xy-plane. Let's examine this case more closely:

$$g(x, y, z) = f(x, y) - z \Rightarrow \nabla g = \frac{\partial f}{\partial x}\hat{\boldsymbol{i}} + \frac{\partial f}{\partial y}\hat{\boldsymbol{j}} - \hat{\boldsymbol{k}}$$

$$|\nabla g| = \sqrt{\left(\frac{\partial f}{\partial x}\right)^2 + \left(\frac{\partial f}{\partial y}\right)^2 + (-1)^2} = \sqrt{\left(\frac{\partial z}{\partial x}\right)^2 + \left(\frac{\partial z}{\partial y}\right)^2 + 1}$$

$$\begin{aligned} |\nabla g \cdot \boldsymbol{p}| = |\nabla g \cdot \hat{\boldsymbol{k}}| &= \left|\left\langle \frac{\partial f}{\partial x}, \frac{\partial f}{\partial y}, -1 \right\rangle \cdot \langle 0, 0, 1 \rangle\right| \\ &= \left|\frac{\partial f}{\partial x} \times 0 + \frac{\partial f}{\partial y} \times 0 + -1 \times 1\right| \\ &= |0 + 0 - 1| = |-1| \\ &= 1 \end{aligned}$$

$$\begin{aligned} A(S) &= \iint\limits_{D_{xy}} \frac{|\nabla g|}{|\nabla g \cdot \boldsymbol{p}|} dA = \iint\limits_{D_{xy}} \frac{\sqrt{\left(\frac{\partial z}{\partial x}\right)^2 + \left(\frac{\partial z}{\partial y}\right)^2 + 1}}{1} dA \\ &= \iint\limits_{D_{xy}} \sqrt{\left(\frac{\partial z}{\partial x}\right)^2 + \left(\frac{\partial z}{\partial y}\right)^2 + 1} dA \end{aligned}$$

2.2 Solved Problems

Example 2.1 Find the surface area of the portion of the plane $3x + 2y + z = 6$ that lies in the first octant (where $x \geq 0$, $y \geq 0$, and $z \geq 0$).

Solution The surface is defined by the implicit equation $3x + 2y + z - 6 = 0$. Let:

$$g(x, y, z) = 3x + 2y + z - 6$$

Let's find the partial derivatives of g:

$$\frac{\partial g}{\partial x} = 3$$

$$\frac{\partial g}{\partial y} = 2$$

$$\frac{\partial g}{\partial z} = 1$$

The gradient vector is:

$$\nabla g = \frac{\partial g}{\partial x}\hat{\boldsymbol{\imath}} + \frac{\partial g}{\partial y}\hat{\boldsymbol{\jmath}} + \frac{\partial g}{\partial z}\hat{\boldsymbol{k}} \Rightarrow \nabla g = 3\hat{\boldsymbol{\imath}} + 2\hat{\boldsymbol{\jmath}} + 1\hat{\boldsymbol{k}}$$

The magnitude of the gradient vector is:

$$|\nabla g| = \left|3\hat{\boldsymbol{\imath}} + 2\hat{\boldsymbol{\jmath}} + 1\hat{\boldsymbol{k}}\right| = \sqrt{3^2 + 2^2 + 1^2} \Rightarrow |\nabla g| = \sqrt{14}$$

Since the surface is given by $z = f(x, y)$, the most convenient projection is onto the xy-plane.

For projection onto the xy-plane, the projection vector $\boldsymbol{p}$ is $\hat{\boldsymbol{k}} = \langle 0, 0, 1\rangle$.

The denominator term is:

$$\begin{aligned}|\nabla g \cdot \boldsymbol{p}| = \left|\nabla g \cdot \hat{\boldsymbol{k}}\right| &= |\langle 3, 2, 1\rangle \cdot \langle 0, 0, 1\rangle| \\ &= |3 \times 0 + 2 \times 0 + 1 \times 1| = |0 + 0 + 1| = |1| = 1\end{aligned}$$

The projection region D in the xy-plane is found by setting $z = 0$ in the plane equation:

$$\begin{aligned}3x + 2y + z - 6 = 0 &\overset{z=0}{\Rightarrow} 3x + 2y + 0 - 6 = 0 \\ &\Rightarrow 3x + 2y = 6 \Rightarrow y = 3 - \frac{3}{2}x\end{aligned}$$

The region D is shown in Fig. 2.1.

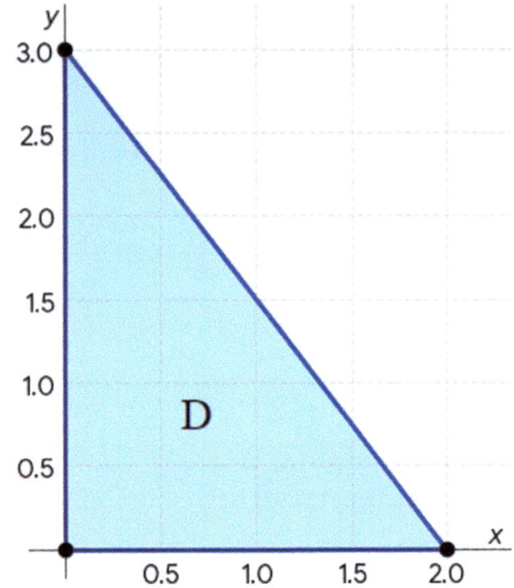

Fig. 2.1 Region D

The limits of integration for this triangular region are:

$$0 \leq x \leq 2, 0 \leq y \leq 3 - \frac{3}{2}x$$

The surface area $A(S)$ is:

$$A(S) = \iint_D \frac{|\nabla g|}{|\nabla g \cdot \boldsymbol{p}|} dA = \iint_D \frac{\sqrt{14}}{1} dA = \sqrt{14} \iint_D dA$$

Since $\iint_D dA$ is simply the area of the region D (a triangle with base 2 and height 3):

$$\begin{aligned} A(S) &= \iint_D \frac{|\nabla g|}{|\nabla g \cdot \boldsymbol{p}|} dA = \sqrt{14} \iint_D dA \\ &= \sqrt{14} \times \left(\frac{1}{2} \times 2 \times 3 \right) = 3\sqrt{14} \approx 11.2250 \end{aligned}$$

The surface area is $3\sqrt{14} \approx 11.2250$ square units.

Example 2.2 Find the surface area of the portion of the plane $3x + 2y + z = 6$ that lies in the first octant (where $x \geq 0$, $y \geq 0$, and $z \geq 0$).

Solution Although this problem was solved in the preceding example, a geometric method will be employed here to confirm the previously obtained result.

The plane $3x + 2y + z = 6$ intersects the x-axis at $x_0 = 2, 0, 0$, the y-axis at $y_0 = 0, 3, 0$, and the z-axis at $z_0 = 0, 0, 6$. The area of the triangle formed by these three intercept points can be found using the Heron's formula.

Heron's Formula is a method used to find the area of a triangle when the lengths of all three sides are known. If a triangle has side lengths a, b, and c, then the area A is given by:

$$A = \sqrt{p(p-a)(p-b)(p-c)}$$

where p is the semi-perimeter (half the perimeter): $p = \frac{a+b+c}{2}$. The side lengths of the triangle $x_0y_0z_0$ are as follows:

$$\begin{aligned} a &= |x_0y_0| = |\langle 2,0,0\rangle - \langle 0,3,0\rangle| = |\langle 2,-3,0\rangle| \\ &= \sqrt{2^2 + (-3)^2 + 0^2} = \sqrt{13} \approx 3.6056 \end{aligned}$$

$$\begin{aligned} b &= |x_0z_0| = |\langle 2,0,0\rangle - \langle 0,0,6\rangle| = |\langle 2,0,-6\rangle| \\ &= \sqrt{2^2 + 0^2 + (-6)^2} = \sqrt{40} \approx 6.3246 \end{aligned}$$

$$\begin{aligned} c &= |y_0z_0| = |\langle 0,3,0\rangle - \langle 0,0,6\rangle| = |\langle 0,3,-6\rangle| \\ &= \sqrt{0^2 + 3^2 + (-6)^2} = \sqrt{45} \approx 6.7082 \end{aligned}$$

Therefore,

$$p = \frac{a+b+c}{2} = \frac{\sqrt{13}+\sqrt{40}+\sqrt{45}}{2} \approx 8.3192$$

$$\begin{aligned} A &= \sqrt{p(p-a)(p-b)(p-c)} \\ &\approx \sqrt{8.3192(8.3192-3.6056)(8.3192-6.3246)(8.3192-6.7082)} \\ &\approx 11.2250 \end{aligned}$$

This result matches the result obtained in the previous example.

Example 2.3 Find the surface area of the portion of the surface $z = 4 - x^2 - y^2$ lying above the disk $x^2 + y^2 \le 4$ (with $z \ge 0$).

Solution $z = 4 - x^2 - y^2 \Rightarrow 4 - x^2 - y^2 - z = 0$.

Let $g(x, y, z) = 4 - x^2 - y^2 - z$. The partial derivatives of g are:

$$\frac{\partial g}{\partial x} = -2x$$

$$\frac{\partial g}{\partial y} = -2y$$

$$\frac{\partial g}{\partial z} = -1$$

The gradient vector is:

$$\nabla g = \frac{\partial g}{\partial x}\hat{\boldsymbol{i}} + \frac{\partial g}{\partial y}\hat{\boldsymbol{j}} + \frac{\partial g}{\partial z}\hat{\boldsymbol{k}} \Rightarrow \nabla g = -2x\hat{\boldsymbol{i}} - 2y\hat{\boldsymbol{j}} - \hat{\boldsymbol{k}}$$

The magnitude of the gradient vector is:

$$|\nabla g| = \left|-2x\hat{\boldsymbol{i}} - 2y\hat{\boldsymbol{j}} - \hat{\boldsymbol{k}}\right| = \sqrt{(-2x)^2 + (-2y)^2 + (-1)^2}$$
$$\Rightarrow |\nabla g| = \sqrt{4\left(x^2 + y^2\right) + 1}$$

Since the surface is given by $z = f(x, y)$, the most convenient projection is onto the xy-plane.

For projection onto the xy-plane, the projection vector $\boldsymbol{p}$ is $\hat{\boldsymbol{k}} = \langle 0, 0, 1 \rangle$.

The denominator term is:

$$\begin{aligned} |\nabla g \cdot \boldsymbol{p}| = \left|\nabla g \cdot \hat{\boldsymbol{k}}\right| &= |\langle -2x, -2y, -1 \rangle \cdot \langle 0, 0, 1 \rangle| \\ &= |-2x \times 0 - 2y \times 0 + (-1) \times 1| \\ &= |0 + 0 - 1| = |-1| = 1 \end{aligned}$$

The region D is shown in Fig. 2.2.

Fig. 2.2 Region D

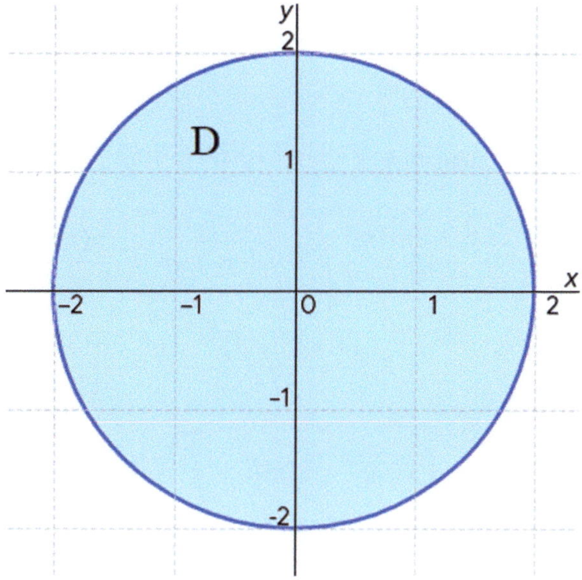

The limits of integration for this region are:

$$-2 \leq x \leq 2, -\sqrt{4-x^2} \leq y \leq \sqrt{4-x^2}$$

The surface area $A(S)$ is:

$$A(S) = \iint_D \frac{|\nabla g|}{|\nabla g \cdot \boldsymbol{p}|} dA = \iint_D \sqrt{4(x^2+y^2)+1} dA$$

$$= \int_{-2}^{2} \int_{-\sqrt{4-x^2}}^{+\sqrt{4-x^2}} \sqrt{4(x^2+y^2)+1} dydx$$

The integral is straightforward to solve using polar coordinates. In polar coordinates, the region D (Fig. 2.2) is described as $D = \{(r, \theta)|0 \leq r \leq 2, 0 \leq \theta \leq 2\pi\}$. Thus,

$$A(S) = \iint_D \frac{|\nabla g|}{|\nabla g \cdot \boldsymbol{p}|} dA = \int_{-2}^{2} \int_{-\sqrt{4-x^2}}^{+\sqrt{4-x^2}} \sqrt{4(x^2+y^2)+1} dydx$$

$$= \int_0^{2\pi} \int_0^2 \sqrt{4r^2+1} \times r dr d\theta$$

$$= \int_0^{2\pi} \int_0^{2} r\sqrt{4r^2+1}drd\theta$$

The calculation of $\int_0^{2\pi} \int_0^2 r\sqrt{4r^2+1}drd\theta$ is shown below:

$$\int_0^2 r\sqrt{4r^2+1}dr = \frac{1}{8}\int_0^2 8r\sqrt{4r^2+1}dr = \frac{1}{8}\frac{\left(4r^2+1\right)^{\frac{1}{2}+1}}{\frac{1}{2}+1}\Bigg|_0^2 = \frac{\left(4r^2+1\right)^{\frac{3}{2}}}{12}\Bigg|_0^2$$

$$= \frac{\left(4\times 2^2+1\right)^{\frac{3}{2}}}{12} - \frac{\left(4\times 0^2+1\right)^{\frac{3}{2}}}{12} = \frac{17^{\frac{3}{2}}}{12} - \frac{1^{\frac{3}{2}}}{12} = \frac{17\sqrt{17}-1}{12}$$

$$\int_0^{2\pi}\int_0^2 r\sqrt{4r^2+1}drd\theta = \int_0^{2\pi}\frac{17\sqrt{17}-1}{12}d\theta = \frac{17\sqrt{17}-1}{12}\int_0^{2\pi}d\theta$$

$$= \frac{17\sqrt{17}-1}{12}\times(2\pi-0)$$

$$= \frac{17\sqrt{17}-1}{6}\pi \approx 36.1769$$

The surface area is $\frac{17\sqrt{17}-1}{6}\pi \approx 36.1769$ square units.

Example 2.4 Calculate the surface area of the upper hemisphere defined by the equation: $x^2+y^2+z^2=9$.

Solution The equation $x^2+y^2+z^2=9$ represents a sphere centered at the origin with a radius of 3. Therefore, we expect the area of upper hemisphere to be $\frac{1}{2}\times 4\pi R^2 = \frac{1}{2}\times 4\pi\times 3^2 = 18\pi \approx 56.5487$.

Let's solve the sphere equation for z:

$$z = f(x,y) = \sqrt{9-\left(x^2+y^2\right)}$$

Since we are considering the upper hemisphere ($z \geq 0$), we use the positive root.

$$z = \sqrt{9-\left(x^2+y^2\right)} \Rightarrow z - \sqrt{9-\left(x^2+y^2\right)} = 0$$

Let $g(x,y,z) = z - \sqrt{9-\left(x^2+y^2\right)}$. The partial derivatives of g are:

$$\frac{\partial g}{\partial x} = -\frac{1}{2}\times -2x\times\frac{1}{\sqrt{9-\left(x^2+y^2\right)}} = \frac{x}{\sqrt{9-\left(x^2+y^2\right)}}$$

$$\frac{\partial g}{\partial y} = -\frac{1}{2} \times -2y \times \frac{1}{\sqrt{9-(x^2+y^2)}} = \frac{y}{\sqrt{9-(x^2+y^2)}}$$

$$\frac{\partial g}{\partial z} = 1$$

The gradient vector is:

$$\nabla g = \frac{\partial g}{\partial x}\hat{\boldsymbol{i}} + \frac{\partial g}{\partial y}\hat{\boldsymbol{j}} + \frac{\partial g}{\partial z}\hat{\boldsymbol{k}} \Rightarrow \nabla g = \frac{x}{\sqrt{9-(x^2+y^2)}}\hat{\boldsymbol{i}} + \frac{y}{\sqrt{9-(x^2+y^2)}}\hat{\boldsymbol{j}} + \hat{\boldsymbol{k}}$$

The magnitude of the gradient vector is:

$$\begin{aligned}
|\nabla g| &= \left|\frac{x}{\sqrt{9-(x^2+y^2)}}\hat{\boldsymbol{i}} + \frac{y}{\sqrt{9-(x^2+y^2)}}\hat{\boldsymbol{j}} + \hat{\boldsymbol{k}}\right| \\
&= \sqrt{\left(\frac{x}{\sqrt{9-(x^2+y^2)}}\right)^2 + \left(\frac{y}{\sqrt{9-(x^2+y^2)}}\right)^2} + 1^2 \Rightarrow |\nabla g| \\
&= \sqrt{\frac{x^2}{9-(x^2+y^2)} + \frac{y^2}{9-(x^2+y^2)} + 1} = \sqrt{\frac{9}{9-(x^2+y^2)}} \\
&= \frac{3}{\sqrt{9-(x^2+y^2)}}
\end{aligned}$$

Since the surface is given by $z = f(x, y)$, the most convenient projection is onto the xy-plane.

For projection onto the xy-plane, the projection vector $\boldsymbol{p}$ is $\hat{\boldsymbol{k}} = \langle 0, 0, 1\rangle$.

The denominator term is:

$$\begin{aligned}
|\nabla g \cdot \boldsymbol{p}| = \left|\nabla g \cdot \hat{\boldsymbol{k}}\right| &= \left|\frac{x}{\sqrt{9-(x^2+y^2)}}, \frac{y}{\sqrt{9-(x^2+y^2)}}, 1.0, 0, 1\right| \\
&= \left|\frac{x}{\sqrt{9-(x^2+y^2)}} \times 0 + \frac{y}{\sqrt{9-(x^2+y^2)}} \times 0 + 1 \times 1\right| \\
&= |0+0+1| = |1| = 1
\end{aligned}$$

The projection of the upper hemisphere onto the xy-plane (where $z = 0$) is the disk D defined by $x^2 + y^2 \leq 9$ which is a circle centered at the origin with radius 3.

$$x^2 + y^2 + z^2 = 9 \overset{z=0}{\Rightarrow} x^2 + y^2 + 0^2 = 9 \Rightarrow x^2 + y^2 = 9$$

The region D is shown in Fig. 2.3.

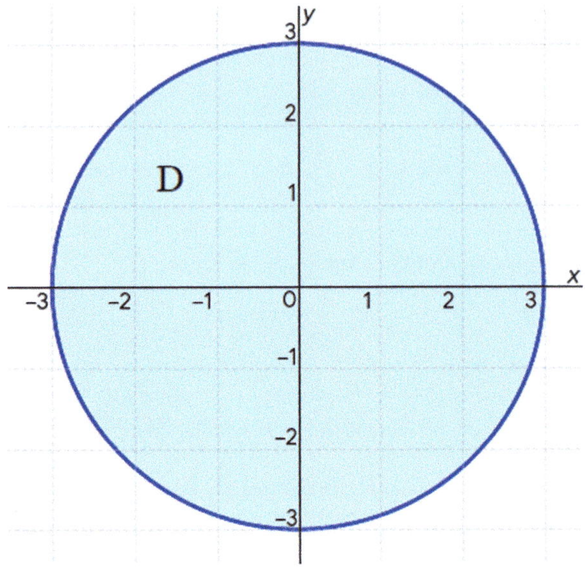

Fig. 2.3 Region D

The limits of integration for this region are:

$$-3 \leq x \leq 3, -\sqrt{9 - x^2} \leq y \leq \sqrt{9 - x^2}$$

The surface area $A(S)$ is:

$$A(S) = \iint_D \frac{|\nabla g|}{|\nabla g \cdot \boldsymbol{p}|} dA = \iint_D \frac{3}{\sqrt{9 - (x^2 + y^2)}} dA$$

$$= \int_{-3}^{3} \int_{-\sqrt{9-x^2}}^{+\sqrt{9-x^2}} \frac{3}{\sqrt{9 - (x^2 + y^2)}} dydx$$

The integral is straightforward to solve using polar coordinates. In polar coordinates, the region D (Fig. 2.3) is described as $D = \{(r, \theta)|0 \leq r \leq 3, 0 \leq \theta \leq 2\pi\}$. Thus,

$$A(S) = \iint_D \frac{|\nabla g|}{|\nabla g \cdot \boldsymbol{p}|} dA = \int_{-3}^{3} \int_{-\sqrt{9-x^2}}^{+\sqrt{9-x^2}} \frac{3}{\sqrt{9 - (x^2 + y^2)}} dydx = \int_0^{2\pi} \int_0^{3} \frac{3}{\sqrt{9 - r^2}} \times rdrd\theta$$

$$= \int_0^{2\pi}\int_0^3 3r\left(9-r^2\right)^{-\frac{1}{2}} dr d\theta$$

The calculation of $\int_0^{2\pi}\int_0^3 3r\left(9-r^2\right)^{-\frac{1}{2}} dr d\theta$ is shown below:

$$\int_0^3 3r\left(9-r^2\right)^{-\frac{1}{2}} dr = \frac{3}{-2}\int_0^3 -2r\left(9-r^2\right)^{-\frac{1}{2}} dr$$

$$= \frac{3}{-2}\frac{\left(9-r^2\right)^{-\frac{1}{2}+1}}{-\frac{1}{2}+1}\Bigg|_0^3 = -3\left(9-r^2\right)^{\frac{1}{2}}\Big|_0^3$$

$$= 3\left(9-r^2\right)^{\frac{1}{2}}\Big|_3^0 = 3\left(9-0^2\right)^{\frac{1}{2}} - 3\left(9-3^2\right)^{\frac{1}{2}}$$

$$= 9 - 0 = 9$$

$$\int_0^{2\pi}\int_0^3 3r\left(9-r^2\right)^{-\frac{1}{2}} dr d\theta = \int_0^{2\pi} 9 d\theta = 9\int_0^{2\pi} d\theta = 9\times 2\pi = 18\pi \approx 56.5487$$

The surface area is $18\pi \approx 56.5487$ square units.

Example 2.5 Calculate the surface area of the upper hemisphere defined by the equation: $x^2+y^2+z^2=9$ that lies above the region $x^2+y^2\le 4$ (Fig. 2.4).

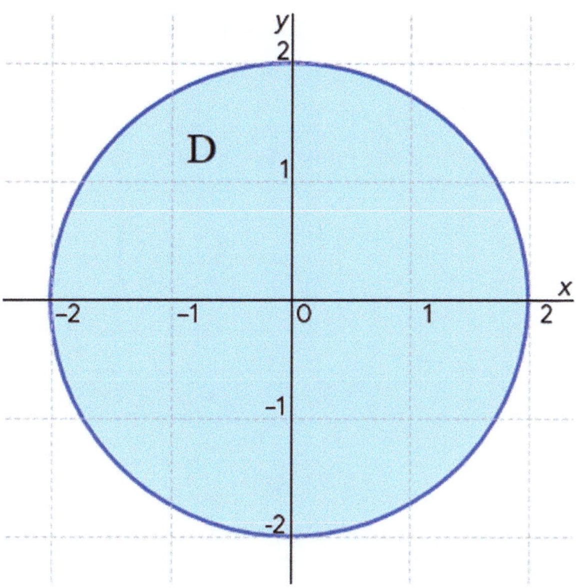

Fig. 2.4 Region $D = \left\{(x,y)|x^2+y^2\le 4\right\}$

Solution Let's solve the sphere equation for z:

$$z = f(x, y) = \sqrt{9 - \left(x^2 + y^2\right)}$$

Since we are considering the upper hemisphere ($z \geq 0$), we use the positive root.

$$z = \sqrt{9 - \left(x^2 + y^2\right)} \Rightarrow z - \sqrt{9 - \left(x^2 + y^2\right)} = 0$$

Let $g(x, y, z) = z - \sqrt{9 - \left(x^2 + y^2\right)}$. The partial derivatives of g are:

$$\frac{\partial g}{\partial x} = -\frac{1}{2} \times -2x \times \frac{1}{\sqrt{9 - \left(x^2 + y^2\right)}} = \frac{x}{\sqrt{9 - \left(x^2 + y^2\right)}}$$

$$\frac{\partial g}{\partial y} = -\frac{1}{2} \times -2y \times \frac{1}{\sqrt{9 - \left(x^2 + y^2\right)}} = \frac{y}{\sqrt{9 - \left(x^2 + y^2\right)}}$$

$$\frac{\partial g}{\partial z} = 1$$

The gradient vector is:

$$\nabla g = \frac{\partial g}{\partial x}\hat{\boldsymbol{i}} + \frac{\partial g}{\partial y}\hat{\boldsymbol{j}} + \frac{\partial g}{\partial z}\hat{\boldsymbol{k}} \Rightarrow \nabla g = \frac{x}{\sqrt{9 - \left(x^2 + y^2\right)}}\hat{\boldsymbol{i}} + \frac{y}{\sqrt{9 - \left(x^2 + y^2\right)}}\hat{\boldsymbol{j}} + \hat{\boldsymbol{k}}$$

The magnitude of the gradient vector is:

$$\begin{aligned}
|\nabla g| &= \left| \frac{x}{\sqrt{9 - \left(x^2 + y^2\right)}}\hat{\boldsymbol{i}} + \frac{y}{\sqrt{9 - \left(x^2 + y^2\right)}}\hat{\boldsymbol{j}} + \hat{\boldsymbol{k}} \right| \\
&= \sqrt{\left(\frac{x}{\sqrt{9 - \left(x^2 + y^2\right)}}\right)^2 + \left(\frac{y}{\sqrt{9 - \left(x^2 + y^2\right)}}\right)^2 + 1^2} \\
\Rightarrow |\nabla g| &= \sqrt{\frac{x^2}{9 - \left(x^2 + y^2\right)} + \frac{y^2}{9 - \left(x^2 + y^2\right)} + 1} \\
&= \sqrt{\frac{9}{9 - \left(x^2 + y^2\right)}} = \frac{3}{\sqrt{9 - \left(x^2 + y^2\right)}}
\end{aligned}$$

Since the surface is given by $z = f(x, y)$, the most convenient projection is onto the xy-plane.

For projection onto the xy-plane, the projection vector $\boldsymbol{p}$ is $\hat{\boldsymbol{k}} = \langle 0, 0, 1\rangle$. The denominator term is:

$$
\begin{aligned}
|\nabla g \cdot \boldsymbol{p}| = \left|\nabla g \cdot \hat{\boldsymbol{k}}\right| &= \left|\left\langle \frac{x}{\sqrt{9-(x^2+y^2)}}, \frac{y}{\sqrt{9-(x^2+y^2)}}, 1\right\rangle \cdot \langle 0, 0, 1\rangle\right| \\
&= \left|\frac{x}{\sqrt{9-(x^2+y^2)}} \times 0 + \frac{y}{\sqrt{9-(x^2+y^2)}} \times 0 + 1 \times 1\right| \\
&= |0+0+1| = |1| = 1
\end{aligned}
$$

The surface area $A(S)$ is:

$$
\begin{aligned}
A(S) &= \iint\limits_D \frac{|\nabla g|}{|\nabla g \cdot \boldsymbol{p}|} dA = \iint\limits_D \frac{3}{\sqrt{9-(x^2+y^2)}} dA \\
&= \int\limits_{-2}^{2} \int\limits_{-\sqrt{4-x^2}}^{+\sqrt{4-x^2}} \frac{3}{\sqrt{9-(x^2+y^2)}} dydx
\end{aligned}
$$

The integral is straightforward to solve using polar coordinates. In polar coordinates, the region D (Fig. 2.4) is described as $D = \{(r, \theta) | 0 \leq r \leq 2, 0 \leq \theta \leq 2\pi\}$. Thus,

$$
\begin{aligned}
A(S) &= \iint\limits_D \frac{|\nabla g|}{|\nabla g \cdot \boldsymbol{p}|} dA \\
&= \int\limits_{-2}^{2} \int\limits_{-\sqrt{4-x^2}}^{+\sqrt{4-x^2}} \frac{3}{\sqrt{9-(x^2+y^2)}} dydx \\
&= \int\limits_{0}^{2\pi} \int\limits_{0}^{2} \frac{3}{\sqrt{9-r^2}} \times r dr d\theta = \int\limits_{0}^{2\pi} \int\limits_{0}^{2} 3r\left(9-r^2\right)^{-\frac{1}{2}} dr d\theta
\end{aligned}
$$

The calculation of $\int_0^{2\pi} \int_0^2 3r\left(9-r^2\right)^{-\frac{1}{2}} dr d\theta$ is shown below:

$$
\begin{aligned}
\int\limits_{0}^{2} 3r\left(9-r^2\right)^{-\frac{1}{2}} dr &= \frac{3}{-2} \int\limits_{0}^{2} -2r\left(9-r^2\right)^{-\frac{1}{2}} dr = \left.\frac{3}{-2} \frac{\left(9-r^2\right)^{-\frac{1}{2}+1}}{-\frac{1}{2}+1}\right|_{0}^{2} \\
&= \left.-3\left(9-r^2\right)^{\frac{1}{2}}\right|_{0}^{2} = \left.3\left(9-r^2\right)^{\frac{1}{2}}\right|_{2}^{0}
\end{aligned}
$$

$$= 3\left(9 - 0^2\right)^{\frac{1}{2}} - 3\left(9 - 2^2\right)^{\frac{1}{2}} = 9 - 3\sqrt{5} \approx 2.2918$$

$$\int_0^{2\pi}\int_0^{2}3r\left(9 - r^2\right)^{-\frac{1}{2}}drd\theta = \int_0^{2\pi}\left(9 - 3\sqrt{5}\right)d\theta = \left(9 - 3\sqrt{5}\right)\int_0^{2\pi}d\theta$$
$$= \left(9 - 3\sqrt{5}\right) \times 2\pi = 2\left(9 - 3\sqrt{5}\right)\pi$$
$$= 6\left(3 - \sqrt{5}\right)\pi \approx 14.3998$$

The surface area is $2\left(9 - 3\sqrt{5}\right)\pi \approx 14.3998$ square units.

Example 2.6 Find the surface area of the portion of the cylinder $x^2 + z^2 = 16$ lying in the region $x \geq 0$ and satisfying $0 \leq y \leq x + 4$.

Solution $x^2 + z^2 = 16 \Rightarrow x = \pm\sqrt{16 - z^2} \overset{x\geq 0}{\Rightarrow} x = \sqrt{16 - z^2} \Rightarrow x = f(y, z) = \sqrt{16 - z^2}$.

Let $g(x, y, z) = x - \sqrt{16 - z^2}$. The partial derivatives of g are:

$$\frac{\partial g}{\partial x} = 1$$

$$\frac{\partial g}{\partial y} = 0$$

$$\frac{\partial g}{\partial z} = -\frac{1}{2} \times -2z \times \frac{1}{\sqrt{16 - z^2}} = \frac{z}{\sqrt{16 - z^2}}$$

The gradient vector is:

$$\nabla g = \frac{\partial g}{\partial x}\hat{\boldsymbol{i}} + \frac{\partial g}{\partial y}\hat{\boldsymbol{j}} + \frac{\partial g}{\partial z}\hat{\boldsymbol{k}} \Rightarrow \nabla g = \hat{\boldsymbol{i}} + \frac{z}{\sqrt{16 - z^2}}\hat{\boldsymbol{k}}$$

The magnitude of the gradient vector is:

$$|\nabla g| = \left|\hat{\boldsymbol{i}} + \frac{z}{\sqrt{16 - z^2}}\hat{\boldsymbol{k}}\right| = \sqrt{1^2 + \left(\frac{z}{\sqrt{16 - z^2}}\right)^2}$$
$$= \sqrt{1 + \frac{z^2}{16 - z^2}} = \sqrt{\frac{16}{16 - z^2}} = \frac{4}{\sqrt{16 - z^2}}$$

Since the surface is given by $x = f(y, z)$, the most convenient projection is onto the yz-plane.

For projection onto the yz-plane, the projection vector $\boldsymbol{p}$ is $\hat{\boldsymbol{i}} = 1, 0, 0$.

The denominator term is:

$$
\begin{aligned}
|\nabla g \cdot \boldsymbol{p}| = \left|\nabla g \cdot \hat{\boldsymbol{\imath}}\right| &= \left|\left\langle 1, 0, \frac{z}{\sqrt{16-z^2}}\right\rangle . \langle 1, 0, 0\rangle\right| \\
&= \left|1 \times 1 + 0 \times 0 + \frac{z}{\sqrt{16-z^2}} \times 0\right| \\
&= |1+0+0| = |1| = 1
\end{aligned}
$$

The region D is defined by the constraints given in the problem, projected onto the yz-plane:

Cylinder: $x^2 + z^2 = 16 \Rightarrow x = \pm\sqrt{16-z^2} \overset{x\geq 0}{\Rightarrow} x = \sqrt{16-z^2}$.

***y* Limits**: $0 \leq y \leq x + 4 \Rightarrow 0 \leq y \leq \sqrt{16-z^2} + 4$.

***z* Limits**: From $x^2 + z^2 = 16$ and $x^2 \geq 0$, we must have $16 - z^2 \geq 0$, so $-4 \leq z \leq 4$.

The region D in the yz-plane is:

$$
D = \left\{ (y, z) | -4 \leq z \leq 4, 0 \leq y \leq \sqrt{16-z^2} + 4 \right\}
$$

The region D is shown in Fig. 2.5.

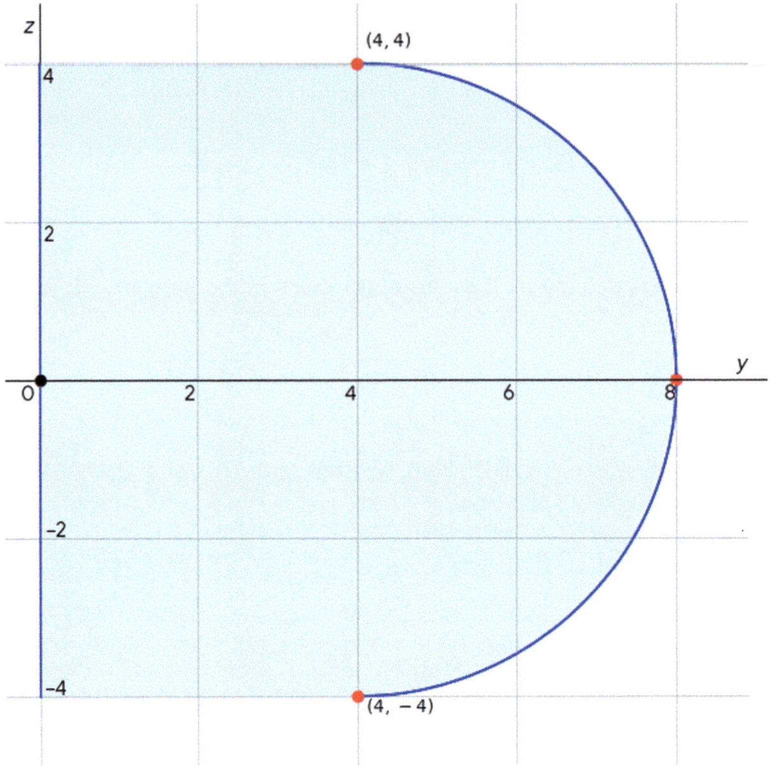

Fig. 2.5 Region $D = \left\{ (y, z) \mid -4 \leq z \leq 4, 0 \leq y \leq \sqrt{16 - z^2} + 4 \right\}$

The surface area $A(S)$ is:

$$A(S) = \iint\limits_D \frac{|\nabla g|}{|\nabla g \cdot \boldsymbol{p}|} dA = \iint\limits_D \frac{\frac{4}{\sqrt{16-z^2}}}{1} dA$$

$$= \iint\limits_D \frac{4}{\sqrt{16 - z^2}} dA = \int\limits_{-4}^{4} \int\limits_{0}^{\sqrt{16-z^2}+4} \frac{4}{\sqrt{16 - z^2}} dydz$$

The calculation of $\int_{-4}^{4} \int_0^{\sqrt{16-z^2}+4} \frac{4}{\sqrt{16-z^2}} dydz$ is shown below:

$$\int\limits_{0}^{\sqrt{16-z^2}+4} \frac{4}{\sqrt{16 - z^2}} dy = \frac{4}{\sqrt{16 - z^2}} \int\limits_{0}^{\sqrt{16-z^2}+4} dy$$

$$= \frac{4}{\sqrt{16 - z^2}} \left(\sqrt{16 - z^2} + 4 - 0 \right)$$

$$= 4 + \frac{16}{\sqrt{16 - z^2}}$$

$$\int_{-4}^{4} \int_{0}^{\sqrt{16-z^2}+4} \frac{4}{\sqrt{16 - z^2}} dydz = \int_{-4}^{4} \left(4 + \frac{16}{\sqrt{16 - z^2}}\right) dz$$

$$= 2 \int_{0}^{4} \left(4 + \frac{16}{\sqrt{16 - z^2}}\right) dz$$

$$= 2 \int_{0}^{4} 4dz + 2 \int_{0}^{4} \frac{16}{\sqrt{16 - z^2}} dz$$

$$= 8 \int_{0}^{4} dz + 32 \int_{0}^{4} \frac{1}{\sqrt{4^2 - z^2}} dz$$

$$= 8 \times (4 - 0) + 32 \times \sin^{-1}\left(\frac{z}{4}\right)\Big|_{0}^{4}$$

$$= 32 + \left(32 \sin^{-1}\left(\frac{4}{4}\right) - 32 \sin^{-1}\left(\frac{0}{4}\right)\right)$$

$$= 32 + 32 \sin^{-1}(1) = 32 + 32 \times \frac{\pi}{2}$$

$$= 32 + 16\pi \approx 82.2655$$

The surface area is $32 + 16\pi \approx 82.2655$ square units.

Example 2.7 Find the surface area of the portion of the plane $2x + 2y + z = 10$ that lies directly above the circular region R in the xy-plane defined by $x^2 + y^2 \leq 9$.

Solution We define the surface explicitly as a function of x and y:

$$z = f(x, y) = 10 - 2x - 2y$$

Next, we find the partial derivatives:

$$\frac{\partial z}{\partial x} = -2$$

$$\frac{\partial z}{\partial y} = -2$$

The total surface area is given by:

$$A(S) = \iint_D \sqrt{\left(\frac{\partial z}{\partial x}\right)^2 + \left(\frac{\partial z}{\partial y}\right)^2 + 1} dA = \iint_D \sqrt{(-2)^2 + (-2)^2 + 1} dA$$
$$= \iint_D \sqrt{9} dA = 3 \iint_D dA$$

Since $\iint_D dA$ is simply the area of the region D (a circle of radius $\sqrt{9} = 3$):

$$A(S) = 3 \iint_D dA = 3 \times \left(\pi \times 3^2\right) = 27\pi \approx 84.8230$$

The surface area is $27\pi \approx 84.8230$ square units.

Example 2.8 Find the surface area of the portion of the paraboloid $z = x^2 + y^2$ that lies below the plane $z = 9$.

Solution The surface is defined by $z = f(x, y) = x^2 + y^2$. Next, we find the partial derivatives:

$$\frac{\partial z}{\partial x} = 2x$$

$$\frac{\partial z}{\partial y} = 2y$$

The total surface area is given by:

$$A(S) = \iint_D \sqrt{\left(\frac{\partial z}{\partial x}\right)^2 + \left(\frac{\partial z}{\partial y}\right)^2 + 1} dA = \iint_D \sqrt{(2x)^2 + (2y)^2 + 1} dA$$
$$= \iint_D \sqrt{4\left(x^2 + y^2\right) + 1} dA$$

The surface lies below the plane $z = 9$. The boundary of the projection region D in the xy-plane is found by setting $z = 9$:

$$z = x^2 + y^2 \Rightarrow 9 = x^2 + y^2$$

Thus, the region D in the xy-plane is the disk defined by $x^2 + y^2 \leq 9$. The region D is shown in Fig. 2.6.

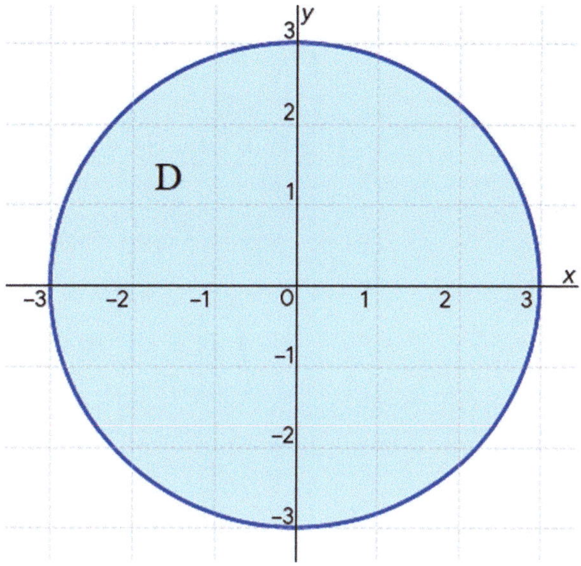

Fig. 2.6 Region D

The limits of integration for this region are:

$$-3 \le x \le 3, -\sqrt{9-x^2} \le y \le \sqrt{9-x^2}$$

Therefore,

$$A(S) = \iint_D \sqrt{4\left(x^2+y^2\right)+1}\,dA = \int_{-3}^{3}\int_{-\sqrt{9-x^2}}^{+\sqrt{9-x^2}} \sqrt{4\left(x^2+y^2\right)+1}\,dydx$$

The integral is straightforward to solve using polar coordinates. In polar coordinates, the region D (Fig. 2.6) is described as $D = \{(r,\theta)|0 \le r \le 3, 0 \le \theta \le 2\pi\}$. Thus,

$$A(S) = \int_{-3}^{3}\int_{-\sqrt{9-x^2}}^{+\sqrt{9-x^2}} \sqrt{4\left(x^2+y^2\right)+1}\,dydx = \int_0^{2\pi}\int_0^3 \sqrt{4r^2+1} \times r\,dr\,d\theta$$

$$= \int_0^{2\pi}\int_0^3 r\sqrt{4r^2+1}\,dr\,d\theta = \int_0^3 r\sqrt{4r^2+1}\,dr \int_0^{2\pi} d\theta$$

Let's calculate each integral separately.

$$\int_0^3 r\sqrt{4r^2+1}dr = \frac{1}{8}\int_0^3 8r\sqrt{4r^2+1}dr = \frac{1}{8}\frac{\left(4r^2+1\right)^{\frac{1}{2}+1}}{\frac{1}{2}+1}\Bigg|_0^3 = \frac{\left(4r^2+1\right)^{\frac{3}{2}}}{12}\Bigg|_0^3$$

$$= \frac{\left(4\times 3^2+1\right)^{\frac{3}{2}}}{12} - \frac{\left(4\times 0^2+1\right)^{\frac{3}{2}}}{12}$$

$$= \frac{37^{\frac{3}{2}}}{12} - \frac{1^{\frac{3}{2}}}{12} = \frac{37^{\frac{3}{2}}-1}{12} = \frac{37\sqrt{37}-1}{12} \approx 18.6719$$

$$\int_0^{2\pi} d\theta = 2\pi - 0 = 2\pi$$

Thus,

$$A(S) = \int_0^3 r\sqrt{4r^2+1}dr\int_0^{2\pi} d\theta = \frac{37\sqrt{37}-1}{12}\times 2\pi = \frac{37\sqrt{37}-1}{6}\pi \approx 117.3187$$

The surface area is $\frac{37\sqrt{37}-1}{6}\pi \approx 117.3187$ square units.

Example 2.9 Find the surface area of the portion of the cone $y=\sqrt{x^2+z^2}$ that lies between the planes $y=1$ and $y=4$.

Solution $y=\sqrt{x^2+z^2} \Rightarrow y-\sqrt{x^2+z^2}=0$.

Let $g(x,y,z) = y-\sqrt{x^2+z^2}$. The partial derivatives of g are:

$$\frac{\partial g}{\partial x} = -\frac{1}{2}\times 2x\times\frac{1}{\sqrt{x^2+z^2}} = -\frac{x}{\sqrt{x^2+z^2}}$$

$$\frac{\partial g}{\partial y} = 1$$

$$\frac{\partial g}{\partial z} = -\frac{1}{2}\times 2z\times\frac{1}{\sqrt{x^2+z^2}} = -\frac{z}{\sqrt{x^2+z^2}}$$

The gradient vector is:

$$\nabla g = \frac{\partial g}{\partial x}\hat{\boldsymbol{\imath}} + \frac{\partial g}{\partial y}\hat{\boldsymbol{\jmath}} + \frac{\partial g}{\partial z}\hat{\boldsymbol{k}} \Rightarrow \nabla g = -\frac{x}{\sqrt{x^2+z^2}}\hat{\boldsymbol{\imath}} + \hat{\boldsymbol{\jmath}} - \frac{z}{\sqrt{x^2+z^2}}\hat{\boldsymbol{k}}$$

The magnitude of the gradient vector is:

$$|\nabla g| = \left|-\frac{x}{\sqrt{x^2+z^2}}\hat{\boldsymbol{\imath}} + \hat{\boldsymbol{\jmath}} - \frac{z}{\sqrt{x^2+z^2}}\hat{\boldsymbol{k}}\right|$$

$$= \sqrt{\left(-\frac{x}{\sqrt{x^2+z^2}}\right)^2 + 1^2 + \left(-\frac{z}{\sqrt{x^2+z^2}}\right)^2}$$

$$= \sqrt{\frac{x^2}{x^2+z^2} + 1 + \frac{z^2}{x^2+z^2}} = \sqrt{1+1} = \sqrt{2}$$

Since the surface is given by $y = f(x, z)$, the most convenient projection is onto the xz-plane.

For projection onto the xz-plane, the projection vector $\boldsymbol{p}$ is $\hat{\boldsymbol{j}} = 0, 1, 0$.

The denominator term is:

$$|\nabla g \cdot \boldsymbol{p}| = \left|\nabla g \cdot \hat{\boldsymbol{j}}\right| = \left|\left\langle -\frac{x}{\sqrt{x^2+z^2}}, 1, -\frac{z}{\sqrt{x^2+z^2}}\right\rangle \cdot \langle 0, 1, 0\rangle\right|$$

$$= \left|-\frac{x}{\sqrt{x^2+z^2}} \times 0 + 1 \times 1 - \frac{z}{\sqrt{x^2+z^2}} \times 0\right|$$

$$= |0 + 1 + 0| = |1| = 1$$

We will now project the surface onto the xz-plane. The projection region D in the xz-plane is determined by the bounds on y:

$$1 \le y \le 4.$$

Substituting the expression for y:

$$1 \le \sqrt{x^2+z^2} \le 4$$

Squaring all parts gives the annular (ring) region D:

$$1 \le x^2 + z^2 \le 16$$

The region D is shown in Fig. 2.7.

Fig. 2.7 Region D

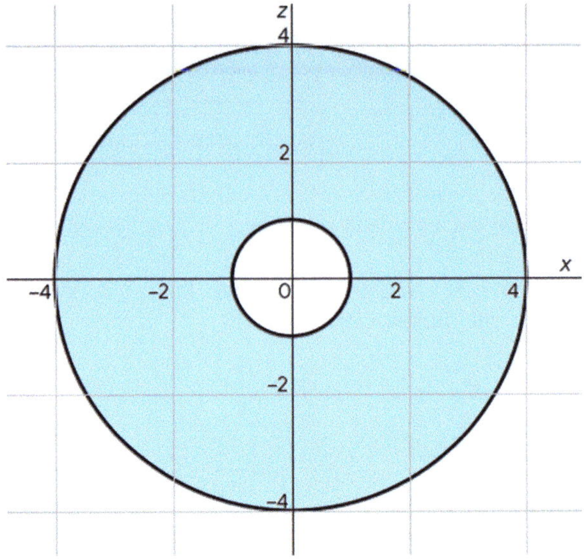

The total surface area is given by:

$$A(S) = \iint\limits_D \frac{|\nabla g|}{|\nabla g \cdot \boldsymbol{p}|} dA = \iint\limits_D \frac{\sqrt{2}}{1} dA = \iint\limits_D \sqrt{2} dA = \sqrt{2} \iint\limits_D dA$$

Since $\iint_D dA$ is simply the area of the region D:

$$\begin{aligned} A(S) &= \sqrt{2} \iint\limits_D dA = \sqrt{2}(\pi \times 4^2 - \pi \times 1^2) \\ &= \sqrt{2}(16\pi - \pi) \\ &= 15\sqrt{2}\pi \approx 66.6432 \end{aligned}$$

The surface area is $15\sqrt{2}\pi \approx 66.6432$ square units.

Example 2.10 Find the surface area of the portion of the cylinder $y^2 + z^2 = 9$ that lies above the triangular region D in the xy-plane with vertices $(0, 0)$, $(3, 0)$, and $(0, 3)$ (Fig. 2.8), assuming $z \geq 0$.

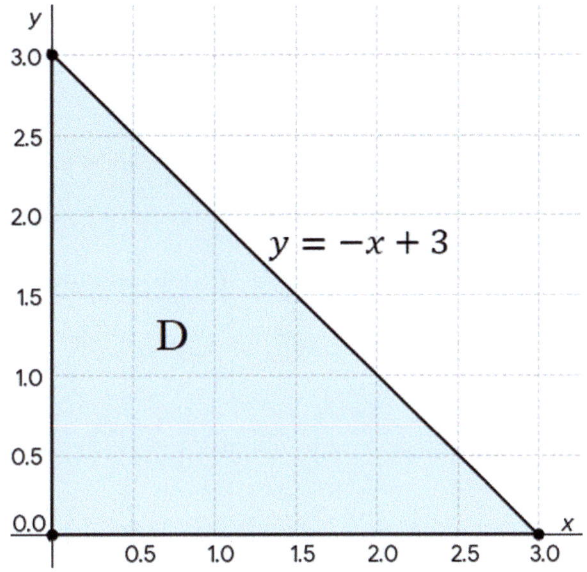

Fig. 2.8 Region D for Example 2.10

Solution $y^2 + z^2 = 9 \Rightarrow z^2 = 9 - y^2 \Rightarrow z = \pm\sqrt{9 - y^2} \overset{z\geq 0}{\Rightarrow} z = \sqrt{9 - y^2}$.

We find the partial derivatives of $z = f(x, y) = \sqrt{9 - y^2}$:

$$\frac{\partial z}{\partial x} = 0$$

$$\frac{\partial z}{\partial y} = \frac{1}{2} \times -2y \times \frac{1}{\sqrt{9 - y^2}} = -\frac{y}{\sqrt{9 - y^2}}$$

The total surface area A is given by:

$$\begin{aligned} A(S) &= \iint_D \sqrt{\left(\frac{\partial z}{\partial x}\right)^2 + \left(\frac{\partial z}{\partial y}\right)^2 + 1dA} \\ &= \iint_D \sqrt{0^2 + \left(-\frac{y}{\sqrt{9 - y^2}}\right)^2 + 1dA} = \iint_D \sqrt{\frac{y^2}{9 - y^2} + 1dA} \\ &= \iint_D \sqrt{\frac{9}{9 - y^2}}dA = \iint_D \frac{3}{\sqrt{9 - y^2}}dA = 3\iint_D \frac{1}{\sqrt{9 - y^2}}dA \end{aligned}$$

The integration region D is shown in Fig. 2.8. Therefore,

$$A(S) = 3\iint_D \frac{1}{\sqrt{9-y^2}}dA = 3\int_0^3\int_0^{-x+3} \frac{1}{\sqrt{9-y^2}}dydx$$

The calculation of $3\int_0^3\int_0^{-x+3} \frac{1}{\sqrt{9-y^2}}dydx$ is shown below:

$$\int_0^{-x+3} \frac{1}{\sqrt{9-y^2}}dy = \int_0^{-x+3} \frac{1}{\sqrt{3^2-y^2}}dy = \sin^{-1}\left(\frac{y}{3}\right)\Big|_0^{-x+3}$$
$$= \sin^{-1}\left(\frac{-x+3}{3}\right) - \sin^{-1}\left(\frac{0}{3}\right)$$
$$= \sin^{-1}\left(\frac{3-x}{3}\right) - 0 = \sin^{-1}\left(\frac{3-x}{3}\right)$$

$$3\int_0^3\int_0^{-x+3} \frac{1}{\sqrt{9-y^2}}dydx = 3\int_0^3 \sin^{-1}\left(\frac{3-x}{3}\right)dx$$
$$= 3\times -3\int_0^3 -\frac{1}{3}\sin^{-1}\left(\frac{3-x}{3}\right)dx$$
$$= 9\int_3^0 -\frac{1}{3}\sin^{-1}\left(\frac{3-x}{3}\right)dx$$

Recall that $\int \sin^{-1}(u)du = u\sin^{-1}(u) + \sqrt{1-u^2} + C$. Thus,

$$3\int_0^3\int_0^{-x+3} \frac{1}{\sqrt{9-y^2}}dydx = 9\int_3^0 -\frac{1}{3}\sin^{-1}\left(\frac{3-x}{3}\right)dx$$
$$= 9\left(\left(\frac{3-x}{3}\right)\sin^{-1}\left(\frac{3-x}{3}\right) + \sqrt{1-\left(\frac{3-x}{3}\right)^2}\right)\Bigg|_3^0$$
$$= 9\left(\left(\frac{3-x}{3}\right)\sin^{-1}\left(\frac{3-x}{3}\right) + \sqrt{\frac{6x-x^2}{9}}\right)\Bigg|_3^0$$
$$= 3\left((3-x)\sin^{-1}\left(\frac{3-x}{3}\right) + \sqrt{6x-x^2}\right)\Bigg|_3^0$$
$$= 3\left((3-0)\sin^{-1}\left(\frac{3-0}{3}\right) + \sqrt{6\times 0-0^2}\right)$$
$$- 3\left((3-3)\sin^{-1}\left(\frac{3-3}{3}\right) + \sqrt{6\times 3-3^2}\right)$$

$$= 9\sin^{-1}(1) - 9 = \frac{9\pi}{2} - 9 \approx 5.1372$$

The surface area is $\frac{9\pi}{2} - 9 \approx 5.1372$ square units.

Example 2.11 Find the surface area of the portion of the paraboloid $z = 1 + x^2 + y^2$ that lies directly above the region D in the xy-plane defined by the constraints:

$$2 \leq r \leq 3, \frac{\pi}{4} \leq \theta \leq \frac{\pi}{2}$$

The region D is shown in Fig. 2.9.

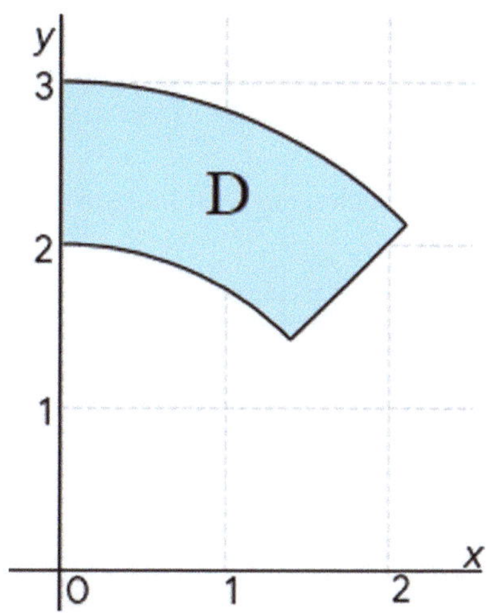

Fig. 2.9 Region D for Example 2.11

Solution $z = 1+x^2+y^2$. We find the partial derivatives of $z = f(x, y) = 1+x^2+y^2$:

$$\frac{\partial z}{\partial x} = 2x$$

$$\frac{\partial z}{\partial y} = 2y$$

The total surface area A is given by:

$$A(S) = \iint\limits_D \sqrt{\left(\frac{\partial z}{\partial x}\right)^2 + \left(\frac{\partial z}{\partial y}\right)^2 + 1dA} = \iint\limits_D \sqrt{(2x)^2 + (2y)^2 + 1dA}$$

$$= \iint\limits_D \sqrt{4(x^2 + y^2) + 1dA}$$

Since the region D is defined by polar coordinates, we must convert the integral into the polar system:

$$x^2 + y^2 = r^2$$

$$dA = rdrd\theta$$

$$A(S) = \iint_D \sqrt{4(x^2 + y^2) + 1dA} = \int_{\frac{\pi}{4}}^{\frac{\pi}{2}} \int_2^3 \sqrt{4r^2 + 1} \times rdrd\theta$$

$$= \int_{\frac{\pi}{4}}^{\frac{\pi}{2}} \int_2^3 r\sqrt{4r^2 + 1}drd\theta = \int_2^3 r\sqrt{4r^2 + 1}dr \int_{\frac{\pi}{4}}^{\frac{\pi}{2}} d\theta$$

Let's calculate each integral separately.

$$\int_2^3 r\sqrt{4r^2 + 1}dr = \frac{1}{8}\int_2^3 8r\sqrt{4r^2 + 1}dr = \frac{1}{8}\frac{\left(4r^2 + 1\right)^{\frac{1}{2}+1}}{\frac{1}{2}+1}\Bigg|_2^3$$

$$= \frac{\left(4r^2 + 1\right)^{\frac{3}{2}}}{12}\Bigg|_2^3 = \frac{\left(4 \times 3^2 + 1\right)^{\frac{3}{2}}}{12} - \frac{\left(4 \times 2^2 + 1\right)^{\frac{3}{2}}}{12}$$

$$= \frac{37^{\frac{3}{2}} - 17^{\frac{3}{2}}}{12} = \frac{37\sqrt{37} - 17\sqrt{17}}{12}$$

$$\int_{\frac{\pi}{4}}^{\frac{\pi}{2}} d\theta = \frac{\pi}{2} - \frac{\pi}{4} = \frac{\pi}{4}$$

Thus,

$$A(S) = \int_2^3 r\sqrt{4r^2 + 1}dr \int_{\frac{\pi}{4}}^{\frac{\pi}{2}} d\theta = \frac{37\sqrt{37} - 17\sqrt{17}}{12}\frac{\pi}{4}$$

$$= \frac{37\sqrt{37} - 17\sqrt{17}}{48}\pi \approx 10.1427$$

The surface area is $\frac{37\sqrt{37}-17\sqrt{17}}{48}\pi \approx 10.1427$ square units.

Example 2.12 Find the surface area of the part of the ellipsoid $x^2 + y^2 + 4z^2 = 10$ that lies above the plane $z = 1$.

Solution If S is the part of the surface $F(x, y, z) = c$ that projects onto a region D in the xy-plane, the surface area $A(S)$ is given by:

$$A(S) = \iint_D \frac{|\nabla F|}{\left|\nabla F \cdot \hat{\boldsymbol{k}}\right|} dA$$

where:

$\nabla F = \langle F_x, F_y, F_z \rangle$ is the gradient vector.

$|\nabla F| = \sqrt{F_x^2 + F_y^2 + F_z^2}$ is the magnitude of the gradient.

$\left|\nabla F \cdot \hat{\boldsymbol{k}}\right| = |F_z|$ is the absolute value of the component of the gradient in the z-direction.

We set the equation to zero to define $F(x, y, z)$:

$$F(x, y, z) = x^2 + y^2 + 4z^2 - 10 = 0$$

Let's calculate the gradient:

$$F_x = \frac{\partial F}{\partial x} = 2x$$

$$F_y = \frac{\partial F}{\partial y} = 2y$$

$$F_z = \frac{\partial F}{\partial z} = 8z$$

$$\nabla F = \langle F_x, F_y, F_z \rangle = \langle 2x, 2y, 8z \rangle$$

The magnitude of the gradient is:

$$|\nabla F| = \sqrt{(2x)^2 + (2y)^2 + (8z)^2} = \sqrt{4x^2 + 4y^2 + 64z^2} = 2\sqrt{x^2 + y^2 + 16z^2}$$

The absolute value of the component of the gradient in the z-direction is:

$$\left|\nabla F \cdot \hat{\boldsymbol{k}}\right| = |F_z| = |8z| = 8|z| \overset{z \geq 1}{\Rightarrow} \left|\nabla F \cdot \hat{\boldsymbol{k}}\right| = 8z$$

The integrand is:

$$\frac{|\nabla F|}{\left|\nabla F \cdot \hat{\boldsymbol{k}}\right|} = \frac{|\nabla F|}{|F_z|} = \frac{2\sqrt{x^2 + y^2 + 16z^2}}{8z} = \frac{\sqrt{x^2 + y^2 + 16z^2}}{4z}$$

Because we are integrating over a region D in the xy-plane, we must replace z using the original surface equation $x^2 + y^2 + 4z^2 = 10$.

$$x^2 + y^2 + 4z^2 = 10 \Rightarrow z^2 = \frac{10 - \left(x^2 + y^2\right)}{4}$$

$$\frac{|\nabla F|}{\left|\nabla F \cdot \hat{\boldsymbol{k}}\right|} = \frac{|\nabla F|}{|F_z|} = \frac{\sqrt{x^2 + y^2 + 16z^2}}{4z} = \frac{\sqrt{x^2 + y^2 + 16 \times \frac{10-(x^2+y^2)}{4}}}{4 \times \sqrt{\frac{10-(x^2+y^2)}{4}}}$$

$$= \frac{\sqrt{x^2 + y^2 + 40 - 4\left(x^2 + y^2\right)}}{2\sqrt{10 - \left(x^2 + y^2\right)}} = \frac{\sqrt{40 - 3\left(x^2 + y^2\right)}}{2\sqrt{10 - \left(x^2 + y^2\right)}}$$

Let's identify the region D. The boundary of the surface is where it intersects $z = 1$.

$$\begin{cases} x^2 + y^2 + 4z^2 = 10 \\ z = 1 \end{cases} \Rightarrow x^2 + y^2 + 4 \times 1^2 = 10 \Rightarrow x^2 + y^2 = 6$$

The projection D is a disk of radius $\sqrt{6}$ in the xy-plane (Fig. 2.10).

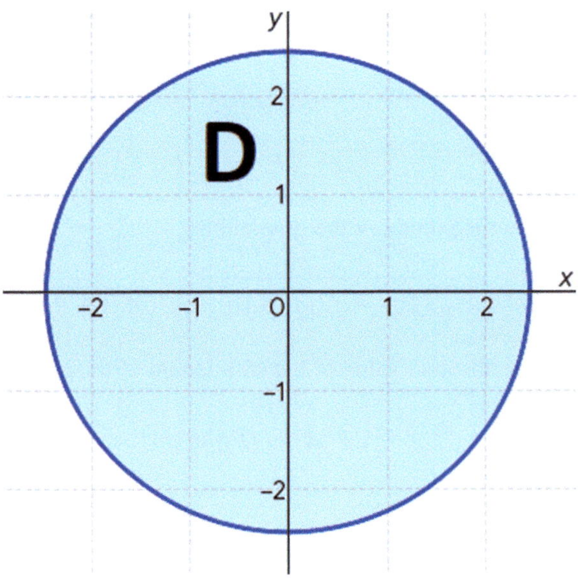

Fig. 2.10 Region D is a disk with a radius of $\sqrt{6}$

Switching to polar coordinates ($x^2 + y^2 = r^2$):

$$\frac{|\nabla F|}{|\nabla F \cdot \hat{\boldsymbol{k}}|} = \frac{\sqrt{40 - 3(x^2 + y^2)}}{2\sqrt{10 - (x^2 + y^2)}} = \frac{\sqrt{40 - 3r^2}}{2\sqrt{10 - r^2}}$$

The integral becomes:

$$A(s) = \int_0^{2\pi}\int_0^{\sqrt{6}} \frac{\sqrt{40 - 3r^2}}{2\sqrt{10 - r^2}} \times r dr d\theta = \frac{1}{2}\int_0^{2\pi} d\theta \int_0^{\sqrt{6}} \frac{r\sqrt{40 - 3r^2}}{\sqrt{10 - r^2}} dr$$

$$= \frac{1}{2} \times (2\pi - 0) \times \int_0^{\sqrt{6}} \frac{r\sqrt{40 - 3r^2}}{\sqrt{10 - r^2}} dr = \pi \int_0^{\sqrt{6}} \frac{r\sqrt{40 - 3r^2}}{\sqrt{10 - r^2}} dr$$

Since performing this integration by hand is computationally difficult, we will evaluate it using MATLAB (Fig. 2.11).

```
Command Window
>> syms r
>> I=eval(int(r*sqrt(40-3*r^2)/sqrt(10-r^2),r,0,sqrt(6)))

I =

    6.3768

fx >>
```

Fig. 2.11 MATLAB code

Therefore,

$$A(s) = \pi \int_0^{\sqrt{6}} \frac{r\sqrt{40 - 3r^2}}{\sqrt{10 - r^2}} dr = \pi \times 6.3768 \approx 20.0333$$

The surface area is 20.0333 square units.

Exercise 2.1 Find the area of the part of the plane $2x + 2y + z = 10$ that lies inside the cylinder $x^2 + y^2 = 9$. (Ans. $27\pi \approx 84.8230$ square units).

Example 2.13 Find the surface area of the helicoid defined by:

$$\boldsymbol{r}(u, v) = \langle u\cos(v), u\sin(v), v\rangle$$

where $0 \leq u \leq 1$ and $0 \leq v \leq \pi$.

Solution To calculate the area of a surface defined parametrically, we use the magnitude of the cross product of the tangent vectors. For a surface defined by the vector-valued function $\boldsymbol{r}(u, v)$ over a parameter domain D, the surface area $A(S)$ is:

$$A(S) = \iint_D |\boldsymbol{r}_u \times \boldsymbol{r}_v| du dv$$

First, we find the partial derivatives of $\boldsymbol{r}$ with respect to $\boldsymbol{u}$ and $\boldsymbol{v}$:

$$\boldsymbol{r}_u = \frac{\partial \boldsymbol{r}}{\partial u} = \langle \cos(v), \sin(v), 0 \rangle$$

$$\boldsymbol{r}_v = \frac{\partial \boldsymbol{r}}{\partial v} = \langle -u\sin(v), u\cos(v), 1 \rangle$$

Next, we find the cross product $\boldsymbol{r}_u \times \boldsymbol{r}_v$:

$$\begin{aligned}
\boldsymbol{r}_u \times \boldsymbol{r}_v &= \begin{vmatrix} \hat{\boldsymbol{i}} & \hat{\boldsymbol{j}} & \hat{\boldsymbol{k}} \\ \cos(v) & \sin(v) & 0 \\ -u\sin(v) & u\cos(v) & 1 \end{vmatrix} \\
&= \hat{\boldsymbol{i}} \begin{vmatrix} \sin(v) & 0 \\ u\cos(v) & 1 \end{vmatrix} - \hat{\boldsymbol{j}} \begin{vmatrix} \cos(v) & 0 \\ -u\sin(v) & 1 \end{vmatrix} + \hat{\boldsymbol{k}} \begin{vmatrix} \cos(v) & \sin(v) \\ -u\sin(v) & u\cos(v) \end{vmatrix} \\
&= \sin(v)\hat{\boldsymbol{i}} - \cos(v)\hat{\boldsymbol{j}} + \left(u\cos^2(v) + u\sin^2(v)\right)\hat{\boldsymbol{k}} \\
&= \sin(v)\hat{\boldsymbol{i}} - \cos(v)\hat{\boldsymbol{j}} + u\hat{\boldsymbol{k}}
\end{aligned}$$

The magnitude of $|\boldsymbol{r}_u \times \boldsymbol{r}_v|$ is:

$$|\boldsymbol{r}_u \times \boldsymbol{r}_v| = \left|\sin(v)\hat{\boldsymbol{i}} - \cos(v)\hat{\boldsymbol{j}} + u\hat{\boldsymbol{k}}\right| = \sqrt{\sin^2(v) + \cos^2(v) + u^2} = \sqrt{1 + u^2}$$

Now, we integrate over the domain D:

$$A(S) = \int_0^{\pi} \int_0^1 \sqrt{1 + u^2} du dv$$

The inner integral $\int_0^1 \sqrt{1 + u^2} du$ is a standard integral (often solved via trigonometric substitution $u = \tan(\theta)$):

$$\int \sqrt{1 + u^2} du = \frac{1}{2}\left(u\sqrt{1 + u^2} + \ln\left(\left|u + \sqrt{1 + u^2}\right|\right)\right) + C$$

$$\int_0^1 \sqrt{1+u^2}du = \frac{1}{2}\left(1 \times \sqrt{1+1^2} + \ln\left(\left|1+\sqrt{1+1^2}\right|\right)\right)$$
$$-\frac{1}{2}\left(0 \times \sqrt{1+0^2} + \ln\left(\left|0+\sqrt{1+0^2}\right|\right)\right)$$
$$= \frac{\sqrt{2}+\ln\left(\left|1+\sqrt{2}\right|\right)}{2} \approx 1.1478$$

Finally, integrate with respect to v:

$$A(S) = \int_0^{\pi}\int_0^1 \sqrt{1+u^2}dudv = \int_0^{\pi} \frac{\sqrt{2}+\ln\left(\left|1+\sqrt{2}\right|\right)}{2}dv$$
$$= \frac{\sqrt{2}+\ln\left(\left|1+\sqrt{2}\right|\right)}{2} \times \pi \approx 3.6059$$

The area is approximately 3.6059 square units.

Exercise 2.2 Find the surface area of a sphere of radius a given by parametric equation:

$$\boldsymbol{r}(\theta, \phi) = a\sin(\phi)\cos(\theta), a\sin(\phi)\sin(\theta), a\cos(\phi)$$

where $0 \le \phi \le \pi$ and $0 \le \theta \le 2\pi$. (Ans. Ans. $A(S) = 4\pi a^2$)

Example 2.14 Use MATLAB to evaluate the function $f(x, y) = x^2\sin(y) + e^{xy}$ at the point $\left(1, \frac{\pi}{6}\right)$.

Solution The required code is shown in Fig. 2.12. Based on the results shown in Fig. 2.12, the value of $f\left(1, \frac{\pi}{6}\right)$ is 2.1881.

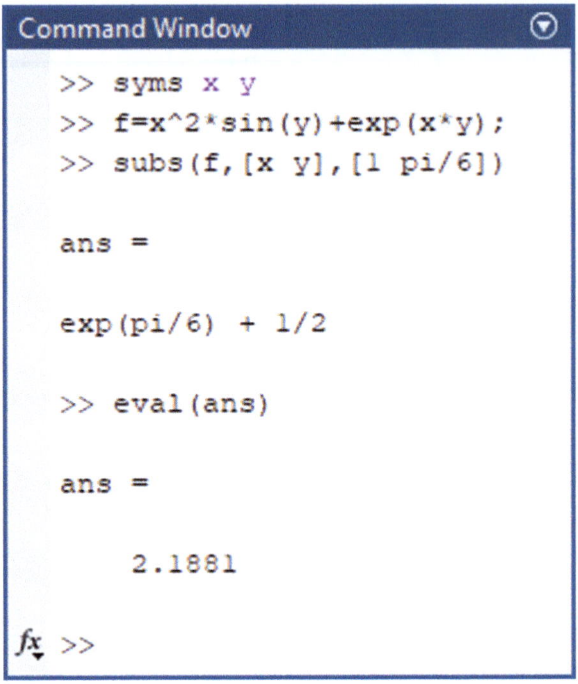

Fig. 2.12 Evaluation of $f(x, y) = x^2 \sin(y) + e^{xy}$ at $\left(1, \frac{\pi}{6}\right)$

Example 2.15 Use MATLAB to determine the partial derivative $\frac{\partial f}{\partial x}$ for the function $f(x, y) = \sqrt{9 - \left(x^2 + y^2\right)}$.

Solution The required code is shown in Fig. 2.13.

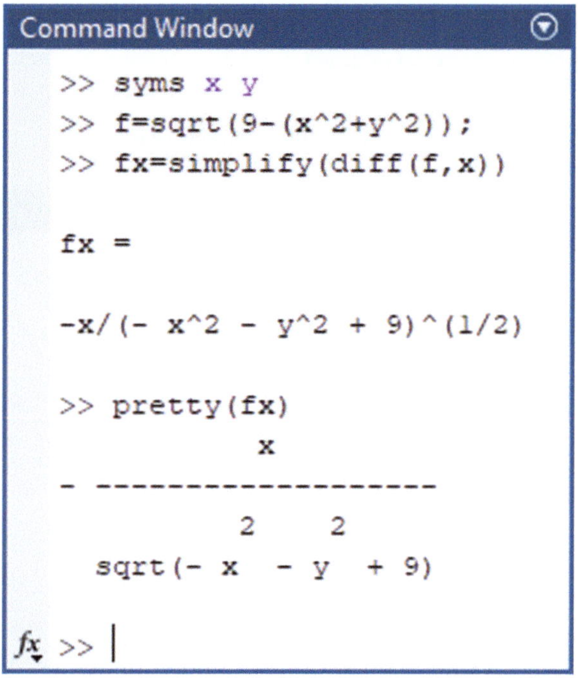

```
Command Window
>> syms x y
>> f=sqrt(9-(x^2+y^2));
>> fx=simplify(diff(f,x))

fx =

-x/(- x^2 - y^2 + 9)^(1/2)

>> pretty(fx)
              x
- -------------------
            2     2
  sqrt(- x  - y  + 9)

>>
```

Fig. 2.13 Calculation of $\frac{\partial f}{\partial x}$ for $f(x, y) = \sqrt{9 - (x^2 + y^2)}$

Example 2.16 Use MATLAB to determine the partial derivative $\frac{\partial f}{\partial y}$ at the point (1, 2) for the function $f(x, y) = \sqrt{9 - \left(x^2 + y^2\right)}$.

Solution The required code is shown in Fig. 2.14.

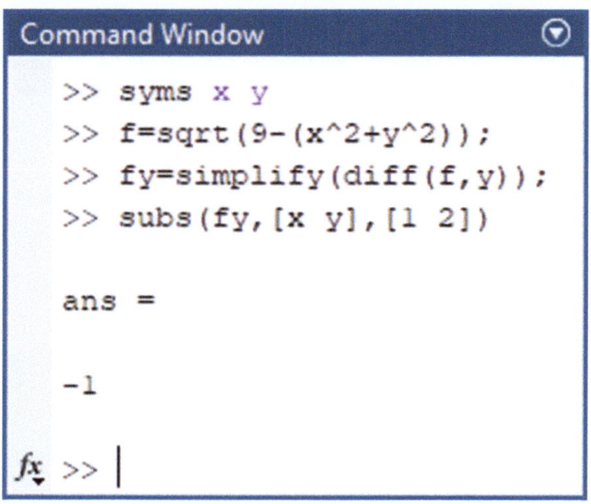

```
Command Window
>> syms x y
>> f=sqrt(9-(x^2+y^2));
>> fy=simplify(diff(f,y));
>> subs(fy,[x y],[1 2])

ans =

-1

>>
```

Fig. 2.14 Calculation of $\frac{\partial f}{\partial y}$ at (1, 2) for $f(x, y) = \sqrt{9 - \left(x^2 + y^2\right)}$

Example 2.17 Use MATLAB to determine the partial derivative $\frac{\partial^2 f}{\partial x^2}$ for the function $f(x, y) = \sqrt{9 - (x^2 + y^2)}$.

Solution The required code is shown in Fig. 2.15.

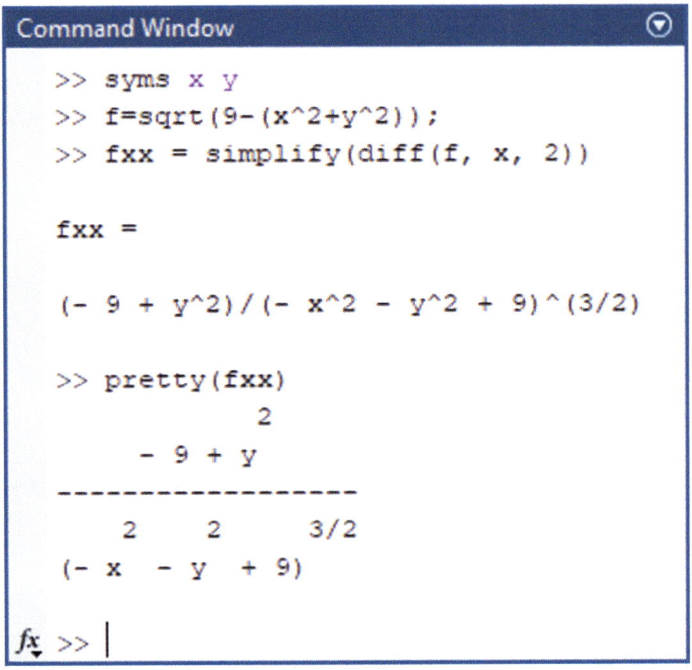

Fig. 2.15 Calculation of $\frac{\partial^2 f}{\partial x^2}$ for $f(x, y) = \sqrt{9 - (x^2 + y^2)}$

Example 2.18 Use MATLAB to determine the partial derivative $\frac{\partial^2 f}{\partial y^2}$ for the function $f(x, y) = \sqrt{9 - (x^2 + y^2)}$.

Solution The required code is shown in Fig. 2.16.

```
Command Window

>> syms x y
>> f=sqrt(9-(x^2+y^2));
>> fyy = simplify(diff(f, y, 2))

fyy =

(- 9 + x^2)/(- x^2 - y^2 + 9)^(3/2)

>> pretty(fyy)
              2
       - 9 + x
------------------
     2     2      3/2
(- x  - y  + 9)

fx >>
```

Fig. 2.16 Calculation of $\frac{\partial^2 f}{\partial y^2}$ for $f(x, y) = \sqrt{9 - (x^2 + y^2)}$

Example 2.19 Use MATLAB to determine the partial derivative $\frac{\partial^2 f}{\partial x \partial y}$ for the function $f(x, y) = \sqrt{9 - (x^2 + y^2)}$.

Solution The required code is shown in Fig. 2.17.

```
Command Window

>> syms x y
>> f=sqrt(9-(x^2+y^2));
>> fxy = simplify(diff(f, x, y))

fxy =

-(x*y)/(- x^2 - y^2 + 9)^(3/2)

>> pretty(fxy)
           x y
- -------------------
       2     2     3/2
  (- x  - y  + 9)

>>
```

Fig. 2.17 Calculation of $\frac{\partial^2 f}{\partial x \partial y}$ for $f(x, y) = \sqrt{9 - (x^2 + y^2)}$

Example 2.20 Use MATLAB to determine the partial derivative $\frac{\partial^3 f}{\partial x \partial y^2}$ for the function $f(x, y, z) = x^3 y^2 z + \sin(xy) + e^z$.

Solution The required code is shown in Fig. 2.18.

```
Command Window

>> syms x y z
>> f = x^3 * y^2 * z + sin(x*y) + exp(z);
>> simplify(diff(f,x,y,y))

ans =

6*x^2*z - 2*x*sin(x*y) - x^2*y*cos(x*y)

>> pretty(ans)
    2                          2
6 x  z - 2 x sin(x y) - x  y cos(x y)

>>
```

Fig. 2.18 Calculation of $\frac{\partial^3 f}{\partial x \partial y^2}$ for $f(x, y, z) = x^3 y^2 z + \sin(xy) + e^z$

Example 2.21 Use MATLAB to determine the partial derivative $\frac{\partial^3 f}{\partial x \partial y \partial z}$ for the function $f(x, y, z) = x^3 y^2 z + \sin(xy) + e^z$ at the point (1, 2, 3).

Solution The required code is shown in Fig. 2.19.

```
Command Window
>> syms x y z
>> f = x^3 * y^2 * z + sin(x*y) + exp(z);
>> fxyz=simplify(diff(f,x,y,z))

fxyz =

6*x^2*y

>> subs(fxyz,[x y z],[1 2 3])

ans =

12

fx >> 
```

Fig. 2.19 Calculation of $\frac{\partial^3 f}{\partial x \partial y \partial z}$ for $f(x, y, z) = x^3 y^2 z + \sin(xy) + e^z$ at (1, 2, 3)

Example 2.22 Use MATLAB to evaluate the double integral $\int_{-3}^{3} \int_{-\sqrt{9-x^2}}^{+\sqrt{9-x^2}} \frac{3}{\sqrt{9-(x^2+y^2)}} dydx$.

Solution The required code is shown in Fig. 2.20.

```
Command Window
>> syms x y
>> f=3/sqrt(9-(x^2+y^2));
>> int(int(f,y,-sqrt(9-x^2),sqrt(9-x^2)),x,-3,3)

ans =

18*pi

fx >> 
```

Fig. 2.20 Calculation of $\int_{-3}^{3} \int_{-\sqrt{9-x^2}}^{+\sqrt{9-x^2}} \frac{3}{\sqrt{9-(x^2+y^2)}} dydx$ using nested int commands

Example 2.23 Use MATLAB to evaluate the double integral $\int_0^3 \int_0^{-x+3} \frac{1}{\sqrt{9-y^2}} dydx$.

Solution The required code is shown in Fig. 2.21.

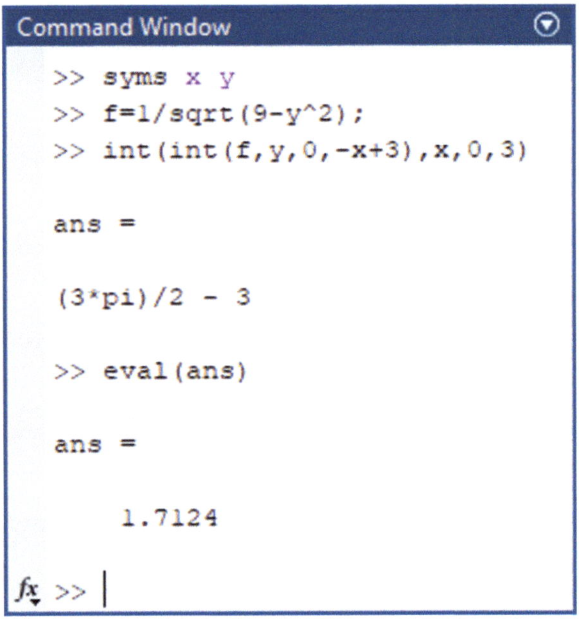

Fig. 2.21 Calculation of $\int_0^3 \int_0^{-x+3} \frac{1}{\sqrt{9-y^2}} dydx$ using nested int commands

Example 2.24 Use MATLAB to evaluate the double integral $\int_{-3}^{3} \int_{-\sqrt{9-x^2}}^{+\sqrt{9-x^2}} \sqrt{4(x^2+y^2)+1} dydx$.

Solution While most double integrals can be evaluated analytically using the `int` command, some cases result in complex symbolic expressions rather than numerical values (Fig. 2.22). For this specific problem, applying nested `int` commands produces an unwieldy result; therefore, using the `integral2` command is more efficient for obtaining a direct numerical solution.

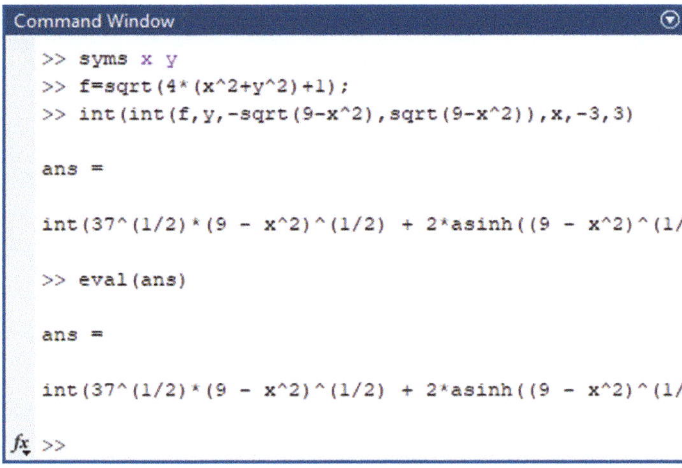

Fig. 2.22 Nested int commands can't evaluate the integral $\int_{-3}^{3}\int_{-\sqrt{9-x^2}}^{+\sqrt{9-x^2}}\sqrt{4(x^2+y^2)+1}dydx$

The script in Fig. 2.23 utilizes the `integral2` command to efficiently compute the numerical value of the given integral. Since constant limits (−3 to 3) belong to x, x must be the first argument in the function handle: @(x, y).

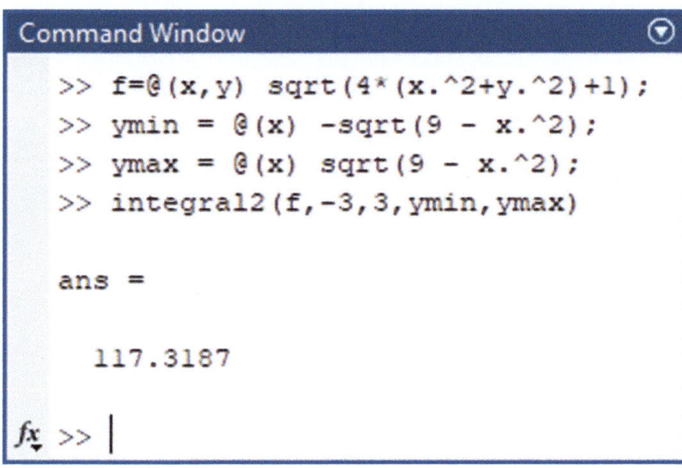

Fig. 2.23 Evaluation of $\int_{-3}^{3}\int_{-\sqrt{9-x^2}}^{+\sqrt{9-x^2}}\sqrt{4(x^2+y^2)+1}dydx$ using the integral2 command

Example 2.25 Use MATLAB to evaluate the double integral $\int_{-2}^{2}\int_{-\sqrt{4-y^2}}^{+\sqrt{4-y^2}}\sqrt{x^2+y^2+1}dxdy$.

Solution The required code is shown in Fig. 2.24. Since constant limits (−2 to 2) belong to y, y must be the first argument in the function handle: @(y, x).

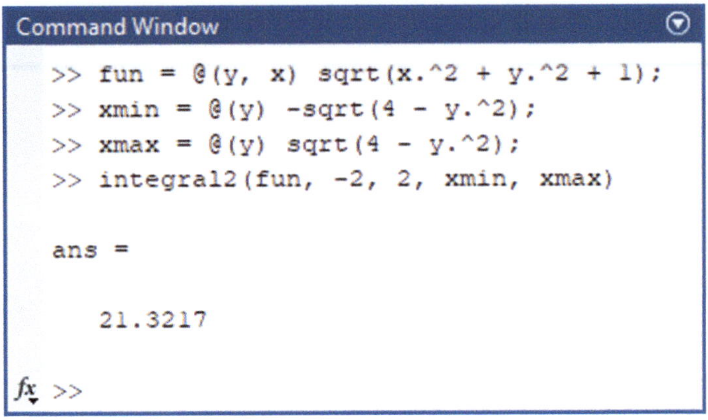

Fig. 2.24 Evaluation of $\int_{-2}^{2}\int_{-\sqrt{4-y^2}}^{+\sqrt{4-y^2}} \sqrt{x^2+y^2+1}dxdy$ using the integral2 command

References

1. Calculus, George Thomas, Maurice Weir, Joel Hass, Pearson, 2014.
2. Calculus, James Stewart, Cengage Learning, 2015.
3. Engineering Mathematics with MATLAB® and Simulink®, Farzin Asadi, Springer, 2025.
4. Applied Numerical Analysis with MATLAB®/Simulink®: For Engineers and Scientists, Farzin Asadi, Springer, 2023.
5. A Problem-Solving Approach to Single-Variable Calculus (Volume I), Farzin Asadi, Springer, 2026.
6. A Problem-Solving Approach to Single-Variable Calculus (Volume II), Farzin Asadi, Springer, 2026.
7. A Problem-Solving Approach to Multi-Variable Calculus (Volume I), Farzin Asadi, Springer, 2026.
8. A Problem-Solving Approach to Multi-Variable Calculus (Volume III), Farzin Asadi, Springer, 2026.
9. A Problem-Solving Approach to Ordinary Differential Equations (Volume I), Farzin Asadi, Springer, 2026.
10. A Problem-Solving Approach to Ordinary Differential Equations (Volume II), Farzin Asadi, Springer, 2026.
11. A Problem-Solving Approach to Ordinary Differential Equations (Volume III), Farzin Asadi, Springer, 2026.
12. A Problem-Solving Approach to Ordinary Differential Equations (Volume IV), Farzin Asadi, Springer, 2026.

Chapter 3
Triple Integrals

3.1 Introduction

Triple integration extends the concept of single and double integrals to functions of three variables and provides a powerful tool for measuring and analyzing quantities distributed throughout three-dimensional regions. Just as single integrals compute areas and double integrals compute volumes or mass over planar regions, triple integrals are used to calculate volumes, masses, moments, and other physical quantities within solid regions of space.

3.1.1 Summary of Integration Formulas

A strong grasp of single-variable calculus is essential for computing triple integrals. To assist you, a review of fundamental integration formulas is provided here for reference.

1. $\int k_1 dx = k_1 x + C$
2. $\int x^n dx = \frac{1}{n+1}x^{n+1} + C \quad n \in \mathbb{R}, \mathrm{n} \neq -1$
3. $\int x^{-1} dx = \int \frac{1}{x} dx = \ln(|x|) + C$
4. $\int \frac{u(x)'}{u(x)} dx = \ln(|u(x)|) + C$
5. $\int \frac{1}{ax+b} dx = \frac{1}{a}\ln(|ax+b|) + C \quad a, b \in \mathbb{R}, \mathrm{a} \neq 0$
6. $\int (ax+b)^n dx = \frac{1}{a}\frac{1}{n+1}(ax+b)^{n+1} + C \quad a, b, n \in \mathbb{R}, \mathrm{a} \neq 0, \mathrm{n} \neq -1$
7. $\int \frac{u'(x)}{(u(x))^2+a^2} dx = \frac{1}{a}\tan^{-1}\left(\frac{u(x)}{a}\right) + C$
8. $\int \frac{1}{x^2+a^2} dx = \frac{1}{a}\tan^{-1}\left(\frac{x}{a}\right) + C$
9. $\int \frac{u'(x)}{\sqrt{a^2-u(x)^2}} dx = \sin^{-1}\left(\frac{u(x)}{a}\right) + C$
10. $\int \frac{u'(x)}{\sqrt{1-u(x)^2}} dx = \sin^{-1}(u(x)) + C$

F. Asadi, *A Problem-Solving Approach to Multi-Variable Calculus*,
https://doi.org/10.1007/978-3-032-21115-6_3

11. $\int \frac{u'(x)}{\sqrt{a^2-u(x)^2}}dx = -\cos^{-1}\left(\frac{u(x)}{a}\right) + C$
12. $\int \frac{u'(x)}{\sqrt{1-u(x)^2}}dx = -\cos^{-1}(u(x)) + C$
13. $\int \frac{u'(x)}{a^2+u(x)^2}dx = \frac{1}{a}\tan^{-1}\left(\frac{u(x)}{a}\right) + C$
14. $\int \frac{u'(x)}{1+u(x)^2}dx = \tan^{-1}(u(x)) + C$
15. $\int \frac{u'(x)}{a^2+u(x)^2}dx = -\frac{1}{a}\cot^{-1}\left(\frac{u(x)}{a}\right) + C$
16. $\int \frac{u'(x)}{1+u(x)^2}dx = -\cot^{-1}(u(x)) + C$
17. $\int \frac{u'(x)}{1-u(x)^2}dx = \frac{1}{2}\ln\left(\left|\frac{1+u(x)}{1-u(x)}\right|\right) + C$
18. $\int \frac{u'(x)}{|u(x)|\sqrt{(u(x))^2-a^2}}dx = \frac{1}{a}sec^{-1}\left(\left|\frac{u(x)}{a}\right|\right) + C$
19. $\int \frac{u'(x)}{|u(x)|\sqrt{u^2(x)-a^2}}dx = -\frac{1}{a}csc^{-1}\left(\left|\frac{u(x)}{a}\right|\right) + C$
20. $\int u'(x)\sin(u(x))dx = -\cos(u(x)) + C$
21. $\int u'(x)\cos(u(x))dx = sin(u(x)) + C$
22. $\int u'(x)\big(1+\tan^2(u(x))\big)dx = \tan(u(x)) + C$
23. $\int u'(x)\big(1+\cot^2(u(x))\big)dx = -\cot(u(x)) + C$
24. $\int \sin(ax+b)dx = -\frac{1}{a}\cos(ax+b) + C \quad a, b \in \mathbb{R}, \mathrm{a} \neq 0$
25. $\int \cos(ax+b)dx = \frac{1}{a}\sin(ax+b) + C \quad a, b \in \mathbb{R}, \mathrm{a} \neq 0$
26. $\int \big(1+\tan^2(ax+b)\big)dx = \frac{1}{a}\tan(ax+b) + C \quad a, b \in \mathbb{R}, \mathrm{a} \neq 0$
27. $\int \big(1+\cot^2(ax+b)\big)dx = -\frac{1}{a}\cot(ax+b) + C \quad a, b \in \mathbb{R}, \mathrm{a} \neq 0$
28. $\int u'(x)e^{u(x)}dx = e^{u(x)} + C$
29. $\int u'(x)(u(x))^n dx = \frac{1}{n+1}u^{n+1}(x) + C$
30. $\int e^{ax}dx = \frac{1}{a}e^{ax} + C$

3.1.2 Wallis' Formula

Wallis' Formula is a highly efficient shortcut in calculus used to evaluate definite integrals of the form $\int_0^{\frac{\pi}{2}} \sin^n(x)dx$ or $\int_0^{\frac{\pi}{2}} \cos^n(x)dx$, where n is a positive integer.

For $n = 1$, $\int_0^{\frac{\pi}{2}} \sin(x)dx = \int_0^{\frac{\pi}{2}} \cos(x)dx = 1$.

For an integer $n \geq 2$, the formula is:

$$\int_0^{\frac{\pi}{2}} \sin^n(x)dx = \int_0^{\frac{\pi}{2}} \cos^n(x)dx = \begin{cases} \frac{n-1}{n} \times \frac{n-3}{n-2} \times \ldots \times \frac{2}{3} \times 1 \ if \ n \ is \ odd \\ \frac{n-1}{n} \times \frac{n-3}{n-2} \times \ldots \times \frac{1}{2} \times \frac{\pi}{2} \ if \ n \ is \ even \end{cases}$$

The "rule of thumb" to remember is to start with $n-1$ in the numerator and n in the denominator, then subtract 2from both until you can't go any further without hitting zero or negative numbers.

If n is odd: The sequence ends at $\frac{2}{3}$ and is multiplied by 1.

If *n* is even: The sequence ends at $\frac{1}{2}$ and is multiplied by $\frac{\pi}{2}$.

For example,

$$\int_0^{\frac{\pi}{2}} \sin^5(x)dx = \frac{4}{5} \times \frac{2}{3} \times 1 = \frac{8}{15}$$

$$\int_0^{\frac{\pi}{2}} \cos^6(x)dx = \frac{5}{6} \times \frac{3}{4} \times \frac{1}{2} \times \frac{\pi}{2} = \frac{15\pi}{96} = \frac{5\pi}{32}$$

These formulas simplify the calculation of integrals on the interval $[0, \pi]$:

$$\forall n \in \mathbb{N} \Rightarrow \int_0^{\pi} \sin^n(x)dx = 2\int_0^{\frac{\pi}{2}} \sin^n(x)dx$$

$$\forall n \in \mathbb{N} \Rightarrow \int_0^{\pi} \cos^n(x)dx = \begin{cases} 2\int_0^{\frac{\pi}{2}} \cos^n(x)dx & if\ n\ is\ even \\ 0 \quad if\ n\ is\ odd \end{cases}$$

3.1.3 The Separability Rule

If $f(x, y, z) = g(x)h(y)i(z)$ and the region of integration is a rectangle $R = \{(x, y, z)|\mathrm{a} \le \mathrm{x} \le \mathrm{b}, \mathrm{c} \le \mathrm{y} \le \mathrm{d}, \mathrm{e} \le \mathrm{z} \le \mathrm{f}\}$, then:

$$\int_e^f \int_c^d \int_a^b f(x, y, z)dxdydz = \int_e^f \int_c^d \int_a^b g(x)h(y)i(z)dxdydz$$

$$= \left(\int_a^b g(x)dx\right)\left(\int_c^d h(y)dy\right)\left(\int_e^f i(z)dz\right)$$

3.1.4 Cylindrical Transformation

A coordinate transformation from cylindrical coordinates (r, θ, z) to Cartesian coordinates (x, y, z) is defined by the following equations:

$$x = r\cos(\theta)$$
$$y = r\sin(\theta)$$
$$z = z$$

Here r is the radial distance from the z-axis ($r \geq 0$), θ is the azimuthal angle measured counterclockwise from the positive x-axis ($0 \leq \theta < 2\pi$), and z is the height along the z-axis ($-\infty \leq z < \infty$) (Fig. 3.1)

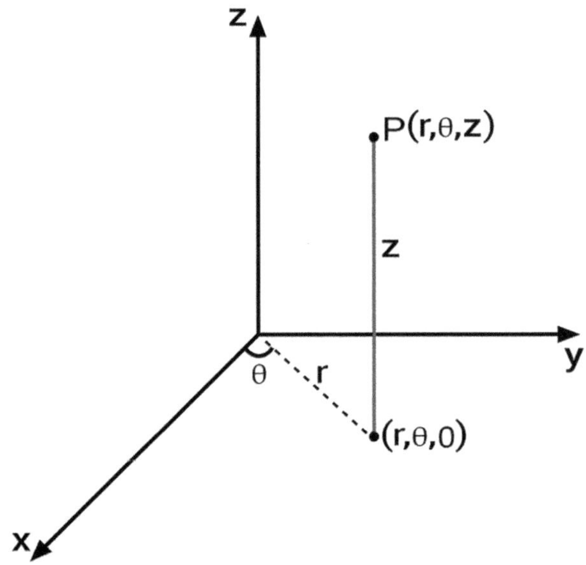

Fig. 3.1 Cylindrical coordinate system

The Jacobian for this transformation is $dV = rdrd\theta dz$.

The following relationship will help you quickly calculate integrals in the cylindrical coordinate system

$$x^2 + y^2 = (r\cos(\theta))^2 + (r\sin(\theta))^2 = r^2\cos^2(\theta) + r^2\sin^2(\theta) = r^2\left(\cos^2(\theta) + \sin^2(\theta)\right)$$
$$= r^2 \times 1 = r^2$$

3.1.5 Spherical Transformation

A coordinate transformation from spherical coordinates (ρ, ϕ, θ) to Cartesian coordinates (x, y, z) is defined by the following equations:

$$x = \rho\sin(\phi)\cos(\theta)$$

$$y = \rho \sin(\phi) \sin(\theta)$$

$$z = \rho \cos(\phi)$$

Here ρ is the distance from the origin ($\rho \geq 0$), ϕ is the polar angle, the angle down from the positive z-axis ($0 \leq \phi \leq \pi$), and θ is the azimuthal angle measured around the z-axis from the positive x-axis ($0 \leq \theta < 2\pi$) (Fig. 3.2).

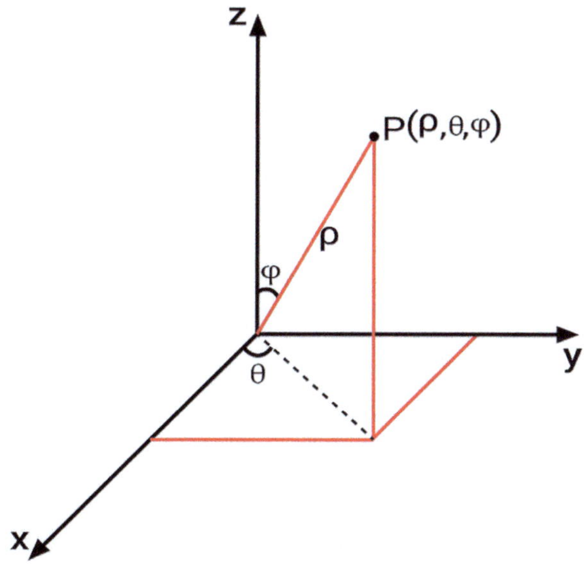

Fig. 3.2 Spherical coordinate system

The Jacobian for this transformation is $dV = \rho^2 \sin(\phi) d\rho d\phi d\theta$.

In general, if the region of integration includes the origin, the lower bound for ρ will always be 0.

The following relationships will help you quickly calculate integrals in the spherical coordinate system:

$$\begin{aligned} x^2 + y^2 &= (\rho \sin(\phi) \cos(\theta))^2 + (\rho \sin(\phi) \sin(\theta))^2 \\ &= \rho^2 \sin^2(\phi) \cos^2(\theta) + \rho^2 \sin^2(\phi) \sin^2(\theta) \\ &= \rho^2 \sin^2(\phi)\big(\cos^2(\theta) + \sin^2(\theta)\big) = \rho^2 \sin^2(\phi) \times 1 = \rho^2 \sin^2(\phi) \end{aligned}$$

$$\begin{aligned} x^2 + y^2 + z^2 &= (\rho \sin(\phi) \cos(\theta))^2 + (\rho \sin(\phi) \sin(\theta))^2 + (\rho \cos(\phi))^2 \\ &= \rho^2 \sin^2(\phi) \cos^2(\theta) + \rho^2 \sin^2(\phi) \sin^2(\theta) + \rho^2 \cos^2(\phi) \\ &= \rho^2 \sin^2(\phi)\big(\cos^2(\theta) + \sin^2(\theta)\big) + \rho^2 \cos^2(\phi) \\ &= \rho^2 \sin^2(\phi) \times 1 + \rho^2 \cos^2(\phi) \end{aligned}$$

$$= \rho^2 \sin^2(\phi) + \rho^2 \cos^2(\phi) = \rho^2\big(\sin^2(\phi) + \cos^2(\phi)\big) = \rho^2 \times 1 = \rho^2$$

In spherical coordinates, z is defined as $z = \rho\cos(\phi)$. Since the radius ρ is always non-negative, the sign of z is entirely determined by $cos(\phi)$.

If $0 \le \phi < \frac{\pi}{2}$, then $\cos(\phi) \ge 0$, which means z is positive.
If $\phi = \frac{\pi}{2}$, then $\cos(\phi) = 0$, which means $z = 0$ (the xy -plane).
If $\frac{\pi}{2} < \phi \le \pi$, then $\cos(\phi) < 0$, which means z is negative.

3.2 Solved Problems

Example 3.1 Calculate $\int_0^1 \int_0^x \int_0^{x^2+y^2} xzdzdydx$.

Solution We start with the innermost integral (with respect to z)

$$\int_0^{x^2+y^2} xzdz = x\int_0^{x^2+y^2} zdz = x\left(\frac{z^2}{2}\bigg|_0^{x^2+y^2}\right)$$
$$= x\left(\frac{\left(x^2+y^2\right)^2}{2} - \frac{0^2}{2}\right) = \frac{x\left(x^2+y^2\right)^2}{2}$$
$$= \frac{x\left(x^4+2x^2y^2+y^4\right)}{2} = \frac{x^5+2x^3y^2+xy^4}{2}$$

The middle integral (with respect to y) is:

$$\int_0^x \int_0^{x^2+y^2} xzdzdy = \int_0^x \frac{x^5+2x^3y^2+xy^4}{2}dy = \int_0^x \left(\frac{x^5}{2} + x^3y^2 + \frac{x}{2}y^4\right)dy$$
$$= \frac{x^5}{2}y + \frac{x^3y^3}{3} + \frac{xy^5}{10}\bigg|_0^x$$
$$= \frac{x^5}{2}\times x + \frac{x^3\times x^3}{3} + \frac{x\times x^5}{10} - \left(\frac{x^5}{2}\times 0 + \frac{x^3\times 0^3}{3} + \frac{x\times 0^5}{10}\right)$$
$$= \frac{x^6}{2} + \frac{x^6}{3} + \frac{x^6}{10} = \frac{28x^6}{30} = \frac{14}{15}x^6$$

The outermost integral (with respect to x) is:

$$\int_0^1 \int_0^x \int_0^{x^2+y^2} xzdzdydx = \int_0^1 \frac{14}{15}x^6dx = \frac{2}{15}x^7\bigg|_0^1 = \frac{2}{15}\times 1^7$$

$$-\frac{2}{15}\times 0^7=\frac{2}{15}\approx 0.1333$$

Example 3.2 Calculate $\int_{-3}^{3}\int_{-2}^{2}\int_{0}^{2-x^2+y^2} dzdydx$.

Solution $\int_0^{2-x^2+y^2} dz = z\Big|_0^{2-x^2+y^2} = 2-x^2+y^2-0 = 2-x^2+y^2$

$$\int_{-2}^{2}\int_{0}^{2-x^2+y^2} dzdy = \int_{-2}^{2}(2-x^2+y^2)dy$$

$$= 2\int_{0}^{2}(2-x^2+y^2)dy = 2\left((2-x^2)y+\frac{y^3}{3}\right)\Big|_0^2$$

$$= 2\left((2-x^2)\times 2+\frac{2^3}{3}\right)-0$$

$$= 4(2-x^2)+\frac{2^4}{3} = 8+\frac{16}{3}-4x^2 = \frac{40}{3}-4x^2$$

$$\int_{-3}^{3}\int_{-2}^{2}\int_{0}^{2-x^2+y^2} dzdydx = \int_{-3}^{3}\left(\frac{40}{3}-4x^2\right)dx$$

$$= 2\int_{0}^{3}\left(\frac{40}{3}-4x^2\right)dx = 2\left(\frac{40}{3}x-\frac{4}{3}x^3\right)\Big|_0^3$$

$$= 2\left(\frac{40}{3}\times 3-\frac{4}{3}\times 3^3\right)-0 = 80-72 = 8$$

Note that for an even function, the integral from $-a$ to a is twice the integral from 0 to a.

Example 3.3 Calculate $\int_{-3}^{3}\int_{4x^2}^{5-x^2}\int_{0}^{1-z} dydzdx$.

Solution $\int_0^{1-z} dy = y\Big|_0^{1-z} = 1-z-0 = 1-z$

$$\int_{4x^2}^{5-x^2}\int_{0}^{1-z} dydz = \int_{4x^2}^{5-x^2}(1-z)dz = \left(z-\frac{z^2}{2}\right)\Big|_{4x^2}^{5-x^2}$$

$$= \left(5-x^2-\frac{(5-x^2)^2}{2}\right)-\left(4x^2-\frac{(4x^2)^2}{2}\right)$$

$$= 5 - 5x^2 + \frac{16x^4 - \left(5 - x^2\right)^2}{2}$$
$$= 5 - 5x^2 + \frac{16x^4 - 25 + 10x^2 - x^4}{2}$$
$$= \frac{15}{2}x^4 - \frac{15}{2} = \frac{15}{2}\left(x^4 - 1\right)$$

$$\int_{-3}^{3}\int_{4x^2}^{5-x^2}\int_{0}^{1-z} dydzdx = \int_{-3}^{3}\frac{15}{2}\left(x^4 - 1\right)dx$$
$$= \frac{15}{2}\int_{-3}^{3}\left(x^4 - 1\right)dx = \frac{15}{2} \times 2\int_{0}^{3}\left(x^4 - 1\right)dx = 15$$
$$\int_{0}^{3}\left(x^4 - 1\right)dx = 15\left(\frac{x^5}{5} - x\right)\Bigg|_0^3 = \left(3x^5 - 15x\right)\Big|_0^3$$
$$= 3 \times 3^5 - 45 - 0 = 3^6 - 45 = 684$$

Example 3.4 Calculate $\int_0^{\frac{\pi}{2}}\int_0^z\int_0^y \cos(x + 2y + 5z)dxdydz$.

Solution We start with the innermost integral (with respect to x)

$$\int_0^y \cos(x + 2y + 5z)dx \qquad = \sin(x + 2y + 5z)\Big|_0^y$$
$$= \sin(y + 2y + 5z) - \sin(0 + 2y + 5z)$$
$$= \sin(3y + 5z) - \sin(2y + 5z)$$

The middle integral (with respect to y) is:

$$\int_0^z\int_0^y \cos(x + 2y + 5z)dxdy = \int_0^z (\sin(3y + 5z) - \sin(2y + 5z))dy$$
$$= \frac{-\cos(3y + 5z)}{3} + \frac{\cos(2y + 5z)}{2}\Bigg|_0^z$$
$$= \left(\frac{-\cos(3z + 5z)}{3} + \frac{\cos(2z + 5z)}{2}\right)$$
$$- \left(\frac{-\cos(3 \times 0 + 5z)}{3} + \frac{\cos(2 \times 0 + 5z)}{2}\right)$$
$$= \left(\frac{-\cos(8z)}{3} + \frac{\cos(7z)}{2}\right) - \left(\frac{-\cos(5z)}{3} + \frac{\cos(5z)}{2}\right)$$

$$= \frac{\cos(7z)}{2} - \frac{\cos(5z)}{6} - \frac{\cos(8z)}{3}$$

The outermost integral (with respect to z) is:

$$\int_0^{\frac{\pi}{2}}\int_0^{z}\int_0^{y} \cos(x+2y+5z)dxdydz = \int_0^{\frac{\pi}{2}} \left(\frac{\cos(7z)}{2} - \frac{\cos(5z)}{6} - \frac{\cos(8z)}{3}\right)dz$$

$$= \frac{\sin(7z)}{14} - \frac{\sin(5z)}{30} - \frac{\sin(8z)}{24}\bigg|_0^{\frac{\pi}{2}}$$

$$= \left(\frac{\sin\left(7\times\frac{\pi}{2}\right)}{14} - \frac{\sin\left(5\times\frac{\pi}{2}\right)}{30} - \frac{\sin\left(8\times\frac{\pi}{2}\right)}{24}\right)$$

$$- \left(\frac{\sin(7\times 0)}{14} - \frac{\sin(5\times 0)}{30} - \frac{\sin(8\times 0)}{24}\right) = \left(\frac{-1}{14} - \frac{1}{30} - 0\right) - (0-0-0)$$

$$= -\frac{11}{105} \approx -0.1048$$

Exercise 3.1 Calculate $\int_0^{\frac{\pi}{2}}\int_0^{z}\int_0^{y} \sin(2x-y+5z)dxdydz$. (Ans. $\frac{1}{30} \approx 0.0333$).

Example 3.5 Calculate $\int_6^7\int_4^5\int_2^3 (x+1)z^2dxdydz$.

Solution Since the limits of integration are all constants and the integrand $f(x,y,z) = (x+1)z^2$ can be separated into a product of functions of single variables $g(x) = (x+1)$, $h(y) = 1$ and $i(z) = z^2$, we can evaluate the triple integral as the product of three single integrals:

$$\int_6^7\int_4^5\int_2^3 (x+1)z^2dxdydz = \int_2^3 (x+1)dx \int_4^5 1dy \int_6^7 z^2dz$$

Let's calculate each integral separately.

$$\int_2^3 ((x+1)dx) = \frac{x^2}{2} + x\bigg|_2^3 = \frac{3^2}{2} + 3 - \left(\frac{2^2}{2} + 2\right) = \frac{5}{2} + 1 = \frac{7}{2} = 3.5$$

$$\int_4^5 1dy = 5 - 4 = 1$$

$$\int_6^7 z^2dz = \frac{z^3}{3}\bigg|_6^7 = \frac{7^3}{3} - \frac{6^3}{3} = \frac{127}{3} \approx 42.3333$$

The value of the triple integral is:

$$\int_6^7\int_4^5\int_2^3 (x+1)z^2 dxdydz = \int_2^3 (x+1)dx \int_4^5 dy \int_6^7 z^2 dz$$
$$= \frac{7}{2} \times 1 \times \frac{127}{3} = \frac{889}{6} \approx 148.1667$$

Example 3.6 Calculate $\int_0^1 \int_0^{\sqrt{1-x^2}} \int_{\sqrt{x^2+y^2}}^{\sqrt{2-x^2-y^2}} xydzdydx$.

Solution We can evaluate this integral more easily by converting it to cylindrical coordinates (r, θ, z).

$$x = r\cos(\theta)$$
$$y = r\sin(\theta)$$
$$z = z$$
$$dV = rdrd\theta dz$$

The bounds for z are:

$$z = \sqrt{x^2+y^2} \Rightarrow z = \sqrt{(r\cos(\theta))^2 + (r\sin(\theta))^2} \Rightarrow z$$
$$= \sqrt{r^2\cos^2(\theta) + r^2\sin^2(\theta)} \Rightarrow z$$
$$= \sqrt{r^2\left(\cos^2(\theta) + \sin^2(\theta)\right)} \Rightarrow z$$
$$= \sqrt{r^2} \Rightarrow z = r$$
$$z = \sqrt{2-x^2-y^2} \Rightarrow z$$
$$= \sqrt{2-\left(x^2+y^2\right)} \Rightarrow z = \sqrt{2-r^2}$$

Therefore, z bounds are $r \le z \le \sqrt{2-r^2}$. The bounds for x and y are:

$$0 \le x \le 1$$
$$0 \le y \le \sqrt{1-x^2} \Rightarrow 0 \le y^2 \le 1-x^2 \Rightarrow x^2+y^2 \le 1$$

This describes the quarter-circle of radius 1 in the first quadrant ($x \ge 0,\ y \ge 0$) (Fig. 3.3). The polar description of the region shown in Fig. 3.3 is: $0 \le r \le 1, 0 \le \theta \le \frac{\pi}{2}$.

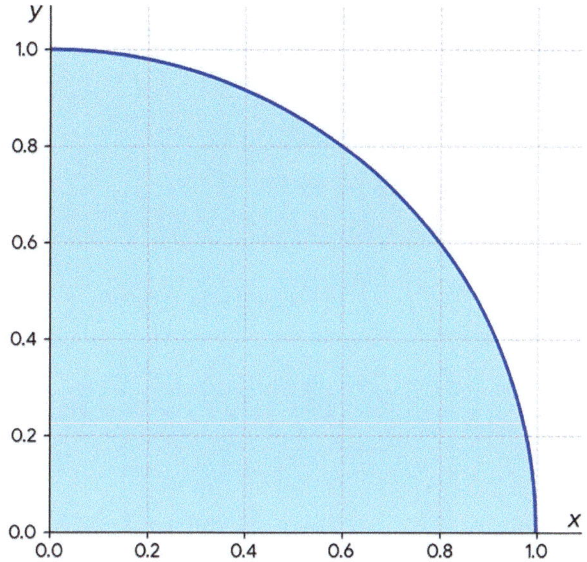

Fig. 3.3 The integration region in the xy-plane

The integrand is:

$$xy = r\cos(\theta) \times r\sin(\theta) = r^2\cos(\theta)\sin(\theta)$$

Substituting the bounds, integrand, and differential element ($dzdydx = rdzdrd\theta$) yields:

$$\int_0^1 \int_0^{\sqrt{1-x^2}} \int_{\sqrt{x^2+y^2}}^{\sqrt{2-x^2-y^2}} xydzdydx = \int_0^{\frac{\pi}{2}} \int_0^1 \int_r^{\sqrt{2-r^2}} \left(r^2\cos(\theta)\sin(\theta)\right) rdzdrd\theta$$

$$= \int_0^{\frac{\pi}{2}} \int_0^1 \int_r^{\sqrt{2-r^2}} \left(r^3\cos(\theta)\sin(\theta)\right) dzdrd\theta$$

The calculation of $\int_0^{\frac{\pi}{2}} \int_0^1 \int_r^{\sqrt{2-r^2}} \left(r^3\cos(\theta)\sin(\theta)\right) dzdrd\theta$ is shown below:

$$\int_r^{\sqrt{2-r^2}} \left(r^3\cos(\theta)\sin(\theta)\right) dz$$

$$= r^3\cos(\theta)\sin(\theta) \int_r^{\sqrt{2-r^2}} dz = r^3\cos(\theta)\sin(\theta) z \Big|_r^{\sqrt{2-r^2}}$$

$$= r^3\cos(\theta)\sin(\theta)\sqrt{2-r^2}$$

$$- r^3 \cos(\theta)\sin(\theta) r$$
$$= r^3 \cos(\theta)\sin(\theta)\left(\sqrt{2-r^2} - r\right) = r^3\left(\sqrt{2-r^2} - r\right)$$
$$\cos(\theta)\sin(\theta) = \left(r^3\sqrt{2-r^2} - r^4\right)$$
$$\cos(\theta)\sin(\theta)$$

$$\int_0^1 \int_r^{\sqrt{2-r^2}} \left(r^3 \cos(\theta)\sin(\theta)\right) dz dr$$
$$= \int_0^1 \left(r^3\sqrt{2-r^2} - r^4\right)\cos(\theta)\sin(\theta) dr$$
$$= \cos(\theta)\sin(\theta)\int_0^1 \left(r^3\sqrt{2-r^2} - r^4\right) dr$$
$$= \cos(\theta)\sin(\theta)\left(\int_3^1 r^3\sqrt{2-r^2} dr - \int_0^1 r^4 dr\right)$$

Let's calculate each integral separately.

$$\int_0^1 r^3\sqrt{2-r^2} dr \overset{u=2-r^2 \Rightarrow du=-2rdr}{\Rightarrow} \frac{1}{-2}\int_0^1 r^2\sqrt{2-r^2}$$
$$\times -2rdr = \frac{1}{-2}\int_2^1 (2-u)\sqrt{u} du$$
$$= \frac{1}{2}\int_1^2 (2-u)\sqrt{u} du = \int_1^2 \sqrt{u} du - \frac{1}{2}\int_1^2 u\sqrt{u} du$$
$$= \int_1^2 u^{\frac{1}{2}} du - \frac{1}{2}\int_1^2 u^{\frac{3}{2}} du$$
$$= \left.\frac{u^{\frac{1}{2}+1}}{\frac{1}{2}+1}\right|_1^2 - \frac{1}{2}\left.\frac{u^{\frac{3}{2}+1}}{\frac{3}{2}+1}\right|_1^2 = \left.\frac{2u^{\frac{3}{2}}}{3}\right|_1^2 - \left.\frac{u^{\frac{5}{2}}}{5}\right|_1^2$$
$$= \left(\frac{2\times 2^{\frac{3}{2}}}{3} - \frac{2\times 1^{\frac{3}{2}}}{3}\right) - \left(\frac{2^{\frac{5}{2}}}{5} - \frac{1^{\frac{5}{2}}}{5}\right)$$
$$= \left(\frac{2^{\frac{5}{2}}}{3} - \frac{2}{3}\right) - \left(\frac{2^{\frac{5}{2}}}{5} - \frac{1}{5}\right) = \left(\frac{5\times 2^{\frac{5}{2}}}{15} - \frac{10}{15}\right)$$

$$-\left(\frac{3\times 2^{\frac{5}{2}}}{15}-\frac{3}{15}\right)=\frac{2\times 2^{\frac{5}{2}}}{15}-\frac{7}{15}$$

$$=\frac{8\sqrt{2}-7}{15}\approx 0.2876$$

$$\int_0^1 r^4 dr=\frac{r^5}{5}\bigg|_0^1=\frac{1^5}{5}-\frac{0^5}{5}=\frac{1}{5}=0.2$$

Therefore,

$$\int_0^1\int_r^{\sqrt{2-r^2}}\left(r^3\cos(\theta)\sin(\theta)\right)dzdr$$

$$=\cos(\theta)\sin(\theta)\left(\int_0^1 r^3\sqrt{2-r^2}dr-\int_0^1 r^4 dr\right)$$

$$=\cos(\theta)\sin(\theta)\left(\frac{8\sqrt{2}-7}{15}-\frac{1}{5}\right)$$

$$=\cos(\theta)\sin(\theta)\left(\frac{8\sqrt{2}-7}{15}-\frac{3}{15}\right)=\frac{8\sqrt{2}-10}{15}\cos(\theta)\sin(\theta)$$

and,

$$\int_0^{\frac{\pi}{2}}\int_0^1\int_r^{\sqrt{2-r^2}}\left(r^3\cos(\theta)\sin(\theta)\right)dzdrd\theta$$

$$=\int_0^{\frac{\pi}{2}}\frac{8\sqrt{2}-10}{15}\cos(\theta)\sin(\theta)d\theta$$

$$=\frac{8\sqrt{2}-10}{15}\int_0^{\frac{\pi}{2}}\cos(\theta)\sin(\theta)d\theta$$

$$=\frac{8\sqrt{2}-10}{15}\frac{\sin^2(\theta)}{2}\bigg|_0^{\frac{\pi}{2}}=\frac{4\sqrt{2}-5}{15}\sin^2(\theta)\bigg|_0^{\frac{\pi}{2}}$$

$$=\frac{4\sqrt{2}-5}{15}\sin^2\left(\frac{\pi}{2}\right)-\frac{4\sqrt{2}-5}{15}\sin^2(0)=\frac{4\sqrt{2}-5}{15}\approx 0.0438$$

The value of the triple integral is:

$$\int_0^1 \int_0^{\sqrt{1-x^2}} \int_{\sqrt{x^2+y^2}}^{\sqrt{2-x^2-y^2}} xydzdydx = \int_0^{\frac{\pi}{2}} \int_0^1 \int_r^{\sqrt{2-r^2}} \left(r^3 \cos(\theta)\sin(\theta)\right) dzdrd\theta$$

$$= \frac{4\sqrt{2}-5}{15} \approx 0.0438$$

Exercise 3.2 Calculate $\int_0^1 \int_0^{\sqrt{1-x^2}} \int_{\sqrt{x^2+y^2}}^{\sqrt{2-x^2-y^2}} xydzdydx$ by converting it to spherical coordinates. (Ans. $\frac{4\sqrt{2}-5}{15} \approx 0.0438$).

Example 3.7 Evaluate $\iiint_E zdV$, where E lies above the paraboloid $z = x^2 + y^2$ and below the plane $z = 2y$.

Solution We will use cylindrical coordinates for the integration, as the region E exhibits circular symmetry when projected onto the xy-plane. The region E is defined by the inequality:

$$x^2 + y^2 \leq z \leq 2y$$

Recall that in cylindrical coordinates:

$$\begin{aligned} x &= r\cos(\theta) \\ y &= r\sin(\theta) \\ z &= z \\ dV &= rdrd\theta dz \end{aligned}$$

Substituting these into the bounds for z:

$$x^2 + y^2 \leq z \leq 2y \Rightarrow r^2 \leq z \leq 2r\sin(\theta)$$

The region R is the projection of E onto the xy-plane, which is defined by where the paraboloid and the plane intersect. Setting their z-values equal:

$$x^2 + y^2 = 2y \Rightarrow x^2 + y^2 - 2y = 0 \Rightarrow x^2 + (y-1)^2 - 1 = 0 \Rightarrow x^2 + (y-1)^2 = 1$$

This is a circle with radius $a = 1$ centered at (0, 1) in the xy-plane. This circle defines the boundary of the projection R (Fig. 3.4).

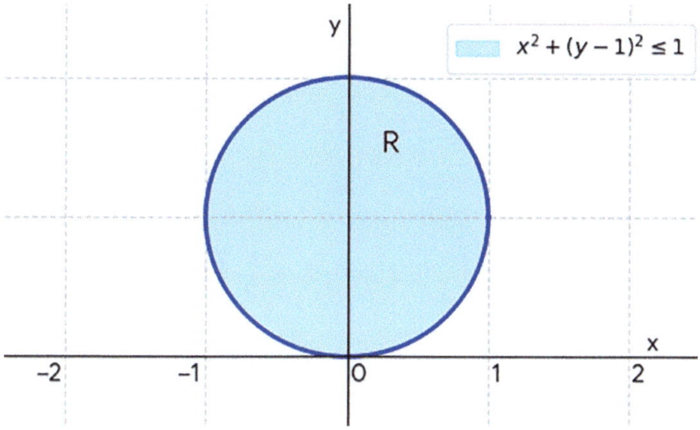

Fig. 3.4 The integration region in the xy-plane

In cylindrical coordinates, the equation of the circle $x^2 + (y-1)^2 = 1$ is:

$$x^2 + (y-1)^2 = 1 \Rightarrow x^2 + y^2 - 2y = 0 \overset{\begin{cases} x = r\cos(\theta) \\ y = r\sin(\theta) \end{cases}}{\Rightarrow} r^2\cos^2(\theta) + r^2\sin^2(\theta) - 2r\sin(\theta) = 0 \Rightarrow r^2\left(\cos^2(\theta) + \sin^2(\theta)\right) = 2r\sin(\theta) \Rightarrow r^2 = 2r\sin(\theta) \overset{assu\min g r\neq 0}{\Rightarrow} r = 2\sin(\theta)$$

The bounds for *r*: r goes from the origin ($r = 0$) out to the boundary of the circle:

$$0 \le r \le 2\sin(\theta)$$

The bounds for *θ*: The circle $r = 2\sin(\theta)$ is centered on the positive y-axis. It covers angles from 0 to π (the upper half-plane, as $\sin(\theta)$ is positive here).

$$0 \le \theta \le \pi$$

The integral $\iiint_E zdV$ in cylindrical coordinates is:

$$\iiint\limits_E zdV = \int_0^{\pi}\int_0^{2\sin(\theta)}\int_{r^2}^{2r\sin(\theta)} z \times rdzdrd\theta = \int_0^{\pi}\int_0^{2\sin(\theta)}\int_{r^2}^{2r\sin(\theta)} zrdzdrd\theta$$

The calculation of $\int_0^{\pi}\int_0^{2\sin(\theta)}\int_{r^2}^{2rsin(\theta)} zrdzdrd\theta$ is shown below:

$$\int_{r^2}^{2r\sin(\theta)} zrdz = r\int_{r^2}^{2r\sin(\theta)} zdz = r\frac{z^2}{2}\Bigg|_{r^2}^{2r\sin(\theta)}$$

$$= r\frac{(2r\sin(\theta))^2}{2} - r\frac{\left(r^2\right)^2}{2} = 2r^3\sin^2(\theta) - \frac{r^5}{2}$$

$$\int_0^{2\sin(\theta)}\int_{r^2}^{2r\sin(\theta)} zrdzdr = \int_0^{2\sin(\theta)}\left(2r^3\sin^2(\theta) - \frac{r^5}{2}\right)dr$$

$$= \left(\frac{r^4}{2}\sin^2(\theta) - \frac{r^6}{12}\right)\Bigg|_0^{2\sin(\theta)}$$

$$= \left(\frac{(2\sin(\theta))^4}{2}\sin^2(\theta) - \frac{(2\sin(\theta))^6}{12}\right)$$

$$-\left(\frac{0^4}{2}\sin^2(\theta) - \frac{0^6}{12}\right) = 8\sin^4(\theta)\sin^2(\theta)$$

$$-\frac{64\sin^6(\theta)}{12} = 8\sin^6(\theta) - \frac{16}{3}\sin^6(\theta) = \frac{8}{3}\sin^6(\theta)$$

$$\int_0^{\pi}\int_0^{2\sin(\theta)}\int_{r^2}^{2r\sin(\theta)} zrdzdrd\theta$$

$$= \int_0^{\pi}\frac{8}{3}\sin^6(\theta)d\theta = \frac{8}{3}\int_0^{\pi}\sin^6(\theta)d\theta$$

Calculating $\int_0^{\pi}\sin^6(\theta)d\theta$ by hand is tedious. Let's use MATLAB (Fig. 3.5).

Fig. 3.5 MATLAB code

```
Command Window

>> syms theta
>> int(sin(theta)^6,theta,0,pi)

ans =

(5*pi)/16

fx >>
```

Based on Fig. 3.5, $\int_0^{\pi}\sin^6(\theta)d\theta = \frac{5\pi}{16}$. Thus,

$$\int_0^{\pi}\int_0^{2\sin(\theta)}\int_{r^2}^{2r\sin(\theta)} zrdzdrd\theta = \int_0^{\pi} \frac{8}{3}\sin^6(\theta)d\theta$$

$$= \frac{8}{3}\int_0^{\pi}\sin^6(\theta)d\theta = \frac{8}{3}\times\frac{5\pi}{16} = \frac{40\pi}{48} = \frac{5\pi}{6}$$

$$\approx 2.6180$$

The value of the triple integral is:

$$\iiint_E zdV = \int_0^{\pi}\int_0^{2\sin(\theta)}\int_{r^2}^{2r\sin(\theta)} zrdzdrd\theta = \frac{5\pi}{6} \approx 2.6180$$

Exercise 3.3 Evaluate $\int_0^{\pi}\sin^6(\theta)d\theta$ using Wallis' Integral Formulas. (Hint: $\int_0^{\pi}\sin^6(\theta)d\theta = 2\int_0^{\frac{\pi}{2}}\sin^6(\theta)d\theta$).

Example 3.8 Determine the volume of the solid region E lying in the first octant, bounded by the surface $y^2 + 4z^2 = 16$ and the plane $x + y = 4$.

Solution The first octant is the region of 3D space where all three coordinates are non-negative:

$$x \geq 0, y \geq 0, z \geq 0$$

The volume V is found by integrating the function $f(x, y, z) = 1$ over the region E:

$$V = \iiint_E 1dV$$

$$y^2 + 4z^2 = 16 \Rightarrow 4z^2 = 16 - y^2 \Rightarrow 2z$$

$$= \sqrt{16 - y^2} \Rightarrow z = \pm\frac{1}{2}\sqrt{16 - y^2}$$

In the first octant $z \geq 0$, therefore, $z = +\frac{1}{2}\sqrt{16 - y^2}$. The projection of E onto the xy -plane is shown in Fig. 3.6.

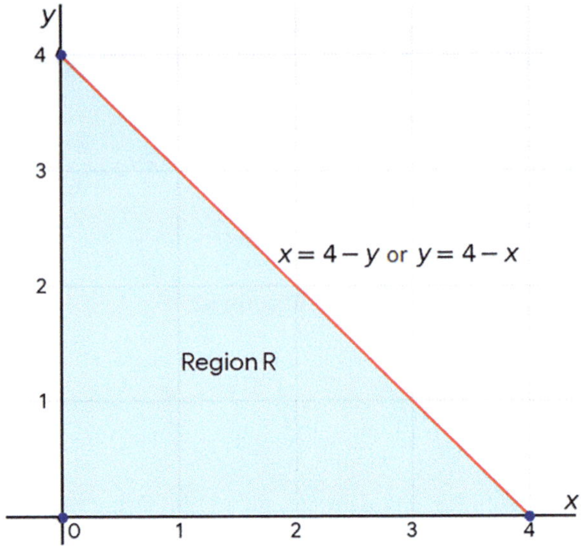

Fig. 3.6 The integration region in the xy -plane

Let's set up the triple integral:

$$V = \int_0^4 \int_0^{4-x} \int_0^{\frac{1}{2}\sqrt{16-y^2}} 1 dzdydx = \int_0^4 \int_0^{4-x} \int_0^{\frac{1}{2}\sqrt{16-y^2}} dzdydx$$

The calculation of $\int_0^4 \int_0^{4-x} \int_0^{\frac{1}{2}\sqrt{16-y^2}} dzdydx$ is shown below:

$$\int_0^{\frac{1}{2}\sqrt{16-y^2}} dz = z\Big|_0^{\frac{1}{2}\sqrt{16-y^2}} = \frac{1}{2}\sqrt{16-y^2} - 0 = \frac{1}{2}\sqrt{16-y^2}$$

$$\int_0^{4-x} \int_0^{\frac{1}{2}\sqrt{16-y^2}} dzdy$$

$$= \int_0^{4-x} \frac{1}{2}\sqrt{16-y^2} dy$$

$$= \frac{1}{2}\int_0^{4-x} \sqrt{16-y^2} dy \begin{cases} y = 4\sin(\theta) \\ dy = 4\cos(\theta)d\theta \end{cases} \Rightarrow \frac{1}{2}$$

$$\int_{0}^{\sin^{-1}\left(1-\frac{x}{4}\right)} \sqrt{16-16\sin^2(\theta)} \times 4\cos(\theta)d\theta$$

$$= 8\int_{0}^{\sin^{-1}\left(1-\frac{x}{4}\right)} \sqrt{1-\sin^2(\theta)}\cos(\theta)d\theta$$

$$= 8\int_{0}^{\sin^{-1}\left(1-\frac{x}{4}\right)} \cos(\theta)\cos(\theta)d\theta$$

$$= 8\int_{0}^{\sin^{-1}\left(1-\frac{x}{4}\right)} \cos^2(\theta)d\theta$$

$$= 8\int_{0}^{\sin^{-1}\left(1-\frac{x}{4}\right)} \frac{1+\cos(2\theta)}{2}d\theta$$

$$= 4\int_{0}^{\sin^{-1}\left(1-\frac{x}{4}\right)} (1+\cos(2\theta))d\theta$$

$$= 4\theta + 2\sin(2\theta)\Big|_{0}^{\sin^{-1}\left(1-\frac{x}{4}\right)} = 4\theta + 2\times 2$$

$$\times \sin(\theta)\times\cos(\theta)\Big|_{0}^{\sin^{-1}\left(1-\frac{x}{4}\right)}$$

$$= 4(\theta+\sin(\theta)\cos(\theta))\Big|_{0}^{\sin^{-1}\left(1-\frac{x}{4}\right)}$$

$$= 4\left(\sin^{-1}\left(1-\frac{x}{4}\right)+\sin\left(\sin^{-1}\left(1-\frac{x}{4}\right)\right)\cos(\sin - 0\right.$$

$$\left.\left(^{-1}\left(1-\frac{x}{4}\right)\right)\right)$$

$$= 4\left(\sin^{-1}\left(1-\frac{x}{4}\right)+\left(1-\frac{x}{4}\right)\sqrt{1-\left(1-\frac{x}{4}\right)^2}\right)$$

$$= 4\left(\sin^{-1}\left(1-\frac{x}{4}\right)+\left(1-\frac{x}{4}\right)\sqrt{\frac{x}{2}-\frac{x^2}{16}}\right)$$

$$= 4\sin^{-1}\left(1-\frac{x}{4}\right)+(4-x)\sqrt{\frac{8x-x^2}{16}}$$

$$= 4\sin^{-1}\left(1-\frac{x}{4}\right)+\left(1-\frac{x}{4}\right)\sqrt{8x-x^2}$$

$$V = \int_0^4 \int_0^{4-x} \int_0^{\frac{1}{2}\sqrt{16-y^2}} dzdydx = \int_0^4 \left(4\sin^{-1}\left(1-\frac{x}{4}\right) + \left(1-\frac{x}{4}\right)\sqrt{8x-x^2}\right)dx$$

$$= \int_0^4 \left(-16 \times -\frac{1}{4}\sin^{-1}\left(1-\frac{x}{4}\right) + \left(\frac{4-x}{4}\right)\sqrt{8x-x^2}\right)dx$$

$$= \int_0^4 \left(-16 \times -\frac{1}{4}\sin^{-1}\left(1-\frac{x}{4}\right) + \left(\frac{8-2x}{8}\right)\sqrt{8x-x^2}\right)dx$$

$$= -16\left(\left(1-\frac{x}{4}\right)\sin^{-1}\left(1-\frac{x}{4}\right) + \sqrt{1-\left(1-\frac{x}{4}\right)^2}\right) + \frac{1}{8} \times \frac{(8x-x^2)^{\frac{3}{2}}}{\frac{3}{2}}\Bigg|_0^4$$

$$= -16\left(\left(1-\frac{x}{4}\right)\sin^{-1}\left(1-\frac{x}{4}\right) + \sqrt{\frac{x}{2}-\frac{x^2}{16}}\right) + \frac{(8x-x^2)^{\frac{3}{2}}}{12}\Bigg|_0^4$$

$$= 16\left(\left(\frac{x}{4}-1\right)\sin^{-1}\left(1-\frac{x}{4}\right) - \sqrt{\frac{x}{2}-\frac{x^2}{16}}\right) + \frac{(8x-x^2)^{\frac{3}{2}}}{12}\Bigg|_0^4$$

$$= 16\left(\left(\frac{4}{4}-1\right)\sin^{-1}\left(1-\frac{4}{4}\right) - \sqrt{\frac{4}{2}-\frac{4^2}{16}}\right) + \frac{(8\times 4-4^2)^{\frac{3}{2}}}{12}$$

$$- 16\left(\left(\frac{0}{4}-1\right)\sin^{-1}\left(1-\frac{0}{4}\right)\right.$$

$$\left.-\sqrt{\frac{0}{2}-\frac{0^2}{16}}\right) + \frac{(8\times 0-0^2)^{\frac{3}{2}}}{12} = -16 + \frac{16}{3} + 16\sin^{-1}(1)$$

$$= -\frac{32}{3} + 16 \times \frac{\pi}{2} = 8\pi - \frac{32}{3} \approx 14.4661$$

The volume is $8\pi - \frac{32}{3} \approx 14.4661$ cubic units.
The following formulas are utilized in the derivation of the solution:

$$\forall x; -1 \le x \le 1 \Rightarrow \sin\left(\sin^{-1}(x)\right) = x$$

$$\forall x; -1 \le x \le 1 \Rightarrow \cos\left(\sin^{-1}(x)\right) = \sqrt{1-x^2}$$

$$\int u' \sin^{-1}(u)dx = u\sin^{-1}(u) + \sqrt{1-u^2} + C$$

Exercise 3.4 Resolve Example 3.8 using the integration order *dxdzdy*. (Ans. $\int_0^4 \int_0^{\sqrt{4-\frac{y^2}{4}}} \int_0^{4-y} dxdzdy = 8\pi - \frac{32}{3} \approx 14.4661$).

Example 3.9 Determine the volume of the solid region E lying in the first octant, bounded by the surface $x^2 + y^2 = 16$ and the plane $x + z = 4$.

Solution The first octant is the region of 3D space where all three coordinates are non-negative:

$$x \geq 0, y \geq 0, z \geq 0$$

The volume V is found by integrating the function $f(x, y, z) = 1$ over the region E:

$$V = \iiint\limits_E 1dV$$

$$x + z = 4 \Rightarrow z = 4 - x$$

$$x^2 + y^2 = 16 \Rightarrow y = \pm\sqrt{16 - x^2}$$

In the first octant $y \geq 0$, therefore, $y = \sqrt{16 - x^2}$. The projection of E onto the xy -plane is shown in Fig. 3.7.

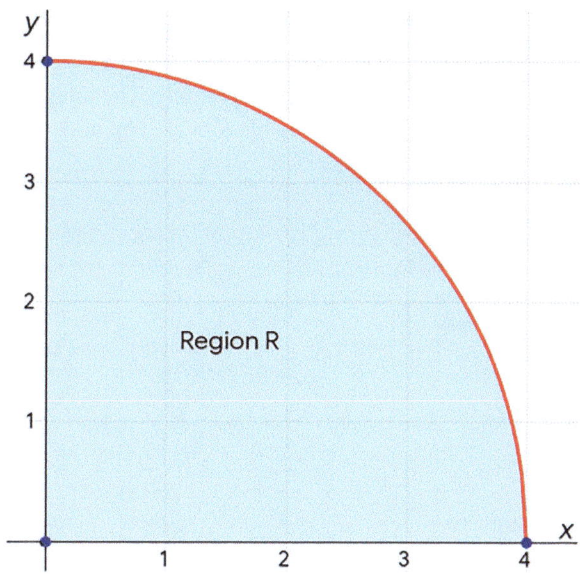

Fig. 3.7 The integration region in the xy -plane

Let's set up the triple integral:

$$V = \int\limits_0^4 \int\limits_0^{\sqrt{16-x^2}} \int\limits_0^{4-x} 1dzdydx = \int\limits_0^4 \int\limits_0^{\sqrt{16-x^2}} \int\limits_0^{4-x} dzdydx$$

The calculation of $\int_0^4 \int_0^{\sqrt{16-x^2}} \int_0^{4-x} dzdydx$ is shown below:

$$\int_0^{4-x} dz = z\Big|_0^{4-x} = 4 - x - 0 = 4 - x$$

$$\int_0^{\sqrt{16-x^2}} \int_0^{4-x} dzdy = \int_0^{\sqrt{16-x^2}} (4-x)dy = (4-x)y\Big|_0^{\sqrt{16-x^2}}$$

$$= (4-x)\sqrt{16-x^2} - 0 = (4-x)\sqrt{16-x^2}$$

$$V = \int_0^4 \int_0^{\sqrt{16-x^2}} \int_0^{4-x} dzdydx = \int_0^4 (4-x)\sqrt{16-x^2}dx$$

$$= \int_0^4 4\sqrt{16-x^2}dx + \int_0^4 -x\sqrt{16-x^2}dx$$

$$= 4\int_0^4 \sqrt{16-x^2}dx + \frac{1}{2}\int_0^4 -2x\sqrt{16-x^2}dx$$

Let's evaluate each integral separately. The integral $\int_0^4 \sqrt{16-x^2}dx$ represents the area of a quarter circle of radius $R = 4$. The area of a full circle is πR^2. Therefore, $\int_0^4 \sqrt{16-x^2}dx = \frac{\pi \times 4^2}{4} = 4\pi$.

$$\frac{1}{2}\int_0^4 -2x\sqrt{16-x^2}dx = \frac{1}{2} \times \frac{(16-x^2)^{\frac{3}{2}}}{\frac{3}{2}}\Bigg|_0^4 = \frac{(16-x^2)^{\frac{3}{2}}}{3}\Bigg|_0^4$$

$$= \frac{(16-4^2)^{\frac{3}{2}}}{3} - \frac{(16-0^2)^{\frac{3}{2}}}{3} = 0 - \frac{16^{\frac{3}{2}}}{3} = 0$$

$$- \frac{(4^2)^{\frac{3}{2}}}{3} = -\frac{4^3}{3} = -\frac{64}{3} \approx 21.3333$$

Therefore,

$$V = \int_0^4 \int_0^{\sqrt{16-x^2}} \int_0^{4-x} dzdydx = \int_0^4 (4-x)\sqrt{16-x^2}dx$$

$$= \int_0^4 4\sqrt{16-x^2}dx + \int_0^4 -x\sqrt{16-x^2}dx$$

$$= 4\int_0^4 \sqrt{16-x^2}dx + \frac{1}{2}\int_0^4 -2x\sqrt{16-x^2}dx$$

$$= 4\times 4\pi - \frac{64}{3} = 16\pi - \frac{64}{3} \approx 28.9321$$

The volume of is $16\pi - \frac{64}{3} \approx 28.9321$ cubic units.

Example 3.10 Determine the volume of the solid region E bounded by the surface $x^2 + 9y^2 = 16$ and the planes $z - x = 4$ and $z = 0$.

Solution The volume V is found by integrating the function $f(x, y, z) = 1$ over the region E:

$$V = \iiint_E 1dV$$

$$z - x = 4 \Rightarrow z = 4 + x$$

$$x^2 + 9y^2 = 16 \Rightarrow y = \pm\sqrt{\frac{16-x^2}{9}} = \pm\frac{\sqrt{16-x^2}}{3}$$

The projection of E onto the xy-plane is shown in Fig. 3.8.

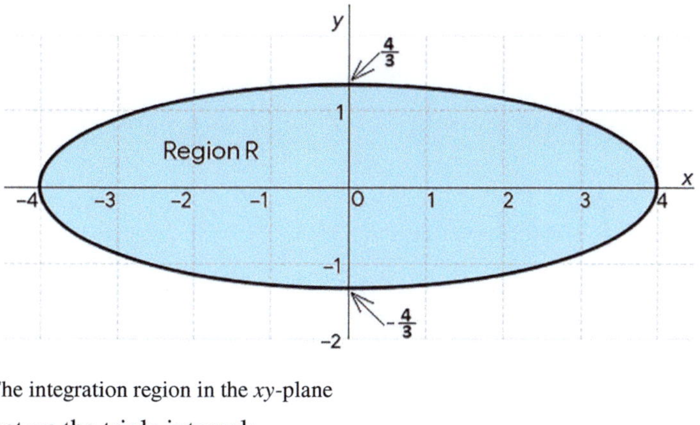

Fig. 3.8 The integration region in the xy-plane

Let's set up the triple integral:

$$V = \int_{-4}^{4}\int_{-\frac{\sqrt{16-x^2}}{3}}^{+\frac{\sqrt{16-x^2}}{3}}\int_0^{4+x} 1dzdydx = \int_{-4}^{4}\int_{-\frac{\sqrt{16-x^2}}{3}}^{+\frac{\sqrt{16-x^2}}{3}}\int_0^{4+x} dzdydx$$

The calculation of $\int_{-4}^{4}\int_{-\frac{\sqrt{16-x^2}}{3}}^{+\frac{\sqrt{16-x^2}}{3}}\int_0^{4+x} dzdydx$ is shown below:

$$\int_{0}^{4+x} dz = z\Big|_{0}^{4+x} = 4 + x - 0 = 4 + x$$

$$\int_{-\frac{\sqrt{16-x^2}}{3}}^{\frac{\sqrt{16-x^2}}{3}} \int_{0}^{4+x} dzdy$$

$$= \int_{-\frac{\sqrt{16-x^2}}{3}}^{\frac{\sqrt{16-x^2}}{3}} (4+x)dy = (4+x)$$

$$\int_{-\frac{\sqrt{16-x^2}}{3}}^{\frac{\sqrt{16-x^2}}{3}} dy = (4+x)\left(\frac{\sqrt{16-x^2}}{3} - \left(-\frac{\sqrt{16-x^2}}{3}\right)\right)$$

$$= (4+x) \times 2\frac{\sqrt{16-x^2}}{3} = \frac{2}{3}(4+x)\sqrt{16-x^2}$$

$$V = \int_{-4}^{4} \int_{-\frac{\sqrt{16-x^2}}{3}}^{+\frac{\sqrt{16-x^2}}{3}} \int_{0}^{4+x} dzdydx$$

$$= \int_{-4}^{4} \frac{2}{3}(4+x)\sqrt{16-x^2}dx = \frac{8}{3}\int_{-4}^{4} \sqrt{16-x^2}dx - \frac{1}{3}$$

$$\int_{-4}^{4} -2x\sqrt{16-x^2}dx$$

Let's evaluate each integral separately. The integral $\int_{-4}^{4} \sqrt{16-x^2}dx$ represents the area of a semicircle of radius $R = 4$. Therefore, $\int_{-4}^{4} \sqrt{16-x^2}dx = \frac{\pi \times 4^2}{2} = 8\pi$.

The function $f(x) = -2x\sqrt{16-x^2}$ is an odd function (since $f(-x) = -2 \times -x\sqrt{16-(-x)^2} = 2x\sqrt{16-x^2} = -f(x)$). Since the limits of integration are symmetric (± 4), the integral of an odd function over a symmetric interval is zero:

$$\frac{1}{3}\int_{-4}^{4} -2x\sqrt{16-x^2}dx = 0$$

Therefore,

$$V = \frac{8}{3}\int_{-4}^{4}\sqrt{16-x^2}dx - \frac{1}{3}\int_{-4}^{4} -2x\sqrt{16-x^2}dx = \frac{8}{3} \times 8\pi - 0 = \frac{64\pi}{3} \approx 67.0206$$

The volume of is $\frac{64\pi}{3} \approx 67.0206$ cubic units.

Example 3.11 Find the volume (V) of the tetrahedron (a four-sided solid) E bounded by the four planes (Fig. 3.9):

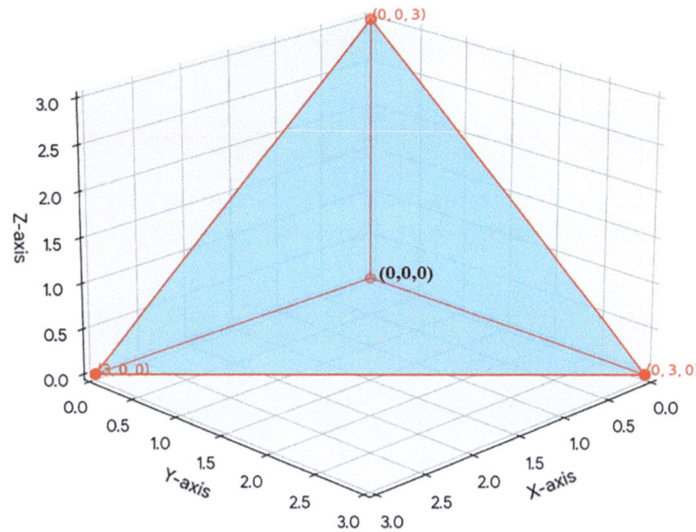

Fig. 3.9 Tetrahedron for Example 3.11

1. $x = 0$ (the yz-plane)
2. $y = 0$ (the xz-plane)
3. $z = 0$ (the xy-plane)
4. $x + y + z = 3$ (a plane intersecting all axes)

Solution This problem was solved in Chapter 10, Volume I (Example 10.50). We will now resolve this problem using triple integrals.

The volume V is found by integrating the function $f(x, y, z) = 1$ over the region E:

$$V = \iiint_E 1dV$$

Since the region is bounded by $x \geq 0$, $y \geq 0$, and $z \geq 0$, we are in the first octant. We need to determine the limits for the z, y, and x variables from the defining plane $x + y + z = 3$.

A. **z-limits (Height)**: The solid is bounded below by the xy-plane ($z = 0$) and above by the plane $z = 3 - x - y$. Thus, $0 \leq z \leq 3 - x - y$.
B. **Projection onto the xy-plane (R)**: The two surfaces ($z = 0$and $z = 3 - x - y$) intersect when:

$$0 = 3 - x - y \Rightarrow y = 3 - x$$

The projection R is a triangular region in the xy-plane bounded by $x = 0$, $y = 0$, and the line $y = 3 - x$.

C. **y-limits**: The projection is bounded below by $y = 0$ and above by the line $y = 3 - x$. Therefore,

$$0 \leq y \leq 3 - x$$

D. **x-limits**: The triangular region spans from $x = 0$ to the intersection of the line $y = 3 - x$ with the x-axis ($y = 0$). Therefore,

$$0 \leq x \leq 3$$

Let's set up the triple integral:

$$V = \int_0^3 \int_0^{3-x} \int_0^{3-x-y} 1dzdydx$$

The calculation of $V = \int_0^3 \int_0^{3-x} \int_0^{3-x-y} 1dzdydx$ is shown below:

$$\int_0^{3-x-y} 1dz = z \Big|_0^{3-x-y} = 3 - x - y - 0 = 3 - x - y$$

$$\int_0^{3-x} \int_0^{3-x-y} 1dzdy = \int_0^{3-x} (3 - x - y)dy = 3y - xy - \frac{y^2}{2} \Big|_0^{3-x}$$

$$= \left(3(3 - x) - x(3 - x) - \frac{(3 - x)^2}{2}\right)$$

$$- \left(3 \times 0 - x \times 0 - \frac{0^2}{2}\right) = 9 - 3x - 3x + x^2$$

$$- \frac{9 - 6x + x^2}{2} = \frac{x^2}{2} - 3x + \frac{9}{2}$$

$$V = \int_0^3 \int_0^{3-x} \int_0^{3-x-y} 1dzdydx = \int_0^3 \left(\frac{x^2}{2} - 3x + \frac{9}{2}\right)dx$$
$$= \frac{x^3}{6} - \frac{3}{2}x^2 + \frac{9}{2}x\Big|_0^3 = \left(\frac{3^3}{6} - \frac{3}{2} \times 3^2 + \frac{9}{2} \times 3\right)$$
$$- \left(\frac{0^3}{6} - \frac{3}{2} \times 0^2 + \frac{9}{2} \times 0\right) = \frac{27}{6} - \frac{27}{2} + \frac{27}{2} - 0 = \frac{27}{6} = \frac{9}{2} = 4.5$$

The volume of the tetrahedron is $\frac{9}{2}$ cubic units. This value is identical to the result from Example 10.50.

Example 3.12 Calculate the volume (V) of the solid shown in Fig. 3.10.

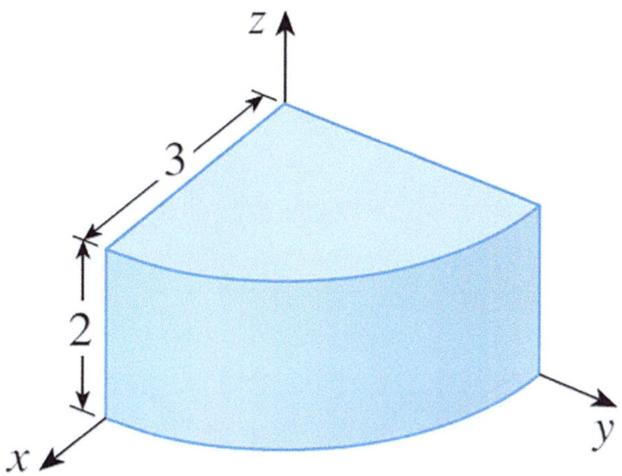

Fig. 3.10 A quarter-cylinder

Solution This problem was solved in Chapter 10, Volume I (Example 10.51). We will now resolve this problem using triple integrals.

The volume V is found by integrating the function $f(x, y, z) = 1$ over the region E shown in Fig. 3.10:

$$V = \iiint_E 1dV$$

Let's set up the triple integral (in cylindrical coordinates $dV = rdrd\theta dz$):

$$V = \iiint_E 1dV = \int_0^2 \int_0^{\frac{\pi}{2}} \int_0^3 rdrd\theta dz$$

The calculation of $V = \int_0^2 \int_0^{\frac{\pi}{2}} \int_0^3 rdrd\theta dz$ is shown below:

$$V = \int_0^2 \int_0^{\frac{\pi}{2}} \int_0^3 r dr d\theta dz = \int_0^3 r dr \times \int_0^{\frac{\pi}{2}} d\theta \times \int_0^2 dz$$

$$= \left(\frac{r^2}{2} \bigg|_0^3 \right) \left(\frac{\pi}{2} - 0 \right) (2 - 0) = \left(\frac{3^2}{2} - \frac{0^2}{2} \right) \times \pi = \frac{9\pi}{2} \approx 14.1372$$

The volume of the shown solid is $\frac{9\pi}{2} \approx 14.1372$ cubic units. This value is identical to the result from Example 10.51.

Example 3.13 Calculate the volume (V) of the solid shown in Fig. 3.11.

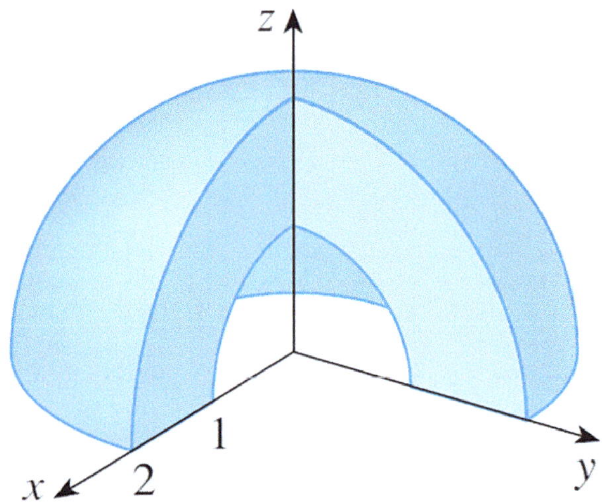

Fig. 3.11 Solid region for Example 3.13

Solution This problem was solved in Chapter 10, Volume I (Example 10.52). We will now resolve this problem using triple integrals.

The volume V is found by integrating the function $f(x, y, z) = 1$ over the region E shown in Fig. 3.11:

$$V = \iiint_E 1 dV$$

Since the solid has spherical symmetry, the volume is most easily calculated using spherical coordinates (ρ, ϕ, θ). The differential volume element is $dV = \rho^2 \sin(\phi) d\rho d\phi d\theta$.

Let's calculate the triple integral bounds:

Radial Distance (ρ): The solid is bounded by the hemisphere of radius 1 and the hemisphere of radius 2. Therefore, $1 \leq \rho \leq 2$.

Angle ϕ (from the positive z-axis): The solid is only the upper half-space (($z \geq 0$)). This means ϕ ranges from the positive z-axis ($\phi = 0$) to the xy-plane $\left(\phi = \frac{\pi}{2}\right)$. Therefore, $0 \leq \phi \leq \frac{\pi}{2}$.

Angle θ (in the xy-plane): The first quadrant in the xy-plane corresponds to $0 \leq \theta \leq \frac{\pi}{2}$. Since this region is excluded, θ must cover the remaining three quadrants. Therefore, the correct range is $\frac{\pi}{2} \leq \theta \leq 2\pi$.

The volume V is given by the triple integral:

$$\int_{\frac{\pi}{2}}^{2\pi}\int_{0}^{\frac{\pi}{2}}\int_{1}^{2} \rho^2 \sin(\phi) d\rho d\phi d\theta = \int_{1}^{2} \rho^2 d\rho \int_{0}^{\frac{\pi}{2}} \sin(\phi) d\phi \int_{\frac{\pi}{2}}^{2\pi} d\theta$$

$$= \left(\frac{\rho^3}{3}\Big|_1^2\right)\left(-\cos(\phi)\Big|_0^{\frac{\pi}{2}}\right)\left(\theta\Big|_{\frac{\pi}{2}}^{2\pi}\right)$$

$$= \left(\frac{\rho^3}{3}\Big|_1^2\right)\left(\cos(\phi)\Big|_{\frac{\pi}{2}}^{0}\right)\left(\theta\Big|_{\frac{\pi}{2}}^{2\pi}\right)$$

$$= \left(\frac{2^3}{3} - \frac{1^3}{3}\right)\left(\cos(0) - \cos\left(\frac{\pi}{2}\right)\right)\left(2\pi - \frac{\pi}{2}\right)$$

$$= \frac{7}{3} \times 1 \times \frac{3\pi}{2} = \frac{7\pi}{2} \approx 10.9956$$

The volume of the shown solid is $\frac{7\pi}{2} \approx 10.9956$ cubic units. This value is identical to the result from Example 10.52.

Example 3.14 A solid object, E, is defined by the region inside the sphere $x^2 + y^2 + z^2 = 4$ and above the xy-plane ($z \geq 0$). The mass density of the object at any point (x, y, z) is given by the function $\rho(x, y, z) = z$.

(A) Find the volume of the solid E.
(B) Find the total mass of the solid E.

Solution The solid E is a hemisphere of radius 2.

ρ (Radius): The solid is inside the sphere of radius 2, so $0 \leq \rho \leq 2$.

ϕ (Polar Angle): The solid is above the xy-plane, meaning the angle from the positive z-axis goes from $\phi = 0$ to $\phi = \frac{\pi}{2}$ (a hemisphere). So, $0 \leq \phi \leq \frac{\pi}{2}$.

θ (Azimuthal Angle): The solid covers the entire circular base, so $0 \leq \theta \leq 2\pi$.

The region E in spherical coordinates is:

$$E = \left\{(\rho, \phi, \theta) | 0 \leq \rho \leq 2, 0 \leq \phi \leq \frac{\pi}{2}, 0 \leq \theta \leq 2\pi\right\}$$

(A) The volume is calculated by integrating the function $f(x, y, z) = 1$ (the integrand for volume) over the region E.

$$\iiint\limits_V 1dV = \int_0^{2\pi}\int_0^{\frac{\pi}{2}}\int_0^2 \rho^2\sin(\phi)d\rho d\phi d\theta$$
$$= \left(\int_0^2 \rho^2 d\rho\right)\left(\int_0^{\frac{\pi}{2}} \sin(\phi)d\phi\right)\left(\int_0^{2\pi} 1d\theta\right)$$

Let's calculate each integral separately.

$$\int_0^2 \rho^2 d\rho = \frac{\rho^3}{3}\bigg|_0^2 = \frac{2^3}{3} - \frac{0^3}{3} = \frac{8}{3}$$
$$\int_0^{\frac{\pi}{2}} \sin(\phi)d\phi = -\cos(\phi)\bigg|_0^{\frac{\pi}{2}} = \cos(\phi)\bigg|_{\frac{\pi}{2}}^0 = \cos(0) - \cos\left(\frac{\pi}{2}\right) = 1$$
$$\int_0^{2\pi} 1d\theta = 2\pi - 0 = 2\pi$$

Therefore,

$$\iiint\limits_V 1dV = \int_0^{2\pi}\int_0^{\frac{\pi}{2}}\int_0^2 \rho^2\sin(\phi)d\rho d\phi d\theta$$
$$= \left(\int_0^2 \rho^2 d\rho\right)\left(\int_0^{\frac{\pi}{2}} \sin(\phi)d\phi\right)\left(\int_0^{2\pi} 1d\theta\right) = \frac{8}{3} \times 1 \times 2\pi$$
$$= \frac{16\pi}{3} \approx 16.7552$$

The volume of the hemisphere is $\frac{16\pi}{3} \approx 16.7552$ cubic units. (This matches the geometric formulas $V_{sphere} = \frac{4}{3}\pi r^3$ and $V_{halfsphere} = \frac{2}{3}\pi r^3$ for $r = 2$).

(B) The mass is calculated by integrating the density function $\rho(x, y, z) = z$ over the region E.

$$M = \iiint\limits_E \rho(x, y, z)dV = \iiint\limits_E zdV$$

First, substitute $z = \rho\cos(\phi)$ and $dV = \rho^2\sin(\phi)d\rho d\phi d\theta$ into the integral:

$$M = \iiint\limits_E \rho(x, y, z)dV = \iiint\limits_E zdV = \int_0^{2\pi}\int_0^{\frac{\pi}{2}}\int_0^2 \rho\cos(\phi)$$
$$\times \rho^2 \sin(\phi)d\rho d\phi d\theta$$
$$= \int_0^{2\pi}\int_0^{\frac{\pi}{2}}\int_0^2 \rho^3 \sin(\phi)\cos(\phi)d\rho d\phi d\theta$$

The calculation of $\int_0^{2\pi}\int_0^{\frac{\pi}{2}}\int_0^2 \rho^3 \sin(\phi)\cos(\phi)d\rho d\phi d\theta$ is shown below:

$$\int_0^{2\pi}\int_0^{\frac{\pi}{2}}\int_0^2 \rho^3 \sin(\phi)\cos(\phi)d\rho d\phi d\theta$$
$$= \left(\int_0^2 \rho^3 d\rho\right)\left(\int_0^{\frac{\pi}{2}} \sin(\phi)\cos(\phi)d\phi\right)\left(\int_0^{2\pi} 1d\theta\right)$$

Let's calculate each integral separately.

$$\int_0^2 \rho^3 d\rho = \frac{\rho^4}{4}\bigg|_0^2 = \frac{2^4}{4} - \frac{0^4}{4} = 4$$
$$\int_0^{\frac{\pi}{2}} \sin(\phi)\cos(\phi)d\phi = \frac{\sin^2(\phi)}{2}\bigg|_0^{\frac{\pi}{2}} = \frac{\sin^2(\frac{\pi}{2})}{2} - \frac{\sin^2(0)}{2} = \frac{1}{2}$$
$$\int_0^{2\pi} 1d\theta = \theta\bigg|_0^{2\pi} = 2\pi - 0 = 2\pi$$

Therefore,

$$M = \iiint\limits_E \rho(x, y, z)dV = \int_0^{2\pi}\int_0^{\frac{\pi}{2}}\int_0^2 \rho^3 \sin(\phi)\cos(\phi)d\rho d\phi d\theta$$
$$= \left(\int_0^2 \rho^3 d\rho\right)\left(\int_0^{\frac{\pi}{2}} \sin(\phi)\cos(\phi)d\phi\right)$$

$$\left(\int_0^{2\pi} 1 d\theta\right) = 4 \times \frac{1}{2} \times 2\pi = 4\pi$$

The total mass of the solid is 4π mass units.

Example 3.15 Calculate the volume of the solid bounded by the surfaces $z = 0$, $y = 0$, $x = 2y^2$, and $x + 2y + z = 4$ in the region $y \geq 0$.

Solution This problem was solved in Chapter 10, Volume I (Example 10.54). We will now resolve this problem using triple integrals.

The volume (V) of the solid is given by $V = \iiint_E 1dV$. The solid E is defined by the following inequalities derived from the surfaces:

1. ***z* Bounds (Inner Integral)**: The solid extends from the xy-plane ($z = 0$) up to the plane $x + 2y + z = 4$.

$$0 \leq z \leq 4 - x - 2y$$

2. ***x* Bounds (Middle Integral)**: The projection onto the xy-plane (R) is bounded by the parabola $x = 2y^2$ and the line $x = 4 - 2y$ (where the top plane $x + 2y + z = 4$ intersects $z = 0$) (Fig. 3.12).

$$2y^2 \leq x \leq 4 - 2y$$

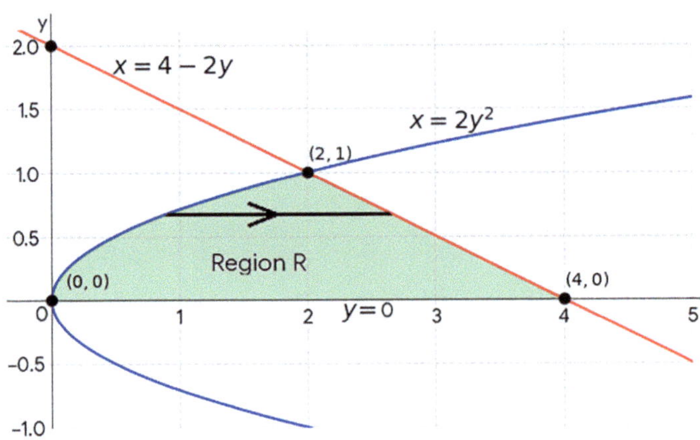

Fig. 3.12 The integration region in the xy-plane

3. ***y* Bounds (Outer Integral)**: The curves intersect at $y = 1$ and the region is bounded by $y = 0$ and the constraint $y \geq 0$ (Fig. 3.12).

$$0 \leq y \leq 1$$

The volume integral is set up as:

$$V = \iiint_E 1dV = \int_0^1 \int_{2y^2}^{4-2y} \int_0^{4-x-2y} 1dzdxdy$$

The calculation of $\int_0^1 \int_{2y^2}^{4-2y} \int_0^{4-x-2y} 1dzdxdy$ is shown below:

$$\int_0^{4-x-2y} 1dz = z\Big|_0^{4-x-2y} = (4-x-2y)-(0) = 4-x-2y$$

$$\int_{2y^2}^{4-2y} \int_0^{4-x-2y} 1dzdx = \int_{2y^2}^{4-2y} (4-x-2y)dx = 4x - \frac{x^2}{2} - 2yx\Big|_{2y^2}^{4-2y}$$

$$= \left(4(4-2y) - \frac{(4-2y)^2}{2} - 2y(4-2y)\right)$$
$$-\left(4(2y^2) - \frac{(2y^2)^2}{2} - 2y(2y^2)\right)$$
$$= \left(16-8y-8+8y-2y^2-8y+4y^2\right)$$
$$-\left(8y^2-2y^4-4y^3\right) = \left(8-8y+2y^2\right)$$
$$-\left(8y^2-2y^4-4y^3\right) = 8-8y-6y^2+4y^3+2y^4$$

$$V = \iiint_E 1dV = \int_0^1 \int_{2y^2}^{4-2y} \int_0^{4-x-2y} 1dzdxdy$$
$$= \int_0^1 \left(8-8y-6y^2+4y^3+2y^4\right)dy$$
$$= 8y-4y^2-2y^3+y^4+\frac{2}{5}y^5\Big|_0^1 = \left(8-4-2+1+\frac{2}{5}\right)-0$$
$$= 3+\frac{2}{5} = \frac{17}{5} = 3.4$$

The volume of the solid is $\frac{17}{5}$ or 3.4 cubic units.

Example 3.16 Find the volume of the solid in the first octant bounded by the plane $x+2y+3z=4$.

Solution This problem was solved in Chapter 10, Volume I (Example 10.55). We will now resolve this problem using triple integrals.

The volume (V) of the solid is given by $V = \iiint_E 1dV$. The solid E is defined by the following inequalities derived from the surfaces:

The plane $x + 2y + 3z = 4$ acts as the upper boundary. The coordinate planes provide the lower bounds ($x \geq 0, y \geq 0, z \geq 0$).

Let's calculate the z bounds:

$$x + 2y + 3z = 4 \Rightarrow z = \frac{4 - x - 2y}{3}$$

The limits for z are from the xy-plane ($z = 0$) up to this plane:

$$0 \leq z \leq \frac{4 - x - 2y}{3}$$

The projection of the solid onto the xy-plane is defined by setting $z = 0$ in the plane equation:

$$x + 2y + 3z = 4 \overset{z=0}{\Rightarrow} x + 2y = 4$$

The projection region (R) is bounded by $x + 2y = 4$, $x = 0$, and $y = 0$ (Fig. 3.13).

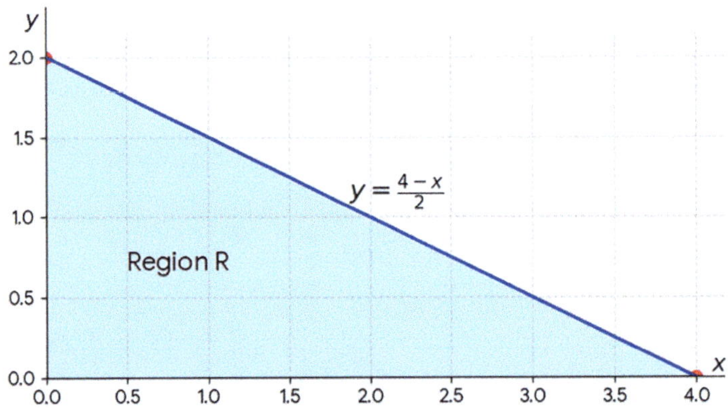

Fig. 3.13 The integration region in the xy-plane

Based on Fig. 3.13,

$$\begin{cases} 0 \leq y \leq \frac{4-x}{2} \\ 0 \leq x \leq 4 \end{cases}$$

The volume integral is set up as:

$$V = \iiint_E 1dV = \int_0^4 \int_0^{\frac{4-x}{2}} \int_0^{\frac{4-x-2y}{3}} 1dzdydx$$

The calculation of $\int_0^4 \int_0^{\frac{4-x}{2}} \int_0^{\frac{4-x-2y}{3}} 1dzdydx$ is shown below:

$$\int_0^{\frac{4-x-2y}{3}} 1dz = z\Big|_0^{\frac{4-x-2y}{3}} = \frac{4-x-2y}{3} - 0 = \frac{4-x-2y}{3}$$

$$\int_0^{\frac{4-x}{2}} \int_0^{\frac{4-x-2y}{3}} 1dzdy = \int_0^{\frac{4-x}{2}} \frac{4-x-2y}{3} dy = \frac{1}{3} \int_0^{\frac{4-x}{2}} (4-x-2y)dy$$

$$= \frac{1}{3}\left((4-x)y - y^2\right)\Big|_0^{\frac{4-x}{2}} = \frac{1}{3}\Bigg((4-x)$$

$$\left(\frac{4-x}{2}\right) - \left(\frac{4-x}{2}\right)^2\Bigg) - 0 = \frac{1}{3}\frac{(4-x)^2}{4} = \frac{1}{12}(4-x)^2$$

$$V = \int_0^4 \int_0^{\frac{4-x}{2}} \int_0^{\frac{4-x-2y}{3}} 1dzdydx = \int_0^4 \frac{1}{12}(4-x)^2 dx$$

$$= \frac{1}{12}\int_0^4 (4-x)^2 dx = -\frac{1}{12}\int_0^4 (4-x)^2 dx$$

$$= -\frac{1}{12}\frac{(4-x)^3}{3}\Big|_0^4 = \frac{1}{12}\frac{(4-x)^3}{3}\Big|_4^0 = \frac{(4-x)^3}{36}\Big|_4^0$$

$$= \frac{(4-0)^3}{36} - \frac{(4-4)^3}{36} = \frac{64}{36} - 0 = \frac{64}{36}$$

$$= \frac{16}{9} \approx 1.7778$$

The volume of the solid is $\frac{16}{9} \approx 1.7778$ cubic units.

Example 3.17 Calculate the volume of the solid E that is bounded above by the paraboloid $z = 3 - x^2 - y^2$ and below by the paraboloid $z = x^2 + y^2$.

Solution This problem was solved in Chapter 10, Volume I (Example 10.56). We will now resolve this problem using triple integrals.

We use cylindrical coordinates due to the symmetry of the surfaces ($x^2+y^2 = r^2$). The volume (V) of the solid is given by $V = \iiint_E 1dV$. The differential volume element in cylindrical coordinates is $dV = rdzdrd\theta$.

The solid is bounded below by $z = x^2 + y^2$ and above by $z = 3 - x^2 - y^2$. Substituting $r^2 = x^2 + y^2$:

$$r^2 \le z \le 3 - r^2$$

The projection region R is a circle of radius $\sqrt{\frac{3}{2}}$ (Fig. 3.14):

$$x^2 + y^2 = 3 - x^2 - y^2 \Rightarrow 2x^2 + 2y^2 = 3 \Rightarrow x^2 + y^2 = \frac{3}{2} \Rightarrow x^2 + y^2 = \left(\sqrt{\frac{3}{2}}\right)^2$$

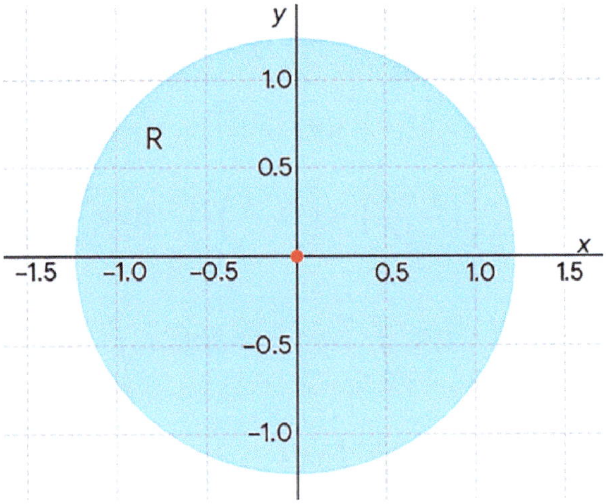

Fig. 3.14 The integration region in the xy-plane

Based on Fig. 3.14,
***r* limits**: $0 \le r \le \sqrt{\frac{3}{2}}$.
***θ* limits**: $0 \le \theta \le 2\pi$.
The volume integral is set up as:

$$V = \iiint\limits_E 1dV = \int_0^{2\pi} \int_0^{\sqrt{\frac{3}{2}}} \int_{r^2}^{3-r^2} 1 \times rdzdrd\theta = \int_0^{2\pi} \int_0^{\sqrt{\frac{3}{2}}} \int_{r^2}^{3-r^2} rdzdrd\theta$$

The calculation of $\int_0^{2\pi} \int_0^{\sqrt{\frac{3}{2}}} \int_{r^2}^{3-r^2} rdzdrd\theta$ is shown below:

$$\int_{r^2}^{3-r^2} rdz = rz\Big|_{r^2}^{3-r^2} = r(3 - r^2) - r \times r^2 = 3r - r^3 - r^3 = 3r - 2r^3$$

$$\int_0^{\sqrt{\frac{3}{2}}} \int_{r^2}^{3-r^2} rdzdr$$

$$= \int_0^{\sqrt{\frac{3}{2}}} (3r - 2r^3)dr = \left(\frac{3}{2}r^2 - \frac{1}{2}r^4\right)\Bigg|_0^{\sqrt{\frac{3}{2}}} = \left(\frac{3}{2}\left(\sqrt{\frac{3}{2}}\right)^2 - \frac{1}{2}\left(\sqrt{\frac{3}{2}}\right)^4\right)$$

$$-0 = \frac{3}{2} \times \frac{3}{2} - \frac{1}{2}\left(\frac{3}{2}\right)^2 = \frac{1}{2}\left(\frac{3}{2}\right)^2 = \frac{9}{8} = 1.125$$

$$V = \int_0^{2\pi} \int_0^{\sqrt{\frac{3}{2}}} \int_{r^2}^{3-r^2} rdzdrd\theta = \int_0^{2\pi} \frac{9}{8} d\theta = \frac{9}{8} \int_0^{2\pi} d\theta$$

$$= \frac{9}{8} \times (2\pi - 0) = \frac{9\pi}{4} \approx 7.0686$$

The volume of the solid E is $\frac{9\pi}{4} \approx 7.0686$ cubic units.

Example 3.18 Calculate the volume of the solid E that is bounded above by the paraboloid $z = 3 - x^2 - y^2$ and below by the paraboloid $z = x^2 + y^2$.

Solution This problem was solved in the previous example. We will now re-evaluate this problem using the Cartesian coordinate system.

$$x^2 + y^2 \le z \le 3 - x^2 - y^2$$

The projection region R is a circle of radius $\sqrt{\frac{3}{2}}$ (Fig. 3.15):

$$x^2 + y^2$$
$$= 3 - x^2 - y^2 \Rightarrow 2x^2 + 2y^2 = 3 \Rightarrow x^2 + y^2$$
$$= \frac{3}{2} \Rightarrow y^2 = \frac{3}{2} - x^2 \Rightarrow y = \pm\sqrt{\frac{3}{2} - x^2} \Rightarrow$$
$$-\sqrt{\frac{3}{2} - x^2} \le y \le \sqrt{\frac{3}{2} - x^2}$$

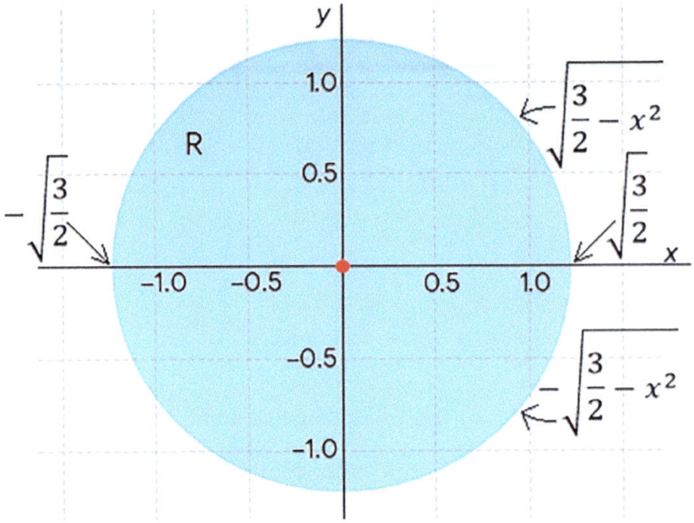

Fig. 3.15 The integration region in the *xy*-plane

The volume integral is set up as:

$$V = \iiint\limits_E 1dV = \int\limits_{-\sqrt{\frac{3}{2}}}^{\sqrt{\frac{3}{2}}} \int\limits_{-\sqrt{\frac{3}{2}-x^2}}^{\sqrt{\frac{3}{2}-x^2}} \int\limits_{x^2+y^2}^{3-x^2-y^2} dzdydx$$

The calculation of $\int_{-\sqrt{\frac{3}{2}}}^{\sqrt{\frac{3}{2}}} \int_{-\sqrt{\frac{3}{2}-x^2}}^{\sqrt{\frac{3}{2}-x^2}} \int_{x^2+y^2}^{3-x^2-y^2} dzdydx$ is shown below:

$$\int\limits_{x^2+y^2}^{3-x^2-y^2} dz = z\Big|_{x^2+y^2}^{3-x^2-y^2} = 3 - x^2 - y^2 - (x^2 + y^2) = 3 - 2x^2 - 2y^2$$

$$\int\limits_{-\sqrt{\frac{3}{2}-x^2}}^{\sqrt{\frac{3}{2}-x^2}} \int\limits_{x^2+y^2}^{3-x^2-y^2} dzdy$$

$$= \int\limits_{-\sqrt{\frac{3}{2}-x^2}}^{\sqrt{\frac{3}{2}-x^2}} (3 - 2x^2 - 2y^2)dy = 2 \int\limits_{0}^{\sqrt{\frac{3}{2}-x^2}} (3 - 2x^2 - 2y^2)dy$$

$$= 2\left((3 - 2x^2)y - \frac{2}{3}y^3\right)\Big|_{0}^{\sqrt{\frac{3}{2}-x^2}}$$

$$= 2\left((3 - 2x^2)\left(\sqrt{\frac{3}{2} - x^2}\right) - \frac{2}{3}\left(\sqrt{\frac{3}{2} - x^2}\right)^3 - 0\right)$$

$$= 2(3 - 2x^2)\left(\sqrt{\frac{3}{2} - x^2}\right) - 2 \times \frac{2}{3}\left(\frac{3}{2} - x^2\right)$$

$$\sqrt{\frac{3}{2} - x^2} = \left(6 - 4x^2 - 2 + \frac{4}{3}x^2\right)\sqrt{\frac{3}{2} - x^2} = \left(4 - \frac{8}{3}x^2\right)\sqrt{\frac{3}{2} - x^2}$$

$$V = \int_{-\sqrt{\frac{3}{2}}}^{\sqrt{\frac{3}{2}}} \int_{-\sqrt{\frac{3}{2}-x^2}}^{\sqrt{\frac{3}{2}-x^2}} \int_{x^2+y^2}^{3-x^2-y^2} dzdydx$$

$$= \int_{-\sqrt{\frac{3}{2}}}^{\sqrt{\frac{3}{2}}} \left(4 - \frac{8}{3}x^2\right)\sqrt{\frac{3}{2} - x^2}dx = 2\int_{0}^{\sqrt{\frac{3}{2}}} \left(4 - \frac{8}{3}x^2\right)\sqrt{\frac{3}{2} - x^2}dx$$

The presence of the term $\sqrt{a^2 - x^2}$ suggests the substitution $x = a\sin(\theta)$. Here, $a^2 = \frac{3}{2}$, so $a = \sqrt{\frac{3}{2}}$. Let:

$$x = \sqrt{\frac{3}{2}}\sin(\theta)$$
$$dx = \sqrt{\frac{3}{2}}\cos(\theta)d\theta$$

Transforming the limits:

$$x = 0 \overset{x=\sqrt{\frac{3}{2}}\sin(\theta)}{\Rightarrow} 0 = \sqrt{\frac{3}{2}}\sin(\theta) \Rightarrow \sin(\theta) = 0 \Rightarrow \theta = 0$$
$$x = \sqrt{\frac{3}{2}} \overset{x=\sqrt{\frac{3}{2}}\sin(\theta)}{\Rightarrow} \sqrt{\frac{3}{2}} = \sqrt{\frac{3}{2}}\sin(\theta) \Rightarrow \sin(\theta) = 1 \Rightarrow \theta = \frac{\pi}{2}$$

Substituting terms:

$$V = 2\int_{0}^{\sqrt{\frac{3}{2}}} \left(4 - \frac{8}{3}x^2\right)\sqrt{\frac{3}{2} - x^2}dx$$

$$= 2\int_{0}^{\frac{\pi}{2}} \left(4 - \frac{8}{3}\left(\frac{3}{2}\sin^2(\theta)\right)\right)\sqrt{\frac{3}{2} - \frac{3}{2}\sin^2(\theta)}\sqrt{\frac{3}{2}}\cos(\theta)d\theta = 12$$

$$\int_0^{\frac{\pi}{2}} (1 - \sin^2(\theta))\sqrt{1 - \sin^2(\theta)} \cos(\theta) d\theta$$

$$= 12 \int_0^{\frac{\pi}{2}} \cos^2(\theta)\sqrt{\cos^2(\theta)} \cos(\theta) d\theta = 12 \int_0^{\frac{\pi}{2}} \cos^4(\theta) d\theta$$

We can use the Wallis' Integral formula for $\int_0^{\frac{\pi}{2}} \cos^n(x)dx$ (for even n):

$$\int_0^{\frac{\pi}{2}} \cos^n(x)dx = \frac{n-1}{n} \times \frac{n-3}{n-2} \times \ldots \frac{1}{2} \times \frac{\pi}{2}$$

For $n = 4$,

$$\int_0^{\frac{\pi}{2}} \cos^4(\theta)d\theta = \frac{3}{4} \times \frac{1}{2} \times \frac{\pi}{2} = \frac{3\pi}{16}$$

Therefore,

$$V = 12 \int_0^{\frac{\pi}{2}} \cos^4(\theta)d\theta = 12 \times \frac{3\pi}{16} = \frac{36\pi}{16} = \frac{18\pi}{8} = \frac{9\pi}{4} \approx 7.0686$$

The volume of the solid E is $\frac{9\pi}{4} \approx 7.0686$ cubic units.

Example 3.19 Find the volume and total mass of the solid region E that lies below the paraboloid $z = 4 - x^2 - y^2$ and above the xy-plane ($z = 0$) (Fig. 3.16). The mass density of the solid is given by the function.

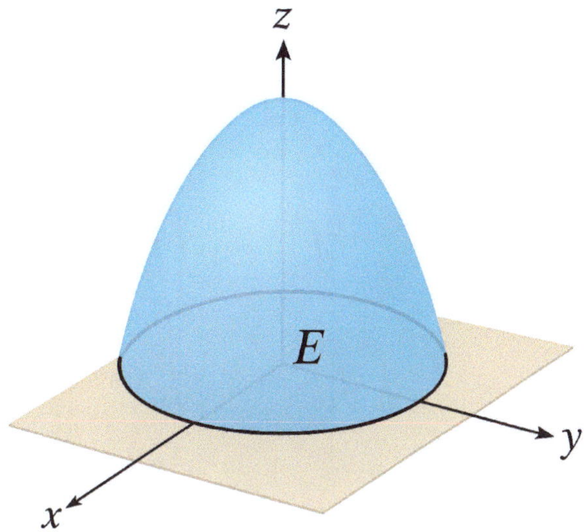

Fig. 3.16 Solid for Example 3.19

Solution The shape of the paraboloid $z = 4 - \left(x^2 + y^2\right)$ strongly suggests using cylindrical coordinates, which are a combination of polar coordinates in the xy-plane and the standard z-coordinate.

The transformation from Cartesian (x, y, z) to Cylindrical (r, θ, z) is:

$$x = r\cos(\theta)$$
$$y = r\sin(\theta)$$
$$z = z$$

The Jacobian for this transformation is $dV = rdrd\theta dz$.

We need to rewrite the bounding surfaces and limits in terms of $r, \theta,$ and z.

Paraboloid: $z = 4 - \left(x^2 + y^2\right)$ becomes $z = 4 - r^2$.

Lower z bound: The solid is above the xy-plane, so $z = 0$. Thus, $0 \leq z \leq 4 - r^2$.

Base (Projection onto the xy-plane): The solid meets the xy-plane ($z = 0$) when:

$$0 = 4 - x^2 - y^2 \Rightarrow x^2 + y^2 = 4$$

This is a circle of radius 2.

***r* (Radius)**: $0 \leq r \leq 2$.

θ (Angle): The region covers the full circle, so $0 \leq \theta \leq 2\pi$.

The region E in cylindrical coordinates is:

$$E = \left\{(r, \theta, z) | 0 \leq r \leq 2, 0 \leq \theta \leq 2\pi, 0 \leq z \leq 4 - r^2\right\}$$

(A) The volume is calculated by integrating $f(x, y, z) = 1$ over E.

$$E = \iiint_V 1dV = \int_0^{2\pi}\int_0^{2}\int_0^{4-r^2} rdzdrd\theta$$

The calculation of $\int_0^{2\pi}\int_0^2\int_0^{4-r^2} rdzdrd\theta$ is shown below:

$$\int_0^{4-r^2} rdz = r\int_0^{4-r^2} dz = rz\Big|_0^{4-r^2}$$
$$= r\left(4-r^2\right) - r\times 0 = 4r - r^3$$
$$\int_0^2\int_0^{4-r^2} rdzdr = \int_0^2 \left(4r-r^3\right)dr = 2r^2 - \frac{r^4}{4}\Big|_0^2$$
$$= 2\times 2^2 - \frac{2^4}{4} - \left(2\times 0^2 - \frac{0^4}{4}\right) = 4$$
$$\int_0^{2\pi}\int_0^2\int_0^{4-r^2} rdzdrd\theta = \int_0^{2\pi} 4d\theta$$
$$= 4\int_0^{2\pi} d\theta = 4\times(2\pi - 0) = 8\pi$$

The volume of the paraboloid is 8π cubic units.

(B) The mass is calculated by integrating the density function $\rho(x,y,z) = \sqrt{x^2+y^2}$ over the region E.

$$M = \iiint_E \rho(x,y,z)dV = \iiint_E \sqrt{x^2+y^2}dV$$

First, substitute the density function: $\rho(x,y,z) = \sqrt{x^2+y^2} = \sqrt{(r\cos(\theta))^2 + (r\sin(\theta))^2}\ \sqrt{r^2cos^2(\theta) + r^2\sin^2(\theta)} = \sqrt{r^2\left(cos^2(\theta)+\sin^2(\theta)\right)} = \sqrt{r^2} = r$ (since $r \ge 0$)

$$M = \iiint_E \rho(x,y,z)dV = \iiint_E \sqrt{x^2+y^2}dV$$
$$= \int_0^{2\pi}\int_0^{2}\int_0^{4-r^2} r\times rdzdrd\theta = \int_0^{2\pi}\int_0^{2}\int_0^{4-r^2} r^2dzdrd\theta$$

The calculation of $\int_0^{2\pi}\int_0^2\int_0^{4-r^2} r\times rdzdrd\theta$ is shown below:

$$\int_0^{4-r^2} r^2 dz = r^2 \int_0^{4-r^2} dz = r^2 z \Big|_0^{4-r^2} = r^2(4-r^2) - r^2 \times 0 = 4r^2 - r^4$$

$$\int_0^2 \int_0^{4-r^2} r^2 dz dr = \int_0^2 (4r^2 - r^4) dr = \frac{4}{3} r^3 - \frac{r^5}{5} \Big|_0^2$$

$$= \left(\frac{4}{3} \times 2^3 - \frac{2^5}{5}\right) - \left(\frac{4}{3} \times 0^3 - \frac{0^5}{5}\right) = \frac{32}{3} - \frac{32}{5} = \frac{64}{15} \approx 4.2667$$

$$\int_0^{2\pi} \int_0^2 \int_0^{4-r^2} r^2 dz dr d\theta = \int_0^{2\pi} \frac{64}{15} d\theta = \frac{64}{15}(2\pi - 0) = \frac{128\pi}{15} \approx 26.8083$$

The total mass of the solid is $\frac{128\pi}{15} \approx 26.8083$ mass unit.

Example 3.20 Evaluate $\iiint_E x^2 dV$, where E is bounded by the xz-plane and the hemisperes $y = \sqrt{9 - x^2 - z^2}$ and $y = \sqrt{16 - x^2 - z^2}$.

Solution This problem asks to evaluate the triple integral $\iiint_E x^2 dV$, where the region E is a spherical shell segment defined by the following boundaries:

The equations $y = \sqrt{R^2 - x^2 - z^2}$ imply $y^2 = R^2 - x^2 - z^2$, or $x^2 + y^2 + z^2 = R^2$, with the restriction $y \geq 0$. This means the E is the region between two hemispheres with radii $R_1 = 3$ and $R_2 = 4$, both restricted to the non-negative y-space.

Due to the spherical nature of the boundaries and the integrand (x^2), the integral is best solved using spherical coordinates.

Let's define the bounds for the region $E = \{(x, y, z) | 9 \leq x^2 + y^2 + z^2 \leq 16, y \geq 0\}$:

1. **Radial Distance (ρ)**: the solid is bounded by spheres of radius $R_1 = 3$ and $R_2 = 4$.

$$3 \leq \rho \leq 4$$

2. **Angle ϕ (Polar Angle)**: Since the region is a shell segment that spans $y \geq 0$, it goes around the z-axis completely. the polar angle ϕ ranges from the positive z-axis ($\phi = 0$) to the negative z-axis ($\phi = \pi$).

$$0 \leq \phi \leq \pi$$

3. **Angle θ (Azimuthal Angle)**: the $y \geq 0$ constraint restricts the region to the upper half of the xz-plane, which corresponds to the first and second quadrants in the standard xy-plane view.

$$0 \leq \theta \leq \pi$$

The integrand is x^2. We substitute the spherical coordinate expression for x:

$$x = \rho \sin(\phi)\cos(\theta) \Rightarrow x^2 = \rho^2 sin^2(\phi) cos^2(\theta)$$

The differential volume element is:
$dV = \rho^2 \sin(\phi) d\rho d\phi d\theta$.
The integral $\iiint_E x^2 dV$ becomes:

$$\iiint\limits_E x^2 dV = \int_0^{\pi}\int_0^{\pi}\int_3^{4} \rho^2 \sin^2(\phi)\cos^2(\theta) \times \rho^2 \sin(\phi) d\rho d\phi d\theta$$

$$= \int_0^{\pi}\int_0^{\pi}\int_3^{4} \rho^4 \sin^3(\phi)\cos^2(\theta) d\rho d\phi d\theta$$

Since the limits are constants and the integrand is a product of functions of single variables, we can separate the integral:

$$\iiint\limits_E x^2 dV = \int_0^{\pi}\int_0^{\pi}\int_3^{4} \rho^2 \sin^2(\phi)\cos^2(\theta) \times \rho^2 \sin(\phi) d\rho d\phi d\theta$$

$$= \int_0^{\pi}\int_0^{\pi}\int_3^{4} \rho^4 \sin^3(\phi)\cos^2(\theta) d\rho d\phi d\theta$$

$$= \int_3^{4} \rho^4 d\rho \int_0^{\pi} \sin^3(\phi) d\phi \int_0^{\pi} \cos^2(\theta) d\theta$$

Let's calculate each integral separately.

$$\int_3^{4} \rho^4 d\rho = \frac{\rho^5}{5}\Big|_3^4 = \frac{4^5}{5} - \frac{3^5}{5} = \frac{4^5}{5} - \frac{3^5}{5} = \frac{1024}{5} - \frac{243}{5} = \frac{781}{5} \approx 156.2$$

$$\int_0^{\pi} \sin^3(\phi) d\phi = \int_0^{\pi} \sin(\phi) sin^2(\phi) d\phi$$

$$= \int_0^{\pi} \sin(\phi)\left(1 - \cos^2(\phi)\right) d\phi = \int_0^{\pi} \sin(\phi) d\phi$$

$$+\int\limits_0^{\pi} -\sin(\phi)\cos^2(\phi)d\phi$$

$$= -\cos(\phi)\Big|_0^{\pi} + \frac{cos^3(\phi)}{3}\Big|_0^{\pi} = \cos(\phi)\Big|_{\pi}^{0} + \frac{cos^3(\phi)}{3}\Big|_0^{\pi}$$

$$= \cos(0) - \cos(\pi) + \frac{cos^3(\pi)}{3} - \frac{cos^3(0)}{3}$$

$$= 1 - (-1) + \frac{-1}{3} - \frac{1}{3} = 2 - \frac{2}{3} = \frac{4}{3} \approx 1.3333$$

$$\int_0^{\pi} cos^2(\theta)d\theta = \int_0^{\pi} \frac{1+\cos(2\theta)}{2}d\theta = \frac{1}{2}\int_0^{\pi} 1d\theta$$

$$+\frac{1}{2}\int_0^{\pi} \cos(2\theta)d\theta = \frac{1}{2}(\pi - 0) + \frac{1}{4}\sin(2\theta)\Big|_0^{\pi} = \frac{\pi}{2}$$

$$+\frac{1}{4}\sin(2\pi) - \frac{1}{4}\sin(0) = \frac{\pi}{2} + 0 - 0 = \frac{\pi}{2} \approx 1.5708$$

Therefore,

$$\iiint\limits_E x^2 dV = \int\limits_3^4 \rho^4 d\rho \int_0^{\pi} \sin^3(\phi)d\phi \int\limits_0^{\pi} \cos^2(\theta)d\theta$$

$$= \frac{781}{5} \times \frac{4}{3} \times \frac{\pi}{2} = \frac{1562\pi}{15} \approx 327.1445$$

Example 3.21 Change the order of integration for $\int\limits_0^2 \int\limits_0^1 \int\limits_0^{1-y} dzdydx$ to $dxdydz$.

Solution The region is $\{(x, y, z)|0 \le z \le 1 - y, 0 \le y \le 1, 0 \le x \le 2\}$. We need to define the limits for x(function of y and z), y(function of z), and z(constants).

$$0 \le x \le 2$$

$$\begin{cases} 0 \le z \le 1 - y \\ 0 \le y \le 1 \end{cases} \Rightarrow \begin{cases} 0 \ge -z \ge -1 + y \\ 0 \le y \le 1 \end{cases}$$

$$\Rightarrow \begin{cases} 1 \ge 1 - z \ge y \\ 0 \le y \le 1 \end{cases} \Rightarrow \begin{cases} y \le 1 - z \le 1 \\ 0 \le y \le 1 \end{cases} \Rightarrow 0 \le y \le 1 - z \le 1$$

$$0 \le z \le 1 - y \overset{0\le y\le 1}{\Rightarrow} 0 \le z \le 1$$

Therefore,

$$\int_0^2 \int_0^1 \int_0^{1-y} dzdydx = \int_0^1 \int_0^{1-z} \int_0^2 dxdydz$$

Example 3.22 Change the order of integration for $\int_0^1 \int_0^x \int_0^y dzdydx$ to $dxdzdy$.

Solution The region is $\{(x, y, z)|0 \le z \le y, 0 \le y \le x, 0 \le x \le 1\}$ or $\{(x, y, z)|0 \le z \le y \le x \le 1\}$. We need to define the limits for x(function of y and z), (function of y), and y(constants). Therefore,

$$\int_0^1 \int_0^x \int_0^y dzdydx = \int_0^1 \int_0^y \int_y^1 dzdydx$$

Example 3.23 Change the order of integration for $\int_0^4 \int_0^{\sqrt{4-x}} \int_0^2 dzdydx$ to $dydzdx$.

Solution The region is $\left\{(x, y, z)|0 \le z \le 2, 0 \le y \le \sqrt{4-x}, 0 \le x \le 4\right\}$. We need to define the limits for y(function of x and z), z(function of x), and x(constants).

$$\begin{aligned} &0 \le y \le \sqrt{4-x} \\ &0 \le z \le 2 \\ &0 \le x \le 4 \end{aligned}$$

Therefore,

$$\int_0^4 \int_0^{\sqrt{4-x}} \int_0^2 dzdydx = \int_0^4 \int_0^{\sqrt{2}} \int_0^{\sqrt{4-x}} dzdydx$$

Example 3.24 Change the order of integration for $\int_0^4 \int_0^{\sqrt{4-x}} \int_0^2 dzdydx$ to $dxdydz$.

Solution The region is $\left\{(x, y, z)|0 \le z \le 2, 0 \le y \le \sqrt{4-x}, 0 \le x \le 4\right\}$. We need to define the limits for x(function of y and z), y(function of z), and z(constants).

$$\begin{cases} 0 \le x \le 4 \\ 0 \le y \le \sqrt{4-x} \end{cases} \Rightarrow \begin{cases} 0 \le x \le 4 \\ 0 \le y^2 \le 4-x \end{cases} \Rightarrow \begin{cases} 0 \le x \le 4 \\ 0 \ge -y^2 \ge x-4 \end{cases}$$

$$\Rightarrow \begin{cases} 0 \le x \le 4 \\ 4 \ge 4-y^2 \ge x \end{cases} \Rightarrow \begin{cases} 0 \le x \le 4 \\ x \le 4-y^2 \le 4 \end{cases} \Rightarrow 0$$

$$\le x \le 4-y^2 \le 4$$

$$0 \le y \le \sqrt{4-x} \overset{0 \le x \le 4}{\Rightarrow} 0 \le y \le \sqrt{4} \Rightarrow 0 \le y \le 2$$

$$0 \le z \le 2$$

Therefore,

$$\int_0^4 \int_0^{\sqrt{4-x}} \int_0^2 dzdydx = \int_0^2 \int_0^2 \int_0^{4-y^2} dxdydz$$

Example 3.25 Find the average value of the temperature function $T(x, y, z) = xz^2$ over the rectangular solid V defined by the region: $V = \{(x, y, z)|0 \leq x \leq 2, 0 \leq y \leq 1, 0 \leq z \leq 3\}$.

Solution The average value of a function $f(x, y, z)$ over a solid region V is given by the formula:

$$f_{avg} = \frac{1}{volume\ of\ V} \iiint\limits_V f(x, y, z)dV$$

Since V is a simple rectangular box (a cuboid), its volume is the product of its side lengths:

$$volume\ of\ V = (2-0) \times (1-0) \times (3-0) = 6$$

Alternatively, the volume is calculated via the triple integral of 1:

$$volume\ of\ V = \int_0^3 \int_0^1 \int_0^2 dxdydz = 6$$

We calculate the triple integral of the temperature function $T(x, y, z) = xz^2$ over the region V:

$$\begin{aligned}
\iiint\limits_V T(x, y, z)dV &= \int_0^3 \int_0^1 \int_0^2 xz^2 dxdydz \\
&= \int_0^3 z^2 dz \int_0^1 dy \int_0^2 xdx = \left.\frac{z^3}{3}\right|_0^3 \times (1-0) \times \left.\frac{x^2}{2}\right|_0^2 \\
&= \left(\frac{3^3}{3} - 0\right) \times 1 \times \left(\frac{2^2}{2} - 0\right) = 9 \times 2 = 18
\end{aligned}$$

Now we apply the average value formula:

$$T_{avg} = \frac{1}{volume\ of\ V} \iiint\limits_V T(x, y, z)dV = \frac{1}{6} \times 18 = 3$$

The average temperature in the solid region V is 3 (units of temperature).

Example 3.26 Find the average density ρ_{avg} of a solid V that is bounded by the plane $x+y+z=3$ and the three coordinate planes ($x=0, y=0, z=0$), given the density function $\rho(x, y, z) = xz$.

Solution The solid V is a tetrahedron in the first octant with intercepts at $(3, 0, 0)$, $(0, 3, 0)$, and $(0, 0, 3)$. We use a triple integral to find the volume:

$$\text{volume of } V = \iiint\limits_V 1dV$$

The z bounds are: $0 \leq z \leq 3 - x - y$.

The projection onto the xy-plane is the triangular region R in the first quadrant bounded by $x+y=3$, $x=0$ and $y=0$ (Fig. 3.17).

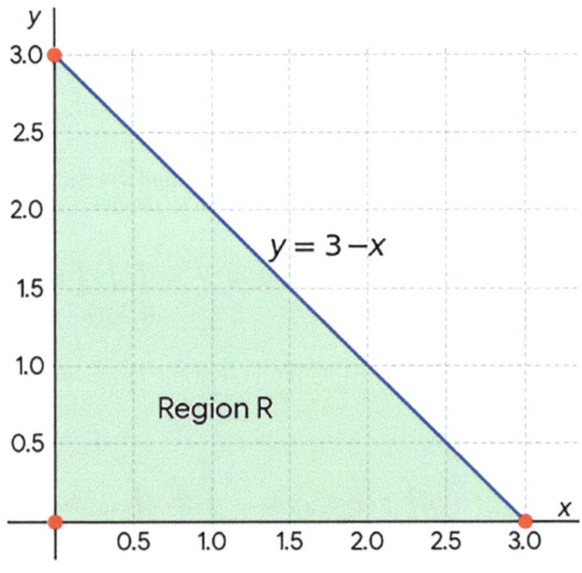

Fig. 3.17 The integration region in the xy-plane

The volume of V is:

$$\text{volume of } V = \int\limits_0^3 \int\limits_0^{3-x} \int\limits_0^{3-x-y} dzdydx$$

The calculation of $V = \int_0^3 \int_0^{3-x} \int_0^{3-x-y} dzdydx$ is shown below:

$$\int_0^{3-x-y} dz = z\Big|_0^{3-x-y} = 3-x-y-0 = 3-x-y$$

$$\int_0^{3-x}\int_0^{3-x-y} dzdy = \int_0^{3-x} (3-x-y)dy$$

$$= \left((3-x)y - \frac{y^2}{2}\right)\Big|_0^{3-x} = (3-x)^2 - \frac{(3-x)^2}{2} - 0 = \frac{(3-x)^2}{2}$$

$$volume\ \ of\ \ V = \int_0^3\int_0^{3-x}\int_0^{3-x-y} dzdydx$$

$$= \int_0^3 \frac{(3-x)^2}{2}dx = \frac{1}{2}\int_0^3 \frac{(3-x)^2}{2}dx = -\frac{1}{2}$$

$$\int_0^3 \frac{(3-x)^2}{2}dx = -\frac{1}{2}\,\frac{(3-x)^3}{3}\Big|_0^3 = -\frac{(3-x)^3}{6}\Big|_0^3 = \frac{(3-x)^3}{6}\Big|_3^0$$

$$= \frac{27}{6} - 0 = \frac{27}{6} = \frac{9}{2} = 4.5$$

Now we integrate the density function $\rho(x, y, z) = xz$:

$$\iiint_V xzdV = \int_0^3\int_0^{3-x}\int_0^{3-x-y} xzdzdydx$$

The calculation of $\int_0^3\int_0^{3-x}\int_0^{3-x-y} xzdzdydx$ is shown below:

$$\int_0^{3-x-y} xzdz = x\frac{z^2}{2}\Big|_0^{3-x-y} = x\frac{(3-x-y)^2}{2} - 0 = \frac{x(3-x-y)^2}{2}$$

$$\int_0^{3-x}\int_0^{3-x-y} xzdzdy$$

$$= \int_0^{3-x} \frac{x(3-x-y)^2}{2}dy = \frac{x}{2}\int_0^{3-x} (3-x-y)^2dy = -\frac{x}{2}\int_0^{3-x} (3-x-y)^2dy$$

$$= -\frac{x}{2}\,\frac{(3-x-y)^3}{3}\Big|_0^{3-x} = -\frac{x(3-x-y)^3}{6}\Big|_0^{3-x} = \frac{x(3-x-y)^3}{6}\Big|_{3-x}^{0}$$

$$= \frac{x(3-x-0)^3}{6} - \frac{x(3-x-(3-x))^3}{6} = \frac{x(3-x)^3}{6}$$

$$\int_0^3\int_0^{3-x}\int_0^{3-x-y} xz\,dz\,dy\,dx = \int_0^3 \frac{x(3-x)^3}{6}dx \overset{\begin{cases}u=3-x\\ du=-dx\end{cases}}{\Rightarrow} \int_3^0 -\frac{(3-u)u^3}{6}du$$

$$= \int_0^3 \frac{(3-u)u^3}{6}du = \int_0^3 \left(\frac{1}{2}u^3 - \frac{1}{6}u^4\right)$$

$$du = \left(\frac{1}{8}u^4 - \frac{1}{30}u^5\right)\Bigg|_0^3 = \frac{1}{8}3^4 - \frac{1}{30}3^5 = \frac{81}{8} - \frac{243}{30} = \frac{81}{8} - \frac{81}{10}$$

$$= \frac{405-324}{40} = \frac{81}{40} \approx 2.0250$$

The average density (ρ_{avg}) is:

$$\rho_{avg} = \frac{1}{volume\ of\ V}\iiint_V \rho(x,y,z)dV = \frac{1}{\frac{9}{2}} \times \frac{81}{40} = \frac{2}{9} \times \frac{81}{40} = \frac{81}{180} = \frac{9}{20} \approx 0.45$$

Exercise 3.5 Find the average value of the function $f(x, y, z) = z$ over the solid region V bounded above by the paraboloid $z = 4 - x^2 - y^2$ and below by the xy-plane ($z = 0$). (Ans. $\frac{1}{8\pi} \times \frac{32\pi}{3} = \frac{4}{3}$).

Example 3.27 Calculate the volume of the sphere $x^2 + y^2 + z^2 = 16$ using a triple integral.

Solution The volume of the sphere $x^2 + y^2 + z^2 = 16$ can be calculated using a triple integral. Since the region is spherical, the most convenient coordinate system is spherical coordinates.

The sphere has a radius of $R = 4$ (since $R^2 = 16$).

Radial Distance (ρ): $0 \le \rho \le 4$.

Polar Angle (ϕ): $0 \le \phi \le \pi$.

Azimuthal Angle (θ): $0 \le \theta \le 2\pi$.

Volume Element (dV): $dV = \rho^2 \sin(\phi)d\rho d\phi d\theta$.

The volume V is given by:

$$V = \iiint_E 1dV = \int_0^{2\pi}\int_0^{\pi}\int_0^{4} \rho^2 \sin(\phi)d\rho d\phi d\theta = \int_0^4 \rho^2 d\rho \int_0^{\pi} \sin(\phi)d\phi \int_0^{2\pi} d\theta$$

Let's calculate each integral separately.

$$\int_0^4 \rho^2 d\rho = \frac{\rho^3}{3}\bigg|_0^4 = \frac{4^3}{3} - \frac{0^3}{3} = \frac{64}{3}$$

$$\int_0^\pi \sin(\phi)d\phi = -\cos(\phi)\bigg|_0^\pi$$

$$= \cos(\phi)\bigg|_\pi^0 = \cos(0) - \cos(\pi) = 1 - (-1) = 2$$

$$\int_0^{2\pi} d\theta = 2\pi - 0 = 2\pi$$

Therefore,

$$V = \int_0^4 \rho^2 d\rho \int_0^\pi \sin(\phi)d\phi \int_0^{2\pi} d\theta = \frac{64}{3} \times 2 \times 2\pi = \frac{256\pi}{3} \approx 268.0826$$

The volume of the solid is $\frac{256\pi}{3} \approx 268.0826$ cubic units.

Example 3.28 Calculate the volume of the sphere $x^2 + y^2 + z^2 = 16$ using a triple integral.

Solution This problem was solved in the previous example. We will now re-evaluate this problem using the cylindrical coordinate system.

In cylindrical coordinates, the transformation is $x^2 + y^2 = r^2$, and the volume element is $dV = rdzdrd\theta$. The equation of the sphere becomes $r^2 + z^2 = 16$.

z **(Height)**: $-\sqrt{16 - r^2} \le z \le \sqrt{16 - r^2}$.

r **(Radius)**: $0 \le r \le 4$.

θ **(Angle)**: $0 \le \theta \le 2\pi$.

The triple integral is:

$$V = \iiint_E 1dV = \int_0^{2\pi}\int_0^4\int_{-\sqrt{16-r^2}}^{+\sqrt{16-r^2}} rdzdrd\theta$$

The calculation of $\int_0^{2\pi}\int_0^4\int_{-\sqrt{16-r^2}}^{+\sqrt{16-r^2}} rdzdrd\theta$ is shown below:

$$\int_{-\sqrt{16-r^2}}^{+\sqrt{16-r^2}} rdz = r\int_{-\sqrt{16-r^2}}^{+\sqrt{16-r^2}} dz = rz\bigg|_{-\sqrt{16-r^2}}^{+\sqrt{16-r^2}}$$

$$= r\left(\sqrt{16-r^2} - \left(-\sqrt{16-r^2}\right)\right) = 2r\sqrt{16-r^2}$$

$$\int_0^4 \int_{-\sqrt{16-r^2}}^{+\sqrt{16-r^2}} rdzdr = \int_0^4 2r\sqrt{16-r^2}dr$$

$$= -\int_0^4 -2r\sqrt{16-r^2}dr = \int_4^0 2r\sqrt{16-r^2}dr$$

$$= \left.\frac{\left(16-r^2\right)^{\frac{3}{2}}}{\frac{3}{2}}\right|_0^0 = \left.\frac{2\left(16-r^2\right)^{\frac{3}{2}}}{3}\right|_4^0$$

$$= \frac{2\left(16-0^2\right)^{\frac{3}{2}}}{3} - \frac{2\left(16-4^2\right)^{\frac{3}{2}}}{3}$$

$$= \frac{2\times 16^{\frac{3}{2}}}{3} - 0 = \frac{2\times \left(4^2\right)^{\frac{3}{2}}}{3} - 0 = \frac{128}{3}$$

Therefore,

$$V = \int_0^{2\pi} \int_0^4 \int_{-\sqrt{16-r^2}}^{+\sqrt{16-r^2}} rdzdrd\theta = \int_0^{2\pi} \frac{128}{3}d\theta$$

$$= \frac{128}{3}\int_0^{2\pi} d\theta = \frac{128}{3} \times (2\pi - 0) = \frac{256\pi}{3} \approx 268.0826$$

The volume of the solid is $\frac{256\pi}{3} \approx 268.0826$ cubic units.

Example 3.29 Calculate the volume (V) of the solid region E that lies below the plane $z = 4$ and above the paraboloid $z = x^2 + y^2$.

Solution We convert the bounding equations to cylindrical coordinates using $x^2 + y^2 = r^2$:

Upper Bound: $z = 4$.

Lower Bound: $z = x^2 + y^2 \Rightarrow z = r^2$.

z **Limits**: $r^2 \leq z \leq 4$.

The projection onto the xy-plane is:

$$\begin{cases} z = 4 \\ z = x^2 + y^2 \end{cases} \Rightarrow x^2 + y^2 = 4$$

The projection onto the xy-plane is shown in Fig. 3.18.

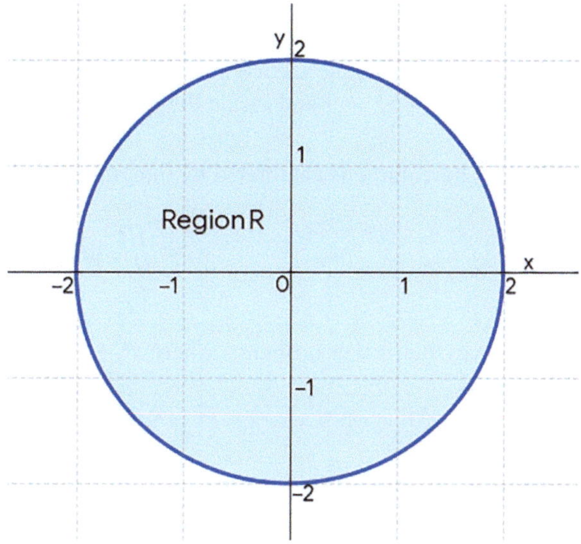

Fig. 3.18 The integration region in the xy-plane

Based on Fig. 3.18,

***r* Limits (Radius)**: $0 \leq r \leq 2$.

***θ* Limits (Angle)**: The full circle is covered, $0 \leq \theta \leq 2\pi$.

The differential volume element in cylindrical coordinates is $dV = rdzdrd\theta$. Therefore,

$$V = \iiint_E 1dV = \int_0^{2\pi}\int_0^2\int_{r^2}^4 rdzdrd\theta$$

The calculation of $\int_0^{2\pi}\int_0^2\int_{r^2}^4 rdzdrd\theta$ is shown below:

$$\int_{r^2}^4 rdz = r\int_{r^2}^4 dz = r(4-r^2) = 4r - r^3$$

$$\int_0^2\int_{r^2}^4 rdzdr = \int_0^2 (4r - r^3)dr = 2r^2 - \frac{r^4}{4}\bigg|_0^2$$

$$= 2\times 2^2 - \frac{2^4}{4} - \left(2\times 0^2 - \frac{0^4}{4}\right) = 4 - 0 = 4$$

$$V = \int_0^{2\pi}\int_0^2\int_{r^2}^4 rdzdrd\theta = \int_0^{2\pi} 4d\theta = 4\int_0^{2\pi} d\theta$$

$$= 4(2\pi - 0) = 8\pi \approx 25.1327$$

The volume of the solid is $8\pi \approx 25.1327$ cubic units.

Example 3.30 Calculate the volume (V) of the solid region E that lies inside the sphere $x^2 + y^2 + z^2 = 9$ and inside the cylinder $x^2 + y^2 = 4$.

Solution The volume is calculated by integrating $f(x, y, z) = 1$ over E.

$$V = \iiint\limits_E 1dV$$

We use cylindrical coordinates where $x^2 + y^2 = r^2$.

$$x^2 + y^2 + z^2 = 9 \Rightarrow r^2 + z^2 = 9 \Rightarrow z^2 = 9 - r^2 \Rightarrow z = \pm\sqrt{9 - r^2}$$

z **Limits**: $-\sqrt{9 - r^2} \le z \le \sqrt{9 - r^2}$.

Since the cylinder $r = 2$ is entirely contained within the sphere (radius 3), the projection onto the xy-plane is the disk $r \le 2$ (Fig. 3.19).

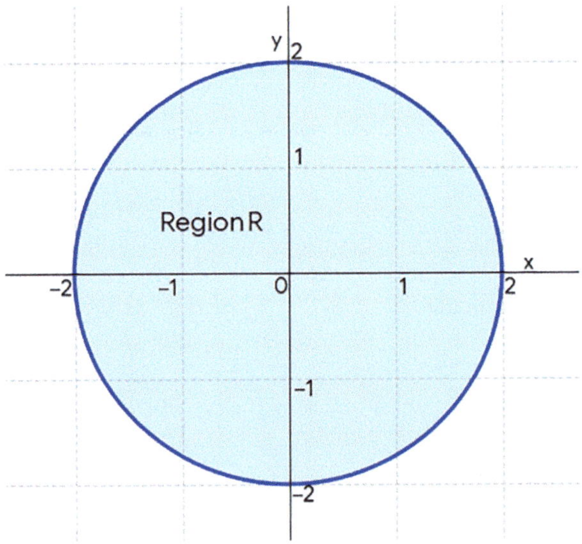

Fig. 3.19 The integration region in the xy-plane

Based on Fig. 3.19,

r **Limits (Radius)**: $0 \le r \le 2$.

θ **Limits (Angle)**: The full circle is covered, $0 \le \theta \le 2\pi$.

The differential volume element in cylindrical coordinates is $dV = rdzdrd\theta$. Therefore,

$$V = \iiint\limits_E 1dV = \int_0^{2\pi} \int_0^2 \int_{-\sqrt{9-r^2}}^{+\sqrt{9-r^2}} rdzdrd\theta$$

The calculation of $\int_0^{2\pi}\int_0^2\int_{-\sqrt{9-r^2}}^{+\sqrt{9-r^2}} rdzdrd\theta$ is shown below:

$$\int_{-\sqrt{9-r^2}}^{+\sqrt{9-r^2}} rdz = r\int_{-\sqrt{9-r^2}}^{+\sqrt{9-r^2}} dz = r\left(\sqrt{9-r^2}-\left(-\sqrt{9-r^2}\right)\right) = 2r\sqrt{9-r^2}$$

$$\int_0^2\int_{-\sqrt{9-r^2}}^{+\sqrt{9-r^2}} rdzdr = \int_0^2 2r\sqrt{9-r^2}dr$$

$$= -\int_0^2 -2r\sqrt{9-r^2}dr = \int_2^0 -2r\sqrt{9-r^2}dr = \left.\frac{(9-r^2)^{\frac{3}{2}}}{\frac{3}{2}}\right|_2^0$$

$$= \left.\frac{2(9-r^2)^{\frac{3}{2}}}{3}\right|_2^0 = \frac{2(9-0^2)^{\frac{3}{2}}}{3} - \frac{2(9-2^2)^{\frac{3}{2}}}{3}$$

$$= \frac{2\times 27}{3} - \frac{2\times\sqrt{5}}{3} = \frac{54-10\sqrt{5}}{3}$$

$$V = \iiint_E 1dV = \int_0^{2\pi}\frac{54-10\sqrt{5}}{3}d\theta = \frac{54-10\sqrt{5}}{3}\int_0^{2\pi}d\theta\frac{54-10\sqrt{5}}{3}$$

$$\times(2\pi-0) = \frac{108-20\sqrt{5}}{3}\pi \approx 66.2652$$

The volume of the solid is $\frac{108-20\sqrt{5}}{3}\pi \approx 66.2652$ cubic units.

Example 3.31 Determine the volume of the solid region that lies above the plane $z = 0$, inside the circular cylinder $r = \cos(\theta)$, and below the surface $z = 3r^2$.

Solution This problem requires calculating the volume of the solid region E using a triple integral in cylindrical coordinates (r, θ, z), as the boundaries are naturally defined by these coordinates.

The volume is given by $V = \iiint_E 1dV$, where $dV = rdzdrd\theta$.

z Limits: $0 \le z \le 3r^2$.

The solid is contained inside the cylinder $r = \cos(\theta)$.

r Limits: The radius extends from the origin ($r = 0$) to the boundary curve $r = \cos(\theta)$.

θ Limits: For the radius r to be non-negative, $\cos(\theta)$ must be non-negative. This traces the full circle defined by the cylinder:

$$-\frac{\pi}{2} \leq \theta \leq \frac{\pi}{2}$$

The volume integral is:

$$V = \int_{-\frac{\pi}{2}}^{\frac{\pi}{2}} \int_{0}^{\cos(\theta)} \int_{0}^{3r^2} rdzdrd\theta$$

The calculation of $V = \int_{-\frac{\pi}{2}}^{\frac{\pi}{2}} \int_{0}^{\cos(\theta)} \int_{0}^{3r^2} rdzdrd\theta$ is shown below:

$$\int_{0}^{3r^2} rdz = r \int_{0}^{3r^2} dz = rz|_0^{3r^2} = r \times 3r^2 - r \times 0 = 3r^3 - 0 = 3r^3$$

$$\int_{0}^{\cos(\theta)} \int_{0}^{3r^2} rdzdr = \int_{0}^{\cos(\theta)} 3r^3 dr = \frac{3}{4} r^4 \Big|_{0}^{\cos(\theta)}$$
$$= \frac{3}{4}(\cos(\theta))^4 - \frac{3}{4}(0)^4 = \frac{3}{4}\cos^4(\theta) - 0 = \frac{3}{4}\cos^4(\theta)$$

$$V = \int_{-\frac{\pi}{2}}^{\frac{\pi}{2}} \int_{0}^{\cos(\theta)} \int_{0}^{3r^2} rdzdrd\theta = \int_{-\frac{\pi}{2}}^{\frac{\pi}{2}} \frac{3}{4}\cos^4(\theta)d\theta$$
$$= 2\int_{0}^{\frac{\pi}{2}} \frac{3}{4}\cos^4(\theta)d\theta = 2 \times \frac{3}{4}\int_{0}^{\frac{\pi}{2}} \cos^4(\theta)d\theta = \frac{3}{2}\int_{0}^{\frac{\pi}{2}} \cos^4(\theta)d\theta$$

The Wallis' formula can be used to calculate $\int_0^{\frac{\pi}{2}} \cos^4(\theta)d\theta$.

$$\int_{0}^{\frac{\pi}{2}} \cos^4(\theta)d\theta = \frac{3}{4} \times \frac{1}{2} \times \frac{\pi}{2} = \frac{3\pi}{16}$$

Therefore,

$$V = \frac{3}{2}\int_{0}^{\frac{\pi}{2}} \cos^4(\theta)d\theta = \frac{3}{2} \times \frac{3\pi}{16} = \frac{9\pi}{32} \approx 0.8836$$

The volume of the solid is $\frac{9\pi}{32} \approx 0.8836$ cubic units.

Exercise 3.6 Determine the volume of the solid region that lies above the plane $z = 0$, inside the circular cylinder $r = 2\cos(\theta)$, and below the surface $z = r^2$. (Ans. $\frac{3\pi}{2} \approx 4.7124$).

Example 3.32 Calculate the volume (V) of the solid region E that lies above the cone $z = \sqrt{x^2 + y^2}$ and below the sphere $x^2 + y^2 + z^2 = 8$.

Solution The volume V is calculated by integrating $f(x, y, z) = 1$ over E. We use cylindrical coordinates where $x^2 + y^2 = r^2$.

$$z = \sqrt{x^2 + y^2} \Rightarrow z = \sqrt{r^2} \Rightarrow z = r$$

$$x^2 + y^2 + z^2 = 8 \Rightarrow r^2 + z^2 = 8 \Rightarrow z^2 = 8 - r^2 \Rightarrow z = \sqrt{8 - r^2}$$

z **Limits**: $r \le z \le \sqrt{8 - r^2}$

$$\begin{cases} z = \sqrt{x^2 + y^2} \\ x^2 + y^2 + z^2 = 8 \end{cases} \Rightarrow x^2 + y^2 + x^2 + y^2 = 8 \Rightarrow 2(x^2 + y^2) = 8 \Rightarrow x^2 + y^2 = 4$$

This is a circle with radius 2. Therefore, the projection onto the xy-plane is the disk $r \le 2$ (Fig. 3.20).

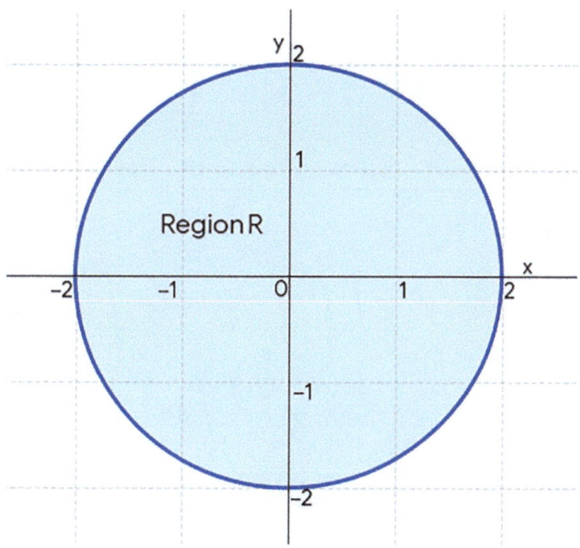

Fig. 3.20 The integration region in the xy-plane

Based on Fig. 3.20,

r **Limits (Radius)**: $0 \le r \le 2$.

θ **Limits (Angle)**: The full circle is covered, $0 \le \theta \le 2\pi$.

The differential volume element in cylindrical coordinates is $dV = rdzdrd\theta$. Therefore,

$$V = \iiint_E 1dV = \int_0^{2\pi}\int_0^2\int_r^{\sqrt{8-r^2}} rdzdrd\theta$$

The calculation of $\int_0^{2\pi}\int_0^2\int_r^{\sqrt{8-r^2}} rdzdrd\theta$ is shown below:

$$\int_r^{\sqrt{8-r^2}} rdz = r\int_r^{\sqrt{8-r^2}} dz = r\left(\sqrt{8-r^2}-r\right) = r\sqrt{8-r^2}-r^2$$

$$\int_0^2\int_r^{\sqrt{8-r^2}} rdzdr = \int_0^2\left(r\sqrt{8-r^2}-r^2\right)dr = \int_0^2 r\sqrt{8-r^2}dr - \int_0^2 r^2dr$$

$$= \frac{1}{-2}\int_0^2 -2r\sqrt{8-r^2}dr - \int_0^2 r^2dr = \frac{1}{2}\int_2^0 -2r\sqrt{8-r^2}dr$$

$$-\int_0^2 r^2dr = \frac{1}{2}\frac{(8-r^2)^{\frac{3}{2}}}{\frac{3}{2}}\Bigg|_2^0 - \frac{r^3}{3}\Bigg|_0^2$$

$$= \frac{(8-r^2)^{\frac{3}{2}}}{3}\Bigg|_2^0 - \frac{r^3}{3}\Bigg|_0^2 = \left(\frac{(8-0^2)^{\frac{3}{2}}}{3} - \frac{(8-2^2)^{\frac{3}{2}}}{3}\right) - \left(\frac{2^3}{3}-\frac{0^3}{3}\right)$$

$$= \frac{8\sqrt{8}-8-8}{3} = \frac{8\sqrt{8}-16}{3} = \frac{16\sqrt{2}-16}{3} = \frac{16\left(\sqrt{2}-1\right)}{3}$$

$$V = \int_0^{2\pi}\int_0^2\int_r^{\sqrt{8-r^2}} rdzdrd\theta = \int_0^{2\pi}\frac{16\left(\sqrt{2}-1\right)}{3}d\theta$$

$$= \frac{16\left(\sqrt{2}-1\right)}{3}\int_0^{2\pi} d\theta = \frac{16\left(\sqrt{2}-1\right)}{3}\times(2\pi-0)$$

$$= \frac{32\left(\sqrt{2}-1\right)}{3}\pi \approx 13.8804$$

The volume of the solid is $\frac{32(\sqrt{2}-1)}{3}\pi \approx 13.8804$ cubic units.

Example 3.33 Determine the volume of the solid region E that lies above the xy-plane ($z = 0$), below the paraboloid $z = 16 - x^2 - y^2$, and outside the cylinder $x^2 + y^2 = 4$.

Solution The volume V is calculated by integrating $f(x, y, z) = 1$ over E. We use cylindrical coordinates where $x^2 + y^2 = r^2$.

$$z = 16 - x^2 - y^2 \Rightarrow z = 16 - r^2$$

z **Limits**: $0 \leq z \leq 16 - r^2$.

The region R is the area outside the cylinder $x^2+y^2 = 4$ and inside the intersection of the paraboloid with the xy-plane ($16 - x^2 - y^2 = 0 \Rightarrow x^2 + y^2 = 16$). This forms an annulus (washer-shaped region).

$$4 \leq x^2 + y^2 \leq 16$$

The region R is shown in Fig. 3.21.

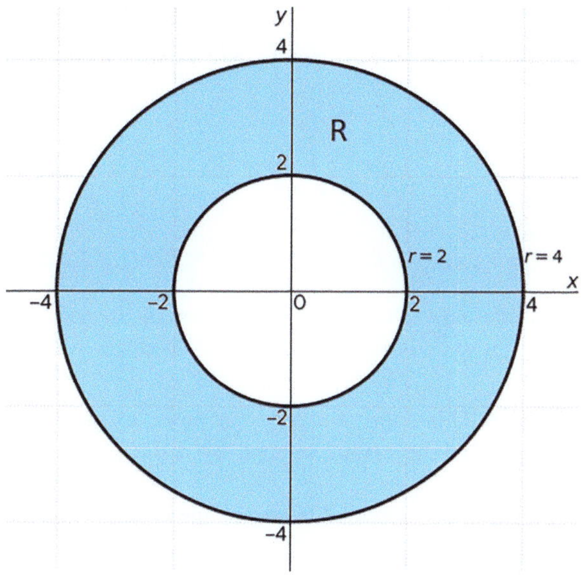

Fig. 3.21 The integration region in the xy-plane

Based on Fig. 3.21,

r **Limits (Radius)**: $2 \leq r \leq 4$.

θ **Limits (Angle)**: The full circle is covered, $0 \leq \theta \leq 2\pi$.

The differential volume element in cylindrical coordinates is $dV = rdzdrd\theta$. Therefore,

$$V = \iiint_E 1dV = \int_0^{2\pi}\int_2^4\int_0^{16-r^2} rdzdrd\theta$$

The calculation of $\int_0^{2\pi}\int_2^4\int_0^{16-r^2} rdzdrd\theta$ is shown below:

$$\int_0^{16-r^2} rdz = r\int_0^{16-r^2} dz = r\left(16-r^2-0\right) = 16r - r^3$$

$$\int_2^4\int_0^{16-r^2} rdzdr = \int_2^4 \left(16r - r^3\right)dr$$

$$= 8r^2 - \frac{r^4}{4}\bigg|_2^4 = \left(8\times 4^2 - \frac{4^4}{4}\right) - \left(8\times 2^2 - \frac{2^4}{4}\right) = (128-64)$$

$$-(32-4) = 64 - 28 = 36$$

$$V = \int_0^{2\pi}\int_2^4\int_0^{16-r^2} rdzdrd\theta = \int_0^{2\pi} 36d\theta$$

$$= 36\int_0^{2\pi} d\theta = 36\times(2\pi - 0) = 72\pi \approx 226.1947$$

The volume of the solid is $72\pi \approx 226.1947$ cubic units.

Example 3.34 Determine the volume of the solid region E bounded below by the cone $z = \sqrt{x^2+y^2}$ and above by the plane $z = 2$.

Solution This solid is a cone (or more accurately, the region inside a cone) with its vertex at the origin, capped by a flat plane. It's best solved using cylindrical coordinates.

The solid is bounded below by the cone $z = \sqrt{x^2+y^2} = \sqrt{r^2} = r$ and above by the plane $z = 2$.

z Limits: $r \le z \le 2$.

The region of integration R (the projection onto the xy-plane) is defined by the intersection of these two surfaces:

$$\begin{cases} z = \sqrt{x^2+y^2} \\ z = 2 \end{cases} \Rightarrow \sqrt{x^2+y^2} = 2 \Rightarrow x^2+y^2 = 4$$

The projection R is a disk of radius $R = 2$ centered at the origin (Fig. 3.22).

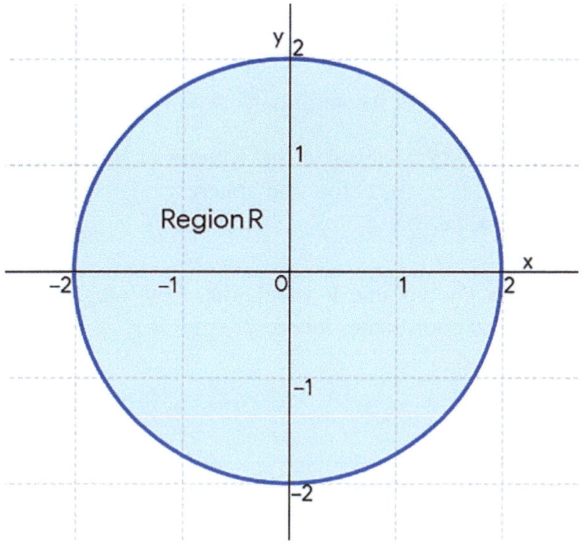

Fig. 3.22 The integration region in the xy-plane

Based on Fig. 3.22,

***r* Limits (Radius)**: $0 \le r \le 2$.

***θ* Limits (Angle)**: The full circle is covered, $0 \le \theta \le 2\pi$.

The differential volume element in cylindrical coordinates is $dV = rdzdrd\theta$. Therefore,

$$V = \iiint_E 1dV = \int_0^{2\pi}\int_0^2\int_r^2 rdzdrd\theta$$

The calculation of $\int_0^{2\pi}\int_0^2\int_r^2 rdzdrd\theta$ is shown below:

$$\int_r^2 rdz = r\int_r^2 dz = r(2-r) = 2r - r^2$$

$$\int_0^2\int_r^2 rdzdr = \int_0^2 (2r - r^2)dr = r^2 - \frac{r^3}{3}\bigg|_0^2$$

$$= 2^2 - \frac{2^3}{3} - \left(0^2 - \frac{0^3}{3}\right) = 4 - \frac{8}{3} - 0 = \frac{4}{3}$$

$$V = \int_0^{2\pi}\int_0^2\int_r^2 rdzdrd\theta = \int_0^{2\pi} \frac{4}{3}d\theta = \frac{4}{3}\int_0^{2\pi} d\theta = \frac{4}{3} \times (2\pi - 0)$$

$$= \frac{8\pi}{3} \approx 8.3776$$

The volume of the solid is $\frac{8\pi}{3} \approx 8.3776$ cubic units.

Example 3.35 Determine the volume of the solid region E E that lies above the xy-plane ($z = 0$), below the sphere $x^2 + y^2 + z^2 = 16$, and inside the cylinder $x^2 + y^2 = 2x$.

Solution The volume V is calculated by integrating $f(x, y, z) = 1$ over E. We use cylindrical coordinates where $x^2 + y^2 = r^2$.

$$x^2 + y^2 + z^2 = 16 \Rightarrow r^2 + z^2 = 16 \Rightarrow z^2 = 16 - r^2 \Rightarrow z = \pm\sqrt{16 - r^2}$$

E lies above the xy-plane ($z = 0$). Thus, $z \geq 0$ and $z = \sqrt{16 - r^2}$.

z **Limits**: $0 \leq z \leq \sqrt{16 - r^2}$.

Let's calculate the projection of E onto the xy-plane.

$$x^2 + y^2 = 2x \Rightarrow x^2 - 2x + y^2 = 0 \Rightarrow (x - 1)^2 - 1 + y^2 = 0 \Rightarrow (x - 1)^2 + y^2 = 1$$

$$\begin{aligned} x^2 + y^2 = 2x &\overset{\begin{cases} x = r\cos(\theta) \\ y = r\sin(\theta) \end{cases}}{\Rightarrow} r^2\cos^2(\theta) + r^2\sin^2(\theta) \\ &= 2r\cos(\theta) \Rightarrow r^2\big(\cos^2(\theta) + \sin^2(\theta)\big) = 2r\cos(\theta) \Rightarrow r^2 \\ &= 2r\cos(\theta) \Rightarrow r = 2\cos(\theta) \end{aligned}$$

The projection R is a disk of radius $R = 1$ centered at the point $(1, 0)$ (Fig. 3.23).

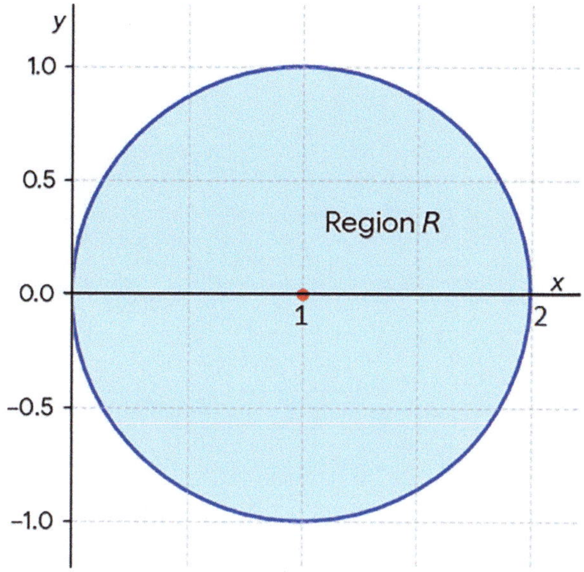

Fig. 3.23 The integration region in the xy-plane

The r and θ limits are:

***r* Limits**: $0 \le r \le 2\cos(\theta)$.

***θ* Limits**: For the radius $r = 2\cos(\theta)$ to be non-negative, $\cos(\theta)$ must be non-negative. This traces the full circle defined by the cylinder:

$$-\frac{\pi}{2} \le \theta \le \frac{\pi}{2}$$

The volume integral is:

$$V = \int_{-\frac{\pi}{2}}^{\frac{\pi}{2}} \int_{0}^{2\cos(\theta)} \int_{0}^{\sqrt{16-r^2}} r dz dr d\theta$$

The calculation of $\int_{-\frac{\pi}{2}}^{\frac{\pi}{2}} \int_{0}^{2\cos(\theta)} \int_{0}^{\sqrt{16-r^2}} r dz dr d\theta$ is shown below:

$$\int_{0}^{\sqrt{16-r^2}} r dz = r \int_{0}^{\sqrt{16-r^2}} dz = rz|_{0}^{\sqrt{16-r^2}} = r\sqrt{16-r^2} - r \times 0 = r\sqrt{16-r^2}$$

$$\int_{0}^{2\cos(\theta)} \int_{0}^{\sqrt{16-r^2}} r dz dr$$

$$= \int_0^{2\cos(\theta)} r\sqrt{16-r^2}dr = \frac{1}{-2}\int_0^{2\cos(\theta)} -2r\sqrt{16-r^2}dr$$

$$= \frac{1}{2}\int_{2\cos(\theta)}^{0} -2r\sqrt{16-r^2}dr$$

$$= \frac{1}{2}\frac{\left(16-r^2\right)^{\frac{3}{2}}}{\frac{3}{2}}\Bigg|_{2\cos(\theta)}^{0}$$

$$= \frac{\left(16-r^2\right)^{\frac{3}{2}}}{3}\Bigg|_{2\cos(\theta)}^{0} = \frac{\left(16-0^2\right)^{\frac{3}{2}}}{3}$$

$$- \frac{\left(16-(2\cos(\theta))^2\right)^{\frac{3}{2}}}{3} = \frac{64}{3} - \frac{8\left(4-\cos^2(\theta)\right)^{\frac{3}{2}}}{3}$$

$$V = \int_{-\frac{\pi}{2}}^{\frac{\pi}{2}}\int_0^{2\cos(\theta)}\int_0^{\sqrt{16-r^2}} rdzdrd\theta = \int_{-\frac{\pi}{2}}^{\frac{\pi}{2}}\left(\frac{64}{3} - \frac{8\left(4-\cos^2(\theta)\right)^{\frac{3}{2}}}{3}\right)d\theta = 2$$

$$\int_0^{\frac{\pi}{2}}\left(\frac{64}{3} - \frac{8\left(4-\cos^2(\theta)\right)^{\frac{3}{2}}}{3}\right)d\theta$$

$$= 2\int_0^{\frac{\pi}{2}}\frac{64}{3}d\theta - 2\int_0^{\frac{\pi}{2}}\frac{8\left(4-\cos^2(\theta)\right)^{\frac{3}{2}}}{3}d\theta = \frac{128}{3}\int_0^{\frac{\pi}{2}}d\theta$$

$$- \frac{16}{3}\int_0^{\frac{\pi}{2}}\left(4-\cos^2(\theta)\right)^{\frac{3}{2}}d\theta$$

$$= \frac{128}{3}\left(\frac{\pi}{2}-0\right) - \frac{16}{3}\int_0^{\frac{\pi}{2}}\left(4-\cos^2(\theta)\right)^{\frac{3}{2}}d\theta$$

$$= \frac{64\pi}{3} - \frac{16}{3}\int_0^{\frac{\pi}{2}}\left(4-\cos^2(\theta)\right)^{\frac{3}{2}}d\theta$$

The last integral is non-elementary (it cannot be expressed in terms of elementary functions). We will use MATLAB to compute its value (Fig. 3.24).

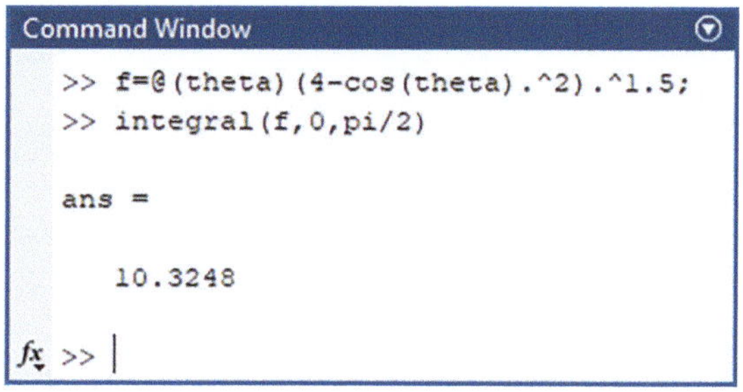

Fig. 3.24 MATLAB code

Based on Fig. 3.24,

$$V = \frac{64\pi}{3} - \frac{16}{3}\int_0^{\frac{\pi}{2}} \left(4 - cos^2(\theta)\right)^{\frac{3}{2}} d\theta = \frac{64\pi}{3} - \frac{16}{3} \times 10.3248 \approx 11.9550$$

The volume of the solid is 11.9550 cubic units.

Example 3.36 Find the integral of $f(x, y, z) = \frac{1}{3\sqrt{x^2+y^2}-(x^2+y^2)}$ over the region E bounded by the surfaces $z = \sqrt{x^2 + y^2}$ and $\frac{1}{2}z = 4 - \sqrt{x^2 + y^2}$.

Solution This problem is most effectively solved using cylindrical coordinates due to the symmetry of the function and the bounding region around the z z-axis. We use the conversion formula:

$$r = \sqrt{x^2 + y^2} \Rightarrow r^2 = x^2 + y^2$$

The integrand becomes:

$$f(x, y, z) = \frac{1}{3\sqrt{x^2 + y^2} - \left(x^2 + y^2\right)} \Rightarrow f(r, \theta, z) = \frac{1}{3\sqrt{r^2} - r^2} \Rightarrow f(r, \theta, z)$$
$$= \frac{1}{3r - r^2} \Rightarrow f(r, \theta, z) = \frac{1}{r(3 - r)}$$

The surfaces become:

$$z = \sqrt{x^2 + y^2} \Rightarrow z = \sqrt{r^2} \Rightarrow z = r$$
$$\frac{1}{2}z = 4 - \sqrt{x^2 + y^2} \Rightarrow z = 8 - 2\sqrt{x^2 + y^2}$$
$$\Rightarrow z = 8 - 2\sqrt{r^2} \Rightarrow z = 8 - 2r$$

The region E is bounded below by the cone $z = r$ and above by the surface $z = 8 - 2r$:

$$r \leq z \leq 8 - 2r$$

The intersection of the two surfaces is:

$$\begin{cases} z=\sqrt{x^2+y^2} \\ \frac{1}{2}z=4-\sqrt{x^2+y^2} \end{cases} \Rightarrow \begin{cases} z=\sqrt{x^2+y^2} \\ z=8-2\sqrt{x^2+y^2} \end{cases} \Rightarrow \sqrt{x^2+y^2}$$
$$=8-2\sqrt{x^2+y^2} \Rightarrow 3\sqrt{x^2+y^2}=8 \Rightarrow \sqrt{x^2+y^2}=\frac{8}{3} \Rightarrow x^2+y^2=\left(\frac{8}{3}\right)^2$$

or,

$$\begin{cases} z=r \\ z=8-2r \end{cases} \Rightarrow r=8-2r \Rightarrow 3r=8 \Rightarrow r=\frac{8}{3}$$

The projection onto the xy-plane (R) is a disk of radius $R=\frac{8}{3}$ centered at the origin (Fig. 3.25).

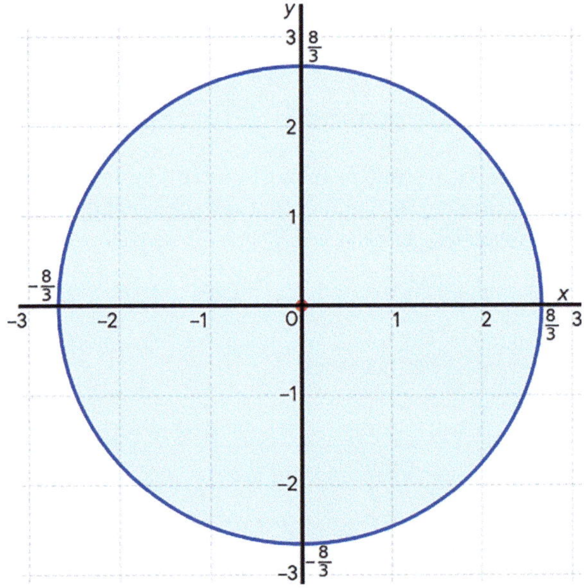

Fig. 3.25 The integration region in the xy-plane

Based on Fig. 3.25,

***r* Limits (Radius)**: $0 \le r \le \frac{8}{3}$.

θ Limits (Angle): The full circle is covered, $0 \le \theta \le 2\pi$.

The differential volume element in cylindrical coordinates is $dV = rdzdrd\theta$. Therefore,

$$\iiint_E f(x,y,z)dV = \int_0^{2\pi}\int_0^{\frac{8}{3}}\int_r^{8-2r} \frac{1}{r(3-r)} \times rdzdrd\theta$$

$$= \int_0^{2\pi} \int_0^{\frac{8}{3}} \int_r^{8-2r} \frac{1}{3-r} dz dr d\theta$$

The calculation of $\int_0^{2\pi} \int_0^{\frac{8}{3}} \int_r^{8-2r} \frac{1}{3-r} dz dr d\theta$ is shown below:

$$\int_r^{8-2r} \frac{1}{3-r} dz = \frac{1}{3-r} \int_r^{8-2r} dz = \frac{1}{3-r}(8-2r-r)$$
$$= \frac{8-3r}{3-r} = \frac{9-3r-1}{3-r} = 3 - \frac{1}{3-r}$$

$$\int_0^{\frac{8}{3}} \int_r^{8-2r} \frac{1}{3-r} dz dr$$
$$= \int_0^{\frac{8}{3}} \left(3 - \frac{1}{3-r}\right) dr = \int_0^{\frac{8}{3}} 3dr + \int_0^{\frac{8}{3}} \frac{-1}{3-r} dr$$
$$= 3 \int_0^{\frac{8}{3}} dr + \int_0^{\frac{8}{3}} \frac{-1}{3-r} dr = 3\left(\frac{8}{3} - 0\right)$$
$$+ \ln(|3-r|) \Big|_0^{\frac{8}{3}} = 8 + \ln\left(\left|3 - \frac{8}{3}\right|\right)$$
$$- \ln(|3-0|) = 8 + \ln\left(\frac{1}{3}\right) - \ln(3) = 8 + \ln(1)$$
$$- \ln(3) - \ln(3) = 8 + 0 - 2\ln(3) = 8 - 2\ln(3)$$

$$\iiint_E f(x, y, z) dV = \int_0^{2\pi} \int_0^{\frac{8}{3}} \int_r^{8-2r} \frac{1}{3-r} dz dr d\theta$$
$$= \int_0^{2\pi} (8 - 2\ln(3)) d\theta = (8 - 2\ln(3)) \int_0^{2\pi} d\theta$$
$$= (8 - 2\ln(3)) \times 2\pi \approx 36.4599$$

Exercise 3.7 Calculate the volume (V) of the solid region E that lies below the paraboloid $z = 9 - x^2 - y^2$, above the xy-plane ($z = 0$), and is contained within the angular sector defined by $-\frac{\pi}{6} \leq \theta \leq \frac{\pi}{6}$. (Ans. $\frac{27\pi}{4} \approx 21.2058$).

Exercise 3.8 Determine the volume of the solid region E enclosed between the two paraboloids defined by the equations:

Upper Paraboloid (Ceiling): $z = 5 - x^2 - y^2$.

Lower Paraboloid (Floor): $z = 4x^2 + 4y^2$

(Ans. $\frac{5\pi}{2} \approx 7.854$)

Exercise 3.9 Find the integral of $f(x, y, z) = \frac{1}{2\sqrt{x^2+y^2-(x^2+y^2)}}$ over the region E bounded by the surfaces $z = \sqrt{x^2 + y^2}$ and $\frac{1}{2}z = 3 - \sqrt{x^2 + y^2}$. (Ans. $12\pi \approx$ 37.6991).

Example 3.37 Calculate the volume of the region of overlap between the spheres $x^2 + y^2 + z^2 = 16\, x^2 + y^2 + z^2 = 16$ and $(x - 6)^2 + y^2 + z^2 = 16$.

Solution The region of overlap is a spherical lens formed by two identical spheres of radius $R = 4$ whose centers are $d = 6$ units apart.

Sphere 1 (S_1): $x^2 + y^2 + z^2 = 16$, centered at $(0, 0, 0)$.

Sphere 2 (S_2): $(x - 6)^2 + y^2 + z^2 = 16$, centered at $(6, 0, 0)$.

The intersection of the two spheres occurs at the plane $x = 3$.

$$\begin{cases} x^2 + y^2 + z^2 = 16 \\ (x-6)^2 + y^2 + z^2 = 16 \end{cases} \Rightarrow x^2 + y^2 + z^2$$
$$= (x-6)^2 + y^2 + z^2 \Rightarrow x^2 = (x-6)^2 \Rightarrow |x|$$
$$= |x-6| \Rightarrow \begin{cases} x = x - 6 \\ x = -(x-6) \end{cases} \Rightarrow \begin{cases} 0 = -6(!) \\ 2x = 6 \end{cases}$$
$$\Rightarrow 2x = 6 \Rightarrow x = 3$$

The total volume of overlap, V, is the sum of the volumes of two identical spherical caps, one from S_1 and one from S_2, both cut by the plane $x = 3$.

$$V = V_1 + V_2 = 2 \times V_1$$

We will calculate the volume V_1 of the cap of S_1 where $x \geq 3$. This cap is defined by the region $3 \leq x \leq 4$ and $y^2 + z^2 \leq 16 - x^2$.

We use a cylindrical system centered on the x-axis (r is the radial distance from the x-axis).

$$x = x$$

$$y = r\cos(\theta)$$

$$z = r\sin(\theta)$$

$$r^2 = y^2 + z^2$$

The volume element is $dV = rdrd\theta dx$.

For the cap V_1 (from $S_1, x \geq 3$):

***x* Limits**: $3 \leq x \leq 4$ (from the intersection plane to the sphere's edge).

***r* Limits**: $0 \leq r \leq \sqrt{16 - x^2}$ (the radius of the cross-sectional disk at a given x x).

***θ* Limits**: $0 \leq \theta \leq 2\pi$ (a full revolution).

$$V_1 = \iiint\limits_{E_1} dV = \int\limits_3^4 \int\limits_0^{2\pi} \int\limits_0^{\sqrt{16-x^2}} rdrd\theta dx$$

The calculation of $\int_3^4 \int_0^{2\pi} \int_0^{\sqrt{16-x^2}} rdrd\theta dx$ is shown below:

$$\int\limits_0^{\sqrt{16-x^2}} rdr = \frac{r^2}{2}\Big|_0^{\sqrt{16-x^2}} = \frac{\left(\sqrt{16 - x^2}\right)^2}{2} - \frac{0}{2} = \frac{16 - x^2}{2}$$

$$\int\limits_0^{2\pi} \int\limits_0^{\sqrt{16-x^2}} rdrd\theta = \int\limits_0^{2\pi} \frac{16 - x^2}{2} d\theta = \frac{16 - x^2}{2} \int\limits_0^{2\pi} d\theta$$

$$= \frac{16 - x^2}{2} \times 2\pi = \left(16 - x^2\right)\pi$$

$$V_1 = \iiint\limits_{E_1} dV = \int\limits_3^4 \int\limits_0^{2\pi} \int\limits_0^{\sqrt{16-x^2}} rdrd\theta dx$$

$$= \int\limits_3^4 \left(16 - x^2\right)\pi dx = \pi \int\limits_3^4 \left(16 - x^2\right)dx = \pi\left(16x - \frac{x^3}{3}\right)\Bigg|$$

$$\begin{matrix}4\\3\end{matrix} = \pi\left(16 \times 4 - \frac{4^3}{3}\right) - \pi\left(16 \times 3 - \frac{3^3}{3}\right)$$

$$= \frac{128\pi}{3} - 39\pi = \frac{11\pi}{3} \approx 11.5192$$

The total volume of overlap V is $2 \times V_1$:

$$V = 2 \times V_1 = 2 \times \frac{11\pi}{3} = \frac{22\pi}{3} \approx 23.0383$$

The volume of is $\frac{22\pi}{3} \approx 23.0383$ cubic units.

Exercise 3.10 Volume of one spherical cap of height h on a sphere of radius R is $V_{cap} = \frac{\pi h^2(3R-h)}{3}$ (Fig. 3.26). So the overlap volume (two caps) is $V = 2V_{cap} = \frac{2\pi h^2(3R-h)}{3}$. Use this formula to verify the result of the previous example, where the overlap volume was found to be $\frac{22\pi}{3} \approx 23.0383$.

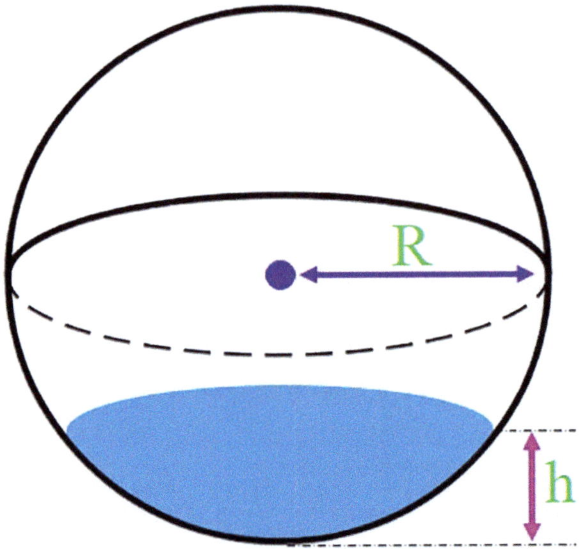

Fig. 3.26 Volume of a spherical cap $(0 \le h \le 2R)$

Example 3.38 Evaluate $\int\limits_{-1}^{1} \int\limits_{-\sqrt{1-x^2}}^{\sqrt{1-x^2}} \int\limits_{x^2+y^2}^{2-x^2-y^2} \sqrt{x^2+y^2}dzdydx$.

Solution This integral is complicated in Cartesian coordinates because of the square root and the parabolic limits. We will convert the integral to cylindrical coordinates (r, θ, z), where $x^2 + y^2 = r^2$ and $dV = rdzdrd\theta$.

The Cartesian integral limits define the region E:

***x* Limits**: $-1 \le x \le 1$.

***y* Limits**: $-\sqrt{1-x^2} \le y \le \sqrt{1-x^2}$. Together with the x limits, this describes the disk $x^2 + y^2 \le 1$ in the xy-plane (Fig. 3.27).

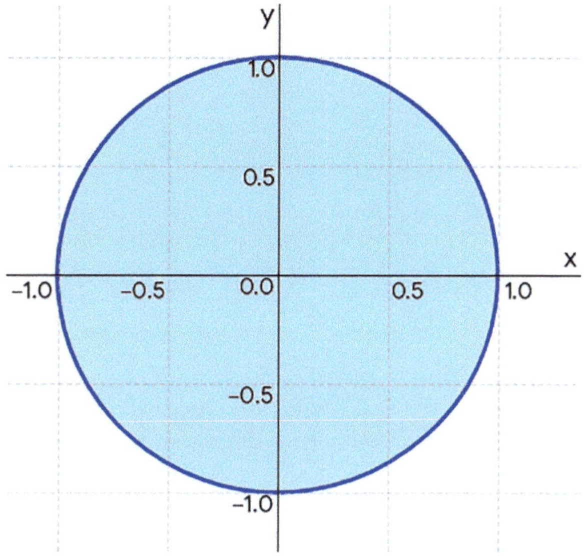

Fig. 3.27 The integration region in the xy-plane

z **Limits**: $x^2 + y^2 \le z \le 2 - \left(x^2 + y^2\right)$. This is the region between the paraboloid $z = x^2 + y^2$ and the paraboloid $z = 2 - x^2 - y^2$.

Let's convert to cylindrical coordinates:

Integrand: $\sqrt{x^2 + y^2} = \sqrt{r^2} = r$.

z **Limits**: $r^2 \le z \le 2 - r^2$.

r **Limits**: $0 \le r \le 1$.

θ **Limits**: $0 \le \theta \le 2\pi$.

The integral transforms to:

$$\int_{-1}^{1} \int_{-\sqrt{1-x^2}}^{\sqrt{1-x^2}} \int_{x^2+y^2}^{2-x^2-y^2} \sqrt{x^2 + y^2} dzdydx = \int_{0}^{2\pi} \int_{0}^{1} \int_{r^2}^{2-r^2} r \times rdzdrd\theta$$

$$= \int_{0}^{2\pi} \int_{0}^{1} \int_{r^2}^{2-r^2} r^2 dzdrd\theta$$

The calculation of $\int_0^{2\pi} \int_0^1 \int_{r^2}^{2-r^2} rdzdrd\theta$ is shown below:

$$\int_{r^2}^{2-r^2} r^2 dz = r^2 \int_{r^2}^{2-r^2} dz = r^2 \times \left(2 - r^2 - r^2\right) = r^2\left(2 - 2r^2\right) = 2\left(r^2 - r^4\right)$$

$$\int_0^1 \int_{r^2}^{2-r^2} r^2 dz dr$$

$$= \int_0^1 2(r^2 - r^4) dr = 2 \int_0^1 (r^2 - r^4) dr$$

$$= 2\left(\frac{r^3}{3} - \frac{r^5}{5}\right)\Big|_0^1 = 2\left(\frac{1^3}{3} - \frac{1^5}{5}\right) - 2\left(\frac{0^3}{3} - \frac{0^5}{5}\right) = \frac{4}{15}$$

$$\approx 0.2667$$

$$\int_0^{2\pi} \int_0^1 \int_{r^2}^{2-r^2} r dz dr d\theta = \int_0^{2\pi} \frac{4}{15} d\theta = \frac{4}{15} \int_0^{2\pi} d\theta = \frac{4}{15} \times (2\pi - 0) = \frac{8\pi}{15} \approx 1.6755$$

Therefore,

$$\int_{-1}^{1} \int_{-\sqrt{1-x^2}}^{\sqrt{1-x^2}} \int_{x^2+y^2}^{2-x^2-y^2} \sqrt{x^2+y^2} dz dy dx = \int_0^{2\pi} \int_0^1 \int_{r^2}^{2-r^2} r^2 dz dr d\theta = \frac{8\pi}{15} \approx 1.6755$$

Example 3.39 Determine the volume of the solid region E that lies above the $z = \sqrt{2(x^2+y^2)}$ $z = \sqrt{2(x^2+y^2)}$ and below the sphere $x^2+y^2+z^2 = 16$ $x^2+y^2+z^2 = 16$.

Solution This problem is best solved using spherical coordinates due to the geometry of the solid, which is bounded by a sphere and a cone centered at the origin.

The spherical coordinates transformation is:

$$\begin{aligned} x &= \rho \sin(\phi) \cos(\theta) \\ y &= \rho \sin(\phi) \sin(\theta) \\ z &= \rho \cos(\phi) \\ x^2 + y^2 &= \rho^2 \sin^2(\phi) \\ x^2 + y^2 + z^2 &= \rho^2 \\ dV &= \rho^2 \sin(\phi) d\rho d\phi d\theta \end{aligned}$$

The solid is bounded above by the sphere of radius 4:

$$x^2 + y^2 + z^2 = 16 \Rightarrow \rho^2 = 16 \Rightarrow \rho = 4$$

The ρ limits are: $0 \le \rho \le 4$.

The solid is bounded by the cone $z = \sqrt{2\left(x^2 + y^2\right)}$. We substitute $z = \rho\cos(\phi)$ and $x^2 + y^2 = \rho^2 sin^2(\phi)$.

$$z = \sqrt{2\left(x^2 + y^2\right)} \Rightarrow \rho\cos(\phi) = \sqrt{2\rho^2 sin^2(\phi)} \Rightarrow \rho\cos(\phi)$$
$$= \sqrt{2}\rho\sin(\phi) \Rightarrow \cos(\phi) = \sqrt{2}\sin(\phi) \Rightarrow \tan(\phi) = \frac{1}{\sqrt{2}}$$

Let $\phi_0 = \tan^{-1}\left(\frac{1}{\sqrt{2}}\right)$. The region E lies above the cone, meaning the angle ϕ starts from the z-axis ($\phi = 0$) and extends to the cone's surface ϕ_0.

The region is symmetric around the z-axis (a full circle). Therefore, θ limits are $0 \le \theta \le 2\pi$.

The volume V is given by the triple integral:

$$V = \iiint\limits_E 1 dV = \int_0^{2\pi}\int_0^{\tan^{-1}\left(\frac{1}{\sqrt{2}}\right)}\int_0^4 1 \times \rho^2\sin(\phi)d\rho d\phi d\theta$$
$$= \int_0^{2\pi}\int_0^{\tan^{-1}\left(\frac{1}{\sqrt{2}}\right)}\int_0^4 \rho^2\sin(\phi)d\rho d\phi d\theta$$

The calculation of $\int_0^{2\pi}\int_0^{\tan^{-1}\left(\frac{1}{\sqrt{2}}\right)}\int_0^4 \rho^2\sin(\phi)d\rho d\phi d\theta$ is shown below:

$$\int_0^4 \rho^2\sin(\phi)d\rho = \sin(\phi)\int_0^4 \rho^2 d\rho = \left.\frac{\sin(\phi)\rho^3}{3}\right|_0^4$$
$$= \frac{\sin(\phi)\times 4^3}{3} - \frac{\sin(\phi)\times 0^3}{3} = \frac{64}{3}\sin(\phi)$$

$$\int_0^{\tan^{-1}\left(\frac{1}{\sqrt{2}}\right)}\int_0^4 \rho^2\sin(\phi)d\rho d\phi$$
$$= \int_0^{\tan^{-1}\left(\frac{1}{\sqrt{2}}\right)} \frac{64}{3}\sin(\phi)d\phi = \frac{64}{3}\int_0^{\tan^{-1}\left(\frac{1}{\sqrt{2}}\right)} \sin(\phi)d\phi$$
$$= -\frac{64}{3}\cos(\phi)\Bigg|_0^{\tan^{-1}\left(\frac{1}{\sqrt{2}}\right)}$$

$$= \frac{64}{3}\cos(\phi)\Big|_{\tan^{-1}\left(\frac{1}{\sqrt{2}}\right)}^{0} = \frac{64}{3}\cos(0) - \frac{64}{3}\cos\left(\tan^{-1}\left(\frac{1}{\sqrt{2}}\right)\right)$$
$$\approx \frac{64}{3} - \frac{64}{3} \times 0.8165$$
$$= 3.9147$$

$$V = \int_0^{2\pi}\int_0^{\tan^{-1}\left(\frac{1}{\sqrt{2}}\right)}\int_0^4 \sin(\phi)d\rho d\phi d\theta$$
$$= \int_0^{2\pi} 3.9147d\theta = 3.9147\int_0^{2\pi} d\theta = 3.9147 \times (2\pi - 0)$$
$$= 24.5966$$

The volume of the solid is 24.5966 cubic units.

Example 3.40 Evaluate the triple integral $\iiint\limits_E x^2dV$, where the region E is the spherical shell bounded by the spheres $x^2 + y^2 + z^2 = 16$ and $x^2 + y^2 + z^2 = 25$.

Solution The region E is the spherical shell between the sphere $x^2 + y^2 + z^2 = 16$ (radius $R_1 = 4$) and the sphere $x^2 + y^2 + z^2 = 25$ (radius $R_2 = 5$). This problem is solved most easily by converting to spherical coordinates.

The conversion to spherical coordinates and the volume element are:

$$x = \rho\sin(\phi)\cos(\theta)$$
$$dV = \rho^2\sin(\phi)d\rho d\phi d\theta$$

The region E is defined by the following bounds:

Radius (ρ): $4 \le \rho \le 5$.

Polar Angle (ϕ): $0 \le \phi \le \pi$.

Azimuthal Angle (θ): $0 \le \theta \le 2\pi$.

The integrand $f(x, y, z) = x^2$ becomes:

$$f(x, y, z) = x^2 \Rightarrow f(r, \theta, \phi) = (\rho\sin(\phi)\cos(\theta))^2 \Rightarrow f(r, \theta, \phi) = \rho^2\sin^2(\phi)\cos^2(\theta)$$

The integral is set up as:

$$\iiint_E x^2 dV = \int_0^{2\pi}\int_0^{\pi}\int_4^{5} \rho^2 \sin^2(\phi)\cos^2(\theta) \times \rho^2 \sin(\phi) d\rho d\phi d\theta$$
$$= \int_0^{2\pi}\int_0^{\pi}\int_4^{5} \rho^4 \sin^3(\phi)\cos^2(\theta) d\rho d\phi d\theta$$

Since the limits are constant and the integrand is separable, we can evaluate the integral as a product of three separate integrals:

$$\iiint_E x^2 dV = \int_0^{2\pi}\int_0^{\pi}\int_4^{5} \rho^4 \sin^3(\phi)\cos^2(\theta) d\rho d\phi d\theta$$
$$= \int_4^{5} \rho^4 d\rho \int_0^{\pi} \sin^3(\phi) d\phi \int_0^{2\pi} \cos^2(\theta) d\theta$$

Let's calculate each integral separately.

$$\int_4^5 \rho^4 d\rho = \frac{\rho^5}{5}\bigg|_4^5 = \frac{5^5}{5} - \frac{4^5}{5} = \frac{2101}{5}$$

$$\int_0^{\pi} \sin^3(\phi)d\phi = \int_0^{\pi} \sin(\phi)\sin^2(\phi)d\phi = \int_0^{\pi} \sin(\phi)\big(1-\cos^2(\phi)\big)d\phi = \int_0^{\pi} \sin(\phi)d\phi$$
$$+ \int_0^{\pi} -\sin(\phi)\cos^2(\phi)d\phi =$$
$$-\cos(\phi)\bigg|_0^{\pi} + \frac{\cos^3(\phi)}{3}\bigg|_0^{\pi} = \cos(\phi)\bigg|_{\pi}^{0} + \frac{\cos^3(\phi)}{3}\bigg|_0^{\pi}$$
$$= \cos(0) - \cos(\pi) + \frac{\cos^3(\pi)}{3}$$
$$- \frac{\cos^3(0)}{3} = 1 - (-1) + \frac{-1}{3} - \frac{1}{3} = 2 - \frac{2}{3} = \frac{4}{3} \approx 1.3333$$

$$\int_0^{2\pi} \cos^2(\theta)d\theta = \int_0^{2\pi} \frac{1+\cos(2\theta)}{2} d\theta$$
$$= \frac{\theta}{2} + \frac{\sin(2\theta)}{4}\bigg|_0^{2\pi} = \frac{2\pi}{2} + \frac{\sin(2\times 2\pi)}{4} - 0 = \pi$$

Therefore,

$$\iiint_E x^2 dV = \int_4^5 \rho^4 d\rho \int_0^\pi \sin^3(\phi) d\phi \int_0^{2\pi} \cos^2(\theta) d\theta$$
$$= \frac{2101}{5} \times \frac{4}{3} \times \pi = \frac{8404\pi}{15} \approx 1760.1$$

Example 3.41 Find the value of $\iiint_E x^2 dV$, where E is the region in the first octant bounded by the spheres $x^2 + y^2 + z^2 = 16$ and $x^2 + y^2 + z^2 = 25$.

Solution This problem is solved most easily by converting to spherical coordinates. The conversion to spherical coordinates and the volume element are:

$$x = \rho \sin(\phi) \cos(\theta)$$
$$dV = \rho^2 \sin(\phi) d\rho d\phi d\theta$$

The radius ρ ranges from $\sqrt{16} = 4$ to $\sqrt{25} = 5$.

Radius (ρ): $4 \le \rho \le 5$.

The first octant ($x \ge 0, y \ge 0, z \ge 0$) restricts the angles ϕ and θ.

Polar Angle (ϕ): $0 \le \phi \le \frac{\pi}{2}$ $0 \le \phi \le \frac{\pi}{2}$.

Azimuthal Angle (θ):$0 \le \theta \le \frac{\pi}{2}$.

The integrand $f(x, y, z) = x^2$ becomes:

$$f(x, y, z) = x^2 \Rightarrow f(r, \theta, \phi) = (\rho \sin(\phi) \cos(\theta))^2 \Rightarrow f(r, \theta, \phi) = \rho^2 sin^2(\phi) cos^2(\theta)$$

The integral is set up as:

$$\iiint_E x^2 dV = \int_0^{\frac{\pi}{2}} \int_0^{\frac{\pi}{2}} \int_4^5 \rho^2 \sin^2(\phi) \cos^2(\theta) \times \rho^2 \sin(\phi) d\rho d\phi d\theta$$
$$= \int_0^{\frac{\pi}{2}} \int_0^{\frac{\pi}{2}} \int_4^5 \rho^4 \sin^3(\phi) \cos^2(\theta) d\rho d\phi d\theta$$

Since the limits are constant and the integrand is separable, we can evaluate the integral as a product of three separate integrals:

$$\iiint_E x^2 dV = \int_0^{\frac{\pi}{2}} \int_0^{\frac{\pi}{2}} \int_4^5 \rho^4 \sin^3(\phi) \cos^2(\theta) d\rho d\phi d\theta$$

$$= \int_4^5 \rho^4 d\rho \int_0^{\frac{\pi}{2}} \sin^3(\phi) d\phi \int_0^{\frac{\pi}{2}} \cos^2(\theta) d\theta$$

Let's calculate each integral separately.

$$\int_4^5 \rho^4 d\rho = \frac{\rho^5}{5}\bigg|_4^5 = \frac{5^5}{5} - \frac{4^5}{5} = \frac{2101}{5}$$

$$\int_0^{\frac{\pi}{2}} \sin^3(\phi) d\phi = \int_0^{\frac{\pi}{2}} \sin(\phi)\sin^2(\phi) d\phi$$

$$= \int_0^{\frac{\pi}{2}} \sin(\phi)\big(1 - \cos^2(\phi)\big) d\phi = \int_0^{\frac{\pi}{2}} \sin(\phi) d\phi$$

$$+ \int_0^{\frac{\pi}{2}} -\sin(\phi)\cos^2(\phi) d\phi = -\cos(\phi)\bigg|_0^{\frac{\pi}{2}} + \frac{\cos^3(\phi)}{3}\bigg|_0^{\frac{\pi}{2}}$$

$$= \cos(\phi)\bigg|_{\frac{\pi}{2}}^{0} + \frac{\cos^3(\phi)}{3}\bigg|_0^{\frac{\pi}{2}}$$

$$= \cos(0) - \cos\left(\frac{\pi}{2}\right) + \frac{\cos^3\left(\frac{\pi}{2}\right)}{3} - \frac{\cos^3(0)}{3}$$

$$= 1 - (0) + \frac{0}{3} - \frac{1}{3} = 1 - \frac{1}{3} = \frac{2}{3} \approx 0.6667$$

$$\int_0^{\frac{\pi}{2}} \cos^2(\theta) d\theta = \int_0^{\frac{\pi}{2}} \frac{1 + \cos(2\theta)}{2} d\theta = \frac{\theta}{2} + \frac{\sin(2\theta)}{4}\bigg|_0^{\frac{\pi}{2}}$$

$$= \frac{\frac{\pi}{2}}{2} + \frac{\sin\left(2 \times \frac{\pi}{2}\right)}{4} - 0 = \frac{\pi}{4}$$

Therefore,

$$\iiint_E x^2 dV = \int_4^5 \rho^4 d\rho \int_0^{\frac{\pi}{2}} \sin^3(\phi) d\phi \int_0^{\frac{\pi}{2}} \cos^2(\theta) d\theta$$

$$= \frac{2101}{5} \times \frac{2}{3} \times \frac{\pi}{4} = \frac{2101\pi}{30} \approx 219.9638$$

Example 3.42 Find the value of $\iiint\limits_E \frac{2}{x^2+y^2+z^2}dV$, where E is the spherical cap defined by the inequalities $x^2 + y^2 + z^2 \leq 16$ and $z \geq 2$.

Solution This problem is best solved using spherical coordinates because the integrand is $\frac{2}{(x^2+y^2+z^2)} = \frac{2}{\rho^2}$ and the boundary $x^2 + y^2 + z^2 = 16$ is $\rho = 4$.

The region E is bounded by:
The lower plane: $z = 2$.
The upper sphere: $x^2 + y^2 + z^2 = 16(\rho = 4)$.
Let's determine the coordinate limits:

***ρ* (Radius)**: The inner boundary is the plane $z = 2$, and the outer boundary is the sphere $\rho = 4$. In spherical coordinates, $z = \rho cos(\phi)$, so the plane $z = 2$ is $\rho cos(\phi) = 2$, or $\rho = \frac{2}{\cos(\phi)} = 2\sec(\phi)$.

$$2\sec(\phi) \leq \rho \leq 4$$

***ϕ* (Polar Angle)**: The cap starts at the North Pole ($\phi = 0$) and extends down to the circle of intersection between $z = 2$ and the sphere $\rho = 4$.

$$z = \rho\cos(\phi) \Rightarrow 2 = 4\cos(\phi_{max}) \Rightarrow \cos(\phi_{max})$$
$$= \frac{2}{4} \Rightarrow \cos(\phi_{max}) = \frac{1}{2} \Rightarrow \phi_{max} = \cos^{-1}\left(\frac{1}{2}\right) \Rightarrow \phi_{max} = \frac{\pi}{3}$$
$$0 \leq \phi \leq \frac{\pi}{3}$$

***θ* (Azimuthal Angle)**: The region covers a full revolution:

$$0 \leq \theta \leq 2\pi$$

The integrand is $\frac{2}{x^2+y^2+z^2} = \frac{2}{\rho^2}$ and the volume element is $dV = \rho^2\sin(\phi)d\rho d\phi d\theta$. Therefore,

$$\iiint\limits_E \frac{2}{x^2+y^2+z^2}dV = \int_0^{2\pi}\int_0^{\frac{\pi}{3}}\int_{2\sec(\phi)}^{4} \frac{2}{\rho^2} \times \rho^2\sin(\phi)d\rho d\phi d\theta$$
$$= \int_0^{2\pi}\int_0^{\frac{\pi}{3}}\int_{2\sec(\phi)}^{4} 2\sin(\phi)d\rho d\phi d\theta = 2\int_0^{2\pi}\int_0^{\frac{\pi}{3}}\int_{2\sec(\phi)}^{4} \sin(\phi)d\rho d\phi d\theta$$

The calculation of $2\int_0^{2\pi}\int_0^{\frac{\pi}{3}}\int_{2\sec(\phi)}^{4}\sin(\phi)d\rho d\phi d\theta$ is shown below:

$$\int_{2\sec(\phi)}^{4} \sin(\phi)d\rho$$

$$= \sin(\phi) \int_{2\sec(\phi)}^{4} d\rho = \sin(\phi)(4 - 2\sec(\phi)) = 4\sin(\phi) - 2\sin(\phi)\sec(\phi) = 4$$

$$\sin(\phi) - 2\sin(\phi) \times \frac{1}{\cos(\phi)} = 4\sin(\phi) - 2\tan(\phi)$$

$$\int_{0}^{\frac{\pi}{3}} \int_{2\sec(\phi)}^{4} \sin(\phi)d\rho d\phi = \int_{0}^{\frac{\pi}{3}} (4\sin(\phi) - 2\tan(\phi))d\phi$$

$$= -4\cos(\phi) + 2\ln(|\cos(\phi)|)\Big|_{0}^{\frac{\pi}{3}} = 2\ln(|\cos(\phi)|) - 4\cos(\phi)\Big|_{0}^{\frac{\pi}{3}}$$

$$= 2\ln\left(\left|\cos\left(\frac{\pi}{3}\right)\right|\right) - 4\cos\left(\frac{\pi}{3}\right) - (2\ln(|\cos(0)|) - 4\cos(0))$$

$$= 2\ln\left(\frac{1}{2}\right) - 4 \times \frac{1}{2} - (2\ln(1) - 4 \times 1)$$

$$= 2(\ln(1) - \ln(2)) - 2 - (2 \times 0 - 4) = 2(0 - \ln(2))$$

$$-2 + 4 = 2 - 2\ln(2) = 2(1 - \ln(2))$$

$$2\int_{0}^{2\pi} \int_{0}^{\frac{\pi}{3}} \int_{2\sec(\phi)}^{4} \sin(\phi)d\rho d\phi d\theta$$

$$= 2\int_{0}^{2\pi} 2(1 - \ln(2))d\theta = 4(1 - \ln(2)) \int_{0}^{2\pi} d\theta = 4(1 - \ln(2)) \times (2\pi - 0)$$

$$= 8(1 - \ln(2))\pi \approx 7.7121$$

Therefore, $\iiint_E \frac{2}{x^2+y^2+z^2} dV = 8(1 - ln(2))\pi \approx 7.7121$.

Example 3.43 Find the value of the integral $\iiint_E \frac{z}{x^2+y^2+z^2} dV$. The region of integration, E, is defined by the inequalities:

$$1 \leq x^2 + y^2 + z^2 \leq 4$$
$$z \geq 0$$
$$y \geq 0$$

Solution The spherical coordinates (ρ, ϕ, θ) relate to rectangular coordinates (x, y, z) as follows:

$$\begin{aligned} x &= \rho \sin(\phi)\cos(\theta) \\ y &= \rho \sin(\phi)\sin(\theta) \\ z &= \rho \cos(\phi) \\ dV &= \rho^2 \sin(\phi) d\rho d\phi d\theta \end{aligned}$$

The integrand is $\frac{z}{x^2+y^2+z^2}$. Substituting the spherical relations:

$$f(x, y, z) = \frac{z}{x^2 + y^2 + z^2} \Rightarrow f(\rho, \phi, \theta) = \frac{\rho \cos(\phi)}{\rho^2} \Rightarrow f(\rho, \phi, \theta) = \frac{\cos(\phi)}{\rho}$$

The region E is defined by: $1 \leq x^2 + y^2 + z^2 \leq 4, z \geq 0$ and $y \geq 0$. Let's determine the coordinate limits:

Radial Distance (ρ): $1 \leq x^2 + y^2 + z^2 \leq 4 \Rightarrow 1 \leq \rho^2 \leq 4 \Rightarrow 1 \leq \rho \leq 2$.

Polar Angle (ϕ): The condition $z \geq 0$ restricts the region to the upper half-space. The angle ϕ is measured from the positive z-axis ($z \geq 0$ means the angle is between the z-axis and the xy-plane).

$$0 \leq \phi \leq \frac{\pi}{2}$$

Azimuthal Angle (θ): The condition $y \geq 0$ restricts the region to the half-space where y is non-negative. Combining this with the entire range of x (implicitly included by the spherical shell), this is the first and second quadrants of the xy-plane.

$$0 \leq \theta \leq \pi$$

The triple integral in spherical coordinates is:

$$\begin{aligned} \iiint_E \frac{z}{x^2 + y^2 + z^2} dV &= \int_0^{\pi} \int_0^{\frac{\pi}{2}} \int_1^2 \frac{\cos(\phi)}{\rho} \times \rho^2 \sin(\phi) d\rho d\phi d\theta \\ &= \int_0^{\pi} \int_0^{\frac{\pi}{2}} \int_1^2 \cos(\phi)\rho \sin(\phi) d\rho d\phi d\theta \\ &= \int_1^2 \rho d\rho \int_0^{\frac{\pi}{2}} \cos(\phi)\sin(\phi) d\phi \int_0^{\pi} d\theta \end{aligned}$$

Let's calculate each integral separately.

$$\int_1^2 \rho d\rho = \frac{\rho^2}{2}\bigg|_1^2 = \frac{2^2}{2} - \frac{1^2}{2} = \frac{3}{2}$$

$$\int_0^{\frac{\pi}{2}} \cos(\phi)\sin(\phi)d\phi = \frac{\sin^2(\phi)}{2}\bigg|_0^{\frac{\pi}{2}}$$

$$= \frac{\sin^2\left(\frac{\pi}{2}\right)}{2} - \frac{\sin^2(0)}{2} = \frac{1}{2} - 0 = \frac{1}{2}$$

$$\int_0^{\pi} d\theta = \pi - 0 = \pi$$

Therefore,

$$\iiint_E \frac{z}{x^2+y^2+z^2} dV = \int_1^2 \rho d\rho \int_0^{\frac{\pi}{2}} \cos(\phi)\sin(\phi)d\phi \int_0^{\pi} d\theta$$

$$= \frac{3}{2} \times \frac{1}{2} \times \pi = \frac{3\pi}{4} \approx 2.3562$$

Example 3.44 Evaluate $\int_0^3 \int_0^{\sqrt{9-x^2}} \int_{\sqrt{x^2+y^2}}^{\sqrt{9-x^2-y^2}} \left(x^2+y^2+z^2\right)^{\frac{3}{2}} dzdydx$.

Solution The complexity of the Cartesian limits and the integrand $\left(x^2+y^2+z^2\right)^{\frac{3}{2}}$ strongly indicate the necessity of converting to spherical coordinates (ρ, ϕ, θ), where $x^2+y^2+z^2 = \rho^2$ and $dV = \rho^2 sin(\phi)d\rho d\phi d\theta$.

The Cartesian limits define the region E:

***x* Limits**: $0 \le x \le 3$.

***y* Limits**: $0 \le y \le \sqrt{9-x^2}$. Together, $x \ge 0$ and $y \ge 0$ define the first quadrant of the disk $x^2+y^2 \le 9$ in the xy-plane (Fig. 3.28).

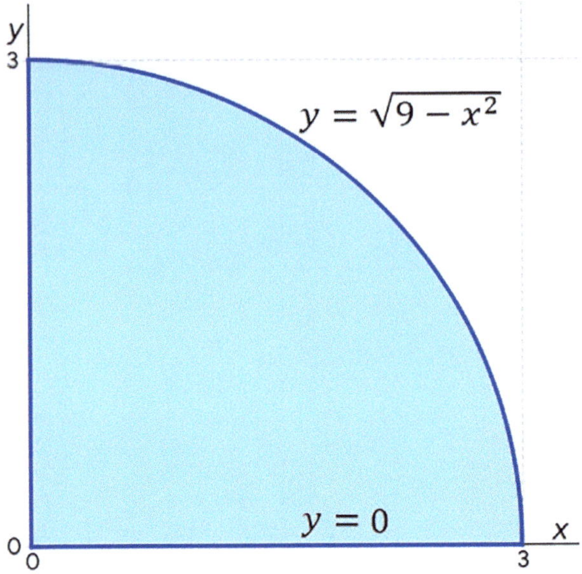

Fig. 3.28 The region defined by $\left\{(x, y)|x \geq 0, y \geq 0, x^2 + y^2 \leq 9\right\}$

z **Limits**: $\sqrt{x^2 + y^2} \leq z \leq \sqrt{9 - x^2 - y^2}$.

$z = \sqrt{x^2 + y^2}$ is the upper half of the cone $z^2 = x^2 + y^2$.

$z = \sqrt{9 - x^2 - y^2}$ is the upper hemisphere of the sphere $x^2 + y^2 + z^2 = 9$.

The region E E is the volume in the first octant bounded below by the cone and above by the sphere.

Let's convert to spherical coordinates:

Integrand: $\left(x^2 + y^2 + z^2\right)^{\frac{3}{2}} = \left(\rho^2\right)^{\frac{3}{2}} = \rho^3$.

ρ **Limits**: Bounded by the origin ($\rho = 0$) and the sphere $\rho^2 = 9$.

$$0 \leq \rho \leq 3$$

ϕ **Limits**: Bounded by the positive z z-axis ($\phi = 0$) and the cone $z = \sqrt{x^2 + y^2}$.

$$\begin{aligned} z = \sqrt{x^2 + y^2} &\Rightarrow z^2 = x^2 + y^2 \Rightarrow (\rho\cos(\phi))^2 \\ &= (\rho\sin(\phi)\cos(\theta))^2 + (\rho\sin(\phi)\sin(\theta))^2 \Rightarrow \rho^2\cos^2(\phi) \\ &= \rho^2\sin^2(\phi)\cos^2(\theta) + \rho^2\sin^2(\phi)\sin^2(\theta) \Rightarrow \rho^2\cos^2(\phi) \\ &= \rho^2\sin^2(\phi)\left(\cos^2(\theta) + \sin^2(\theta)\right) \Rightarrow \rho^2\cos^2(\phi) \\ &= \rho^2\sin^2(\phi) \Rightarrow \cos^2(\phi) = \sin^2(\phi) \Rightarrow \tan^2(\phi) \\ &= 1 \Rightarrow \tan(\phi) = 1 \Rightarrow \phi_{\max} = \tan^{-1}(1) \Rightarrow \phi_{\max} = \frac{\pi}{4} \end{aligned}$$

$$0 \leq \phi \leq \frac{\pi}{4}$$

***θ* Limits**: Since the projection (Fig. 3.28) is a disk quadrant in the *xy*-plane:

$$0 \leq \theta \leq \frac{\pi}{2}$$

The integral transforms to:

$$\int_0^3 \int_0^{\sqrt{9-x^2}} \int_{\sqrt{x^2+y^2}}^{\sqrt{9-x^2-y^2}} \left(x^2+y^2+z^2\right)^{\frac{3}{2}} dzdydx = \int_0^{2\pi} \int_0^{\frac{\pi}{4}} \int_0^3 \rho^3 \times \rho^2 \sin(\phi) d\rho d\phi d\theta$$

$$= \int_0^{\frac{\pi}{2}} \int_0^{\frac{\pi}{4}} \int_0^3 \rho^5 \sin(\phi) d\rho d\phi d\theta = \int_0^3 \rho^5 d\rho \int_0^{\frac{\pi}{4}} \sin(\phi) d\phi \int_0^{\frac{\pi}{2}} d\theta$$

Let's calculate each integral separately.

$$\int_0^3 \rho^5 d\rho = \frac{\rho^6}{6}\bigg|_0^3 = \frac{3^6}{6} - 0 = \frac{729}{6} - 0 = \frac{243}{2} \approx 121.5$$

$$\int_0^{\frac{\pi}{4}} \sin(\phi) d\phi = -\cos(\phi)\bigg|_0^{\frac{\pi}{4}} = \cos(\phi)\bigg|_{\frac{\pi}{4}}^0$$

$$= \cos(0) - \cos\left(\frac{\pi}{4}\right) = 1 - \frac{\sqrt{2}}{2} \approx 0.2929$$

$$\int_0^{\frac{\pi}{2}} d\theta = \frac{\pi}{2} - 0 = \frac{\pi}{2} \approx 1.5708$$

Therefore,

$$\int_0^3 \int_0^{\sqrt{9-x^2}} \int_{\sqrt{x^2+y^2}}^{\sqrt{9-x^2-y^2}} \left(x^2+y^2+z^2\right)^{\frac{3}{2}} dzdydx$$

$$= \int_0^3 \rho^5 d\rho \int_0^{\frac{\pi}{4}} \sin(\phi) d\phi \int_0^{\frac{\pi}{2}} d\theta = \frac{243}{2} \times \left(1 - \frac{\sqrt{2}}{2}\right)$$

$$\times \frac{\pi}{2} = \frac{243\left(2-\sqrt{2}\right)}{8}\pi \approx 55.8992$$

Example 3.45 Evaluate $\int\limits_{-2}^{2}\int\limits_{-\sqrt{4-x^2}}^{\sqrt{4-x^2}}\int\limits_{0}^{\sqrt{4-x^2-y^2}} z^2 dzdydx.$

Solution The integrand z^2 and the spherical limits $\left(\sqrt{4-x^2-y^2}\right)$ make this ideal for conversion to spherical coordinates (ρ, ϕ, θ).

The Cartesian limits define the region E:

***x* Limits**: $-2 \le x \le 2$.

***y* Limits**: $-\sqrt{4-x^2} \le y \le \sqrt{4-x^2}$. Together, this defines the full disk $x^2+y^2 \le 4$ in the xy-plane (Fig. 3.29).

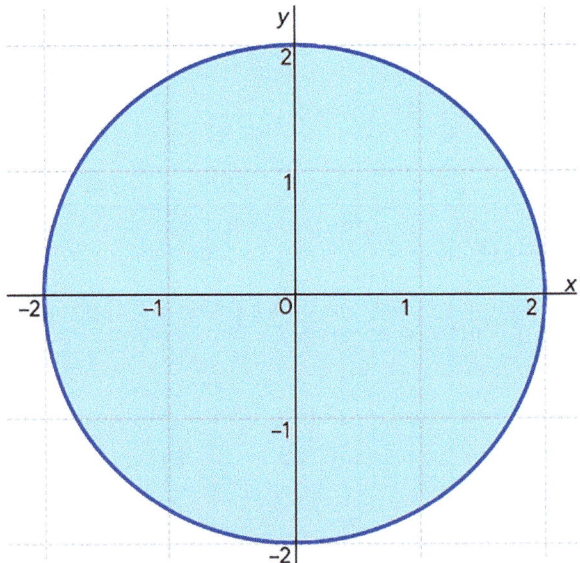

Fig. 3.29 The region defined by $\{(x, y)|x^2+y^2 \le 4\}\{(x, y)|x^2+y^2 \le 4\}$

***z* Limits**: $0 \le z \le \sqrt{4-x^2-y^2}$. This means the region is bounded below by the xy-plane ($z = 0$) and above by the hemisphere $x^2+y^2+z^2 = 4$.

The region E is the upper hemisphere of radius $R = 2$.

Let's convert to spherical coordinates:

Integrand: $z^2 = (\rho\cos(\phi))^2 = \rho^2\cos^2(\phi)$.

***ρ* Limits**: Bounded by the origin ($\rho = 0$) and the sphere $\rho^2 = 4$.

$$0 \le \rho \le 2$$

***ϕ* Limits**: Bounded by the positive z-axis ($\phi = 0$) and the xy-plane ($\phi = \frac{\pi}{2}$).

$$0 \le \phi \le \frac{\pi}{2}$$

***θ* Limits**: Since the projection (Fig. 3.29) is a full disk in the *xy*-plane:

$$0 \le \theta \le 2\pi$$

The integral transforms to:

$$\int_{-2}^{2} \int_{-\sqrt{4-x^2}}^{\sqrt{4-x^2}} \int_{0}^{\sqrt{4-x^2-y^2}} z^2 dz dy dx = \int_0^{2\pi} \int_0^{\frac{\pi}{2}} \int_0^2 \rho^2 \cos^2(\phi)$$
$$\times \rho^2 \sin(\phi) d\rho d\phi d\theta$$
$$= \int_0^{2\pi} \int_0^{\frac{\pi}{2}} \int_0^2 \rho^4 \cos^2(\phi) \sin(\phi) d\rho d\phi d\theta$$
$$= \int_0^2 \rho^4 d\rho \int_0^{\frac{\pi}{2}} \cos^2(\phi) \sin(\phi) d\phi \int_0^{2\pi} d\theta$$

Let's calculate each integral separately.

$$\int_0^2 \rho^4 d\rho = \frac{\rho^5}{5}\Bigg|_0^2 = \frac{2^5}{5} - \frac{0^5}{5} = \frac{32}{5}$$

$$\int_0^{\frac{\pi}{2}} \cos^2(\phi) \sin(\phi) d\phi = -\int_0^{\frac{\pi}{2}} -\sin(\phi) \cos^2(\phi) d\phi$$
$$= \int_{\frac{\pi}{2}}^{0} -\sin(\phi) \cos^2(\phi) d\phi = \frac{\cos^3(\phi)}{3}\Bigg|_{\frac{\pi}{2}}^{0}$$
$$= \frac{\cos^3(0)}{3} - \frac{\cos^3\left(\frac{\pi}{2}\right)}{3} = \frac{1}{3} - \frac{0}{3} = \frac{1}{3}$$

$$\int_0^{2\pi} d\theta = 2\pi - 0 = 2\pi$$

Therefore,

$$\int_{-2}^{2}\int_{-\sqrt{4-x^2}}^{\sqrt{4-x^2}}\int_{0}^{\sqrt{4-x^2-y^2}} z^2 dzdydx$$

$$= \int_0^2 \rho^4 d\rho \int_0^{\frac{\pi}{2}} \cos^2(\phi)\sin(\phi)d\phi \int_0^{2\pi} d\theta$$

$$= \frac{32}{5} \times \frac{1}{3} \times 2\pi = \frac{64\pi}{15} \approx 13.4041$$

Exercise 3.11 Evaluate $\int_{-2}^{2}\int_{-\sqrt{4-x^2}}^{\sqrt{4-x^2}}\int_{0}^{\sqrt{4-x^2-y^2}} (x^2+y^2)dzdydx$. (Ans. $\frac{128\pi}{15} \approx 26.8083$).

Exercise 3.12 Evaluate. $\int_{0}^{1}\int_{0}^{\sqrt{1-x^2}}\int_{\sqrt{x^2+y^2}}^{\sqrt{2(x^2+y^2)}} \frac{1}{\sqrt{x^2+y^2+z^2}}dzdydx$ (Ans. $\frac{\pi(\sqrt{6}-\sqrt{3})}{12} \approx 0.0859$).

Example 3.46 Evaluate $\int_{-1}^{1}\int_{-\sqrt{1-x^2}}^{\sqrt{1-x^2}}\int_{\sqrt{3(x^2+y^2)}}^{\sqrt{4-x^2-y^2}} \sqrt{x^2+y^2+z^2}dzdydx$.

Solution The complexity of the integrand and the conical/spherical limits necessitates conversion to spherical coordinates(ρ, ϕ, θ), where $x^2+y^2+z^2=\rho^2$ and $dV = \rho^2 sin(\phi)d\rho d\phi d\theta$.

The Cartesian limits define the region E:

***x* Limits**: $-1 \le x \le 1$.

***y* Limits**: $-\sqrt{1-x^2} \le y \le \sqrt{1-x^2}$. This is the full disk $x^2+y^2 \le 1$ in the xy-plane (Fig. 3.30).

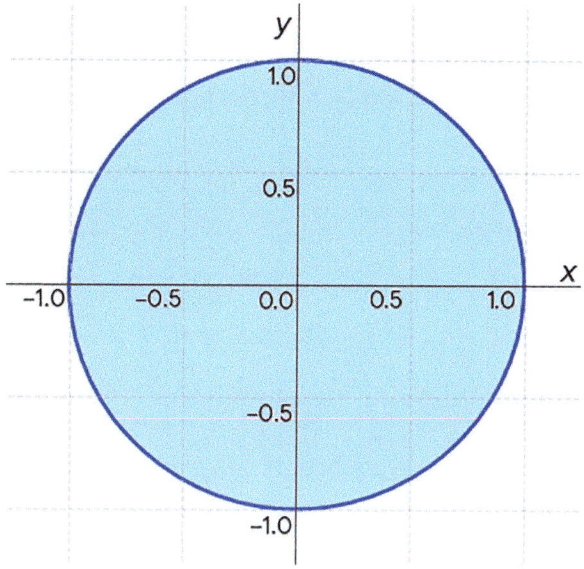

Fig. 3.30 The region defined by $\{(x, y)|x^2 + y^2 \leq 1\}$

z **Limits**: $\sqrt{3(x^2 + y^2)} \leq z \leq \sqrt{4 - x^2 - y^2}$.

$z = \sqrt{3(x^2 + y^2)}$ is the upper half of the cone $z^2 = 3(x^2 + y^2)$.
$z = \sqrt{4 - x^2 - y^2}$ is the upper hemisphere of the sphere $x^2 + y^2 + z^2 = 4$.
The region E is the volume bounded by the sphere above and the cone below.
Let's convert to spherical coordinates:
Integrand: $\sqrt{x^2 + y^2 + z^2} = \sqrt{\rho} = \rho$.
ρ **Limits**: Bounded by the origin ($\rho = 0$) and the sphere $\rho^2 = 4$.

$$0 \leq \rho \leq 2$$

ϕ **Limits**: Bounded by the positive z-axis ($\phi = 0$) and the cone $z = \sqrt{3(x^2 + y^2)}$.

$$z = \sqrt{3(x^2 + y^2)} \Rightarrow z^2 = 3(x^2 + y^2) \Rightarrow (\rho\cos(\phi))^2$$
$$= 3\left((\rho\sin(\phi)\cos(\theta))^2 + (\rho\sin(\phi)\sin(\theta))^2\right) \Rightarrow \rho^2\cos^2(\phi)$$
$$= 3\left(\rho^2\sin^2(\phi)\cos^2(\theta) + \rho^2\sin^2(\phi)\sin^2(\theta)\right) \Rightarrow \rho^2\cos^2(\phi) = 3\rho^2\sin^2(\phi)$$
$$\left(\cos^2(\theta) + \sin^2(\theta)\right) \Rightarrow \rho^2\cos^2(\phi)$$
$$= 3\rho^2\sin^2(\phi) \Rightarrow \cos^2(\phi) = 3\sin^2(\phi) \Rightarrow \tan^2(\phi) = \frac{1}{3} \Rightarrow \tan(\phi) = \frac{1}{\sqrt{3}} \Rightarrow \phi_{\max}$$
$$= \tan^{-1}\left(\frac{1}{\sqrt{3}}\right) \Rightarrow \phi_{\max} = \frac{\pi}{6}$$

$$0 \leq \phi \leq \frac{\pi}{6}$$

***θ* Limits**: Since the projection (Fig. 3.30) is a full disk in the *xy*-plane:

$$0 \leq \theta \leq 2\pi$$

The integral transforms to:

$$\int_{-1}^{1} \int_{-\sqrt{1-x^2}}^{\sqrt{1-x^2}} \int_{\sqrt{3(x^2+y^2)}}^{\sqrt{4-x^2-y^2}} \sqrt{x^2+y^2+z^2}\,dz dy dx = \int_0^{2\pi}\int_0^{\frac{\pi}{6}}\int_0^2 \rho \times \rho^2 \sin(\phi) d\rho d\phi d\theta$$

$$= \int_0^{2\pi}\int_0^{\frac{\pi}{6}}\int_0^2 \rho^3 \sin(\phi) d\rho d\phi d\theta = \int_0^2 \rho^3 d\rho \int_0^{\frac{\pi}{6}} \sin(\phi) d\phi \int_0^{2\pi} d\theta$$

Let's calculate each integral separately.

$$\int_0^2 \rho^3 d\rho = \frac{\rho^4}{4}\bigg|_0^2 = \frac{2^4}{4} - \frac{0^4}{4} = 4$$

$$\int_0^{\frac{\pi}{6}} \sin(\phi) d\phi = -\cos(\phi)\bigg|_0^{\frac{\pi}{6}} = \cos(\phi)\bigg|_{\frac{\pi}{6}}^{0}$$

$$= \cos(0) - \cos\left(\frac{\pi}{6}\right) = 1 - \frac{\sqrt{3}}{2} = \frac{2-\sqrt{3}}{2} \approx 0.1340$$

$$\int_0^{2\pi} d\theta = 2\pi - 0 = 2\pi$$

Therefore,

$$\int_{-1}^{1} \int_{-\sqrt{1-x^2}}^{\sqrt{1-x^2}} \int_{\sqrt{3(x^2+y^2)}}^{\sqrt{4-x^2-y^2}} \sqrt{x^2+y^2+z^2}\,dz dy dx$$

$$= \int_0^2 \rho^3 d\rho \int_0^{\frac{\pi}{6}} \sin(\phi) d\phi \int_0^{2\pi} d\theta = 4 \times \frac{2-\sqrt{3}}{2}$$

$$\times 2\pi = 4\left(2-\sqrt{3}\right)\pi \approx 3.3671$$

Example 3.47 Evaluate $\int\limits_{-2}^{2}\int\limits_{-\sqrt{4-x^2}}^{\sqrt{4-x^2}}\int\limits_{\sqrt{x^2+y^2}}^{2}(x^2+y^2)dzdydx.$

Solution Let's use spherical coordinates(ρ, ϕ, θ), where $x^2 + y^2 = \rho^2 \sin^2(\phi)$ and $dV = \rho^2 \sin(\phi)d\rho d\phi d\theta$.

The Cartesian limits define the region E:

***x* Limits**: $-2 \le x \le 2$.

***y* Limits**: $-\sqrt{4-x^2} \le y \le \sqrt{4-x^2}$. This is the full disk $x^2+y^2 \le 4$ in the xy-plane (Fig. 3.31).

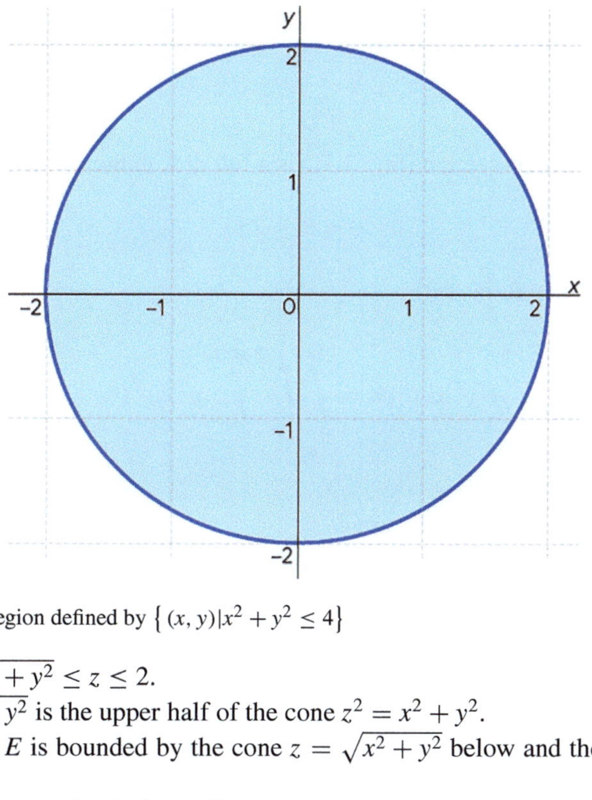

Fig. 3.31 The region defined by $\{(x, y)|x^2 + y^2 \le 4\}$

***z* Limits**: $\sqrt{x^2+y^2} \le z \le 2$.

$z = \sqrt{x^2+y^2}$ is the upper half of the cone $z^2 = x^2 + y^2$.

The region E is bounded by the cone $z = \sqrt{x^2+y^2}$ below and the plane $z = 2$ above.

Let's convert to spherical coordinates:

Integrand: $x^2 + y^2 = \rho^2 \sin^2(\phi)$.

***ρ* Limits**: Bounded by the origin ($\rho = 0$) and the sphere $\rho = 2\sec(\phi)$.

$$z = 2 \Rightarrow \rho\cos(\phi) = 2 \Rightarrow \rho = \frac{2}{\cos(\phi)} \Rightarrow \rho = 2\sec(\phi)$$

$$0 \le \rho \le 2\sec(\phi)$$

ϕ Limits: Bounded by the positive z-axis ($\phi = 0$) and the cone $z = \sqrt{x^2 + y^2}$.

$$\begin{aligned} z &= \sqrt{x^2+y^2} \Rightarrow z^2 = x^2 + y^2 \Rightarrow (\rho\cos(\phi))^2 \\ &= (\rho\sin(\phi)\cos(\theta))^2 + (\rho\sin(\phi)\sin(\theta))^2 \Rightarrow \rho^2\cos^2(\phi) \\ &= \rho^2\sin^2(\phi)\cos^2(\theta) + \rho^2\sin^2(\phi)\sin^2(\theta) \Rightarrow \rho^2\cos^2(\phi) \\ &= \rho^2\sin^2(\phi)\big(\cos^2(\theta)+\sin^2(\theta)\big) \Rightarrow \rho^2\cos^2(\phi) \\ &= \rho^2\sin^2(\phi) \Rightarrow \cos^2(\phi) = \sin^2(\phi) \Rightarrow \tan^2(\phi) \\ &= 1 \Rightarrow \tan(\phi) = 1 \Rightarrow \phi_{\max} = \tan^{-1}(1) \Rightarrow \phi_{\max} = \frac{\pi}{4} \\ 0 &\le \phi \le \frac{\pi}{4} \end{aligned}$$

θ Limits: Since the projection (Fig. 3.31) is a full disk in the xy-plane:

$$0 \le \theta \le 2\pi$$

The integral transforms to:

$$\int_{-2}^{2}\int_{-\sqrt{4-x^2}}^{\sqrt{4-x^2}}\int_{\sqrt{x^2+y^2}}^{2} (x^2+y^2)dzdydx = \int_0^{2\pi}\int_0^{\frac{\pi}{4}}\int_0^{2\sec(\phi)} \rho^2\sin^2(\phi) \times \rho^2\sin(\phi)d\rho d\phi d\theta$$

$$= \int_0^{2\pi}\int_0^{\frac{\pi}{4}}\int_0^{2\sec(\phi)} \rho^4\sin^3(\phi)d\rho d\phi d\theta$$

The calculation of $\int_0^{2\pi}\int_0^{\frac{\pi}{4}}\int_0^{2\sec(\phi)} \rho^4\sin^3(\phi)d\rho d\phi d\theta$ is shown below:

$$\begin{aligned} \int_0^{2\sec(\phi)} \rho^4\sin^3(\phi)d\rho &= \sin^3(\phi)\int_0^{2\sec(\phi)} \rho^4 d\rho \\ &= \sin^3(\phi)\frac{\rho^5}{5}\bigg|_0^{2\sec(\phi)} = \sin^3(\phi)\frac{(2\sec(\phi))^5}{5} \\ &\quad - \sin^3(\phi)\frac{0^5}{5} \\ &= \frac{32}{5}\sin^3(\phi)\sec^5(\phi) = \frac{32\sin^3(\phi)}{5\cos^5(\phi)} \end{aligned}$$

$$= \frac{32\sin^3(\phi)}{5\cos^3(\phi)} \times \frac{1}{\cos^2(\phi)} = \frac{32}{5}\tan^3(\phi)\sec^2(\phi)$$

$$\int_0^{\frac{\pi}{4}} \int_0^{2\sec(\phi)} \rho^4 \sin^3(\phi) d\rho d\phi$$

$$= \int_0^{\frac{\pi}{4}} \frac{32}{5}\tan^3(\phi)\sec^2(\phi) d\phi$$

$$= \frac{32}{5}\int_0^{\frac{\pi}{4}} \sec^2(\phi)\tan^3(\phi) d\phi = \frac{32\tan^4(\phi)}{5}\Big|_0^{\pi} \frac{\pi}{4}$$

$$= \frac{32}{20}\tan^4(\phi)\Big|_0^{\pi} = \frac{32}{20}\tan^4\left(\frac{\pi}{4}\right) - \frac{32}{20}\tan^4(0)$$

$$= \frac{32}{20} \times 1^4 - \frac{32}{20} \times 0^4 = \frac{32}{20} - 0 = \frac{32}{20} = \frac{16}{10}$$

$$= 1.6$$

$$\int_0^{2\pi} \int_0^{\frac{\pi}{4}} \int_0^{2sec(\phi)} \rho^4 \sin^3(\phi) d\rho d\phi d\theta$$

$$= \int_0^{2\pi} \frac{16}{10} d\theta = \frac{16}{10}\int_0^{2\pi} d\theta = \frac{16}{10}(2\pi - 0) = \frac{16\pi}{5} \approx 10.0531$$

Therefore,

$$\int_{-2}^{2} \int_{-\sqrt{4-x^2}}^{\sqrt{4-x^2}} \int_{\sqrt{x^2+y^2}}^{2} (x^2 + y^2) dz dy dx$$

$$= \int_0^{2\pi} \int_0^{\frac{\pi}{4}} \int_0^{2\ \sec(\phi)} \rho^4 \sin^3(\phi) d\rho d\phi d\theta = \frac{16\pi}{5} \approx 10.0531$$

Example 3.48 Evaluate $\int_{-2}^{2} \int_{-\sqrt{4-x^2}}^{\sqrt{4-x^2}} \int_{\sqrt{x^2+y^2}}^{2} (x^2 + y^2) dz dy dx$.

Solution This problem was solved in the previous example. We will now re-evaluate this problem using the cylindrical coordinate system.

In cylindrical coordinates, the transformation is $x^2 + y^2 = r^2$, and the volume element is $dV = rdzdrd\theta$.

The Cartesian limits define the region E:

***x* Limits**: $-2 \leq x \leq 2$.

***y* Limits**: $-\sqrt{4-x^2} \leq y \leq \sqrt{4-x^2}$. This is the full disk $x^2+y^2 \leq 4$ in the xy-plane (Fig. 3.32).

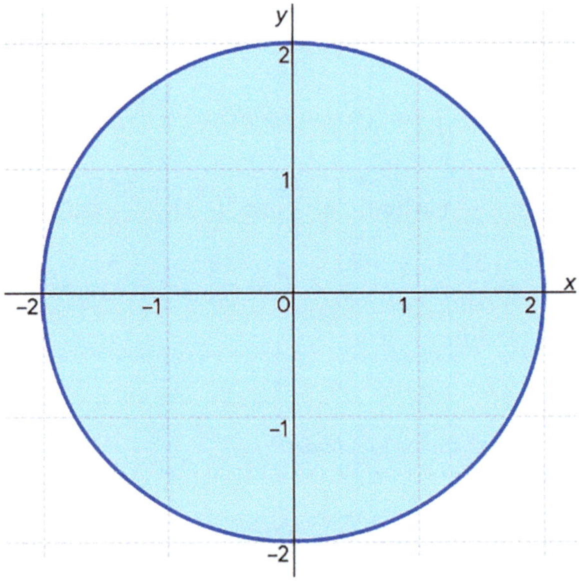

Fig. 3.32 The region defined by $\{(x, y)|x^2 + y^2 \leq 4\}$

Based on Fig. 3.32,

***r* (Radius)**: $0 \leq r \leq 2$.

***θ* (Angle)**: $0 \leq \theta \leq 2\pi$.

The limits for z are:

***z* (Height)**: $\sqrt{x^2+y^2} \leq z \leq 2 \Rightarrow \sqrt{r^2} \leq z \leq 2 \Rightarrow r \leq z \leq 2$.

The integral transforms to:

$$\int_{-2}^{2}\int_{-\sqrt{4-x^2}}^{\sqrt{4-x^2}}\int_{\sqrt{x^2+y^2}}^{2}(x^2+y^2)dzdydx = \int_{0}^{2\pi}\int_{0}^{2}\int_{r}^{2} r^2 \times rdzdrd\theta = \int_{0}^{2\pi}\int_{0}^{2}\int_{r}^{2} r^3dzdrd\theta$$

The calculation of $\int_0^{2\pi}\int_0^2\int_r^2 r^3dzdrd\theta$ is shown below:

$$\int_r^2 r^3 dz = r^3 \int_r^2 dz = r^3(2-r) = 2r^3 - r^4$$

$$\int_0^2 \int_r^2 r^3 dz dr = \int_0^2 (2r^3 - r^4) dr = \frac{r^4}{2} - \frac{r^5}{5}\Big|_0^2$$

$$= \frac{2^4}{2} - \frac{2^5}{5} - \left(\frac{0^4}{2} - \frac{0^5}{5}\right) = 8 - \frac{32}{5} = \frac{8}{5}$$

$$\int_0^{2\pi} \int_0^2 \int_r^2 r^3 dz dr d\theta = \int_0^{2\pi} \frac{8}{5} d\theta = \frac{8}{5} \int_0^{2\pi} d\theta$$

$$= \frac{8}{5} \times (2\pi - 0) = \frac{16\pi}{5} \approx 10.0531$$

Example 3.49 Evaluate $\int_0^{\sqrt{2}} \int_0^{\sqrt{2-x^2}} \int_{\sqrt{x^2+y^2}}^{\sqrt{4-x^2-y^2}} zdzdydx.$

Solution Let's use spherical coordinates(ρ, ϕ, θ), where $x^2 + y^2 = \rho^2 \sin^2(\phi)$ and $dV = \rho^2 sin(\phi) d\rho d\phi d\theta$.

The Cartesian limits define the region E:

***y* Limits**: $0 \le x \le \sqrt{2}$.

***x* Limits:** $0 \le y \le \sqrt{2-x^2}$. Together with $x \ge 0$, this is the portion of the disk $x^2 + y^2 \le 2$ in the first quadrant of the xy-plane (Fig. 3.33).

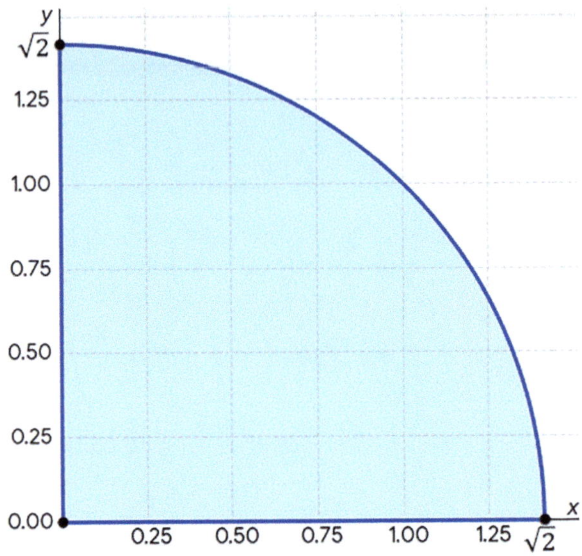

Fig. 3.33 The region defined by $\{(x, y)|x \geq 0, y \geq 0, x^2 + y^2 \leq 2\}$

z Limits: $\sqrt{x^2 + y^2} \leq z \leq \sqrt{4 - (x^2 + y^2)}$.

$z = \sqrt{x^2 + y^2}$ is the upper half of the cone $z^2 = x^2 + y^2$.

$z = \sqrt{4 - x^2 - y^2}$ is the upper hemisphere of the sphere $x^2 + y^2 + z^2 = 4$.

The region E is the volume bounded by the sphere above and the cone below, limited to the first octant ($x \geq 0$, $y \geq 0$ and $z \geq 0$).

Let's convert to spherical coordinates:

Integrand: $z = \rho \cos(\phi)$.

ρ Limits: Bounded by the origin ($\rho = 0$) and the sphere $\rho = 2$.

$$0 \leq \rho \leq 2$$

ϕ Limits: Bounded by the positive z-axis ($\phi = 0$) and the cone $z = \sqrt{x^2 + y^2}$.

$$\begin{aligned}
z &= \sqrt{x^2 + y^2} \Rightarrow z^2 = x^2 + y^2 \Rightarrow (\rho \cos(\phi))^2 \\
&= (\rho \sin(\phi) \cos(\theta))^2 + (\rho \sin(\phi) \sin(\theta))^2 \Rightarrow \rho^2 \cos^2(\phi) \\
&= \rho^2 \sin^2(\phi) \cos^2(\theta) + \rho^2 \sin^2(\phi) \sin^2(\theta) \Rightarrow \rho^2 \cos^2(\phi) \\
&= \rho^2 \sin^2(\phi)\left(\cos^2(\theta) + \sin^2(\theta)\right) \Rightarrow \rho^2 \cos^2(\phi) \\
&= \rho^2 \sin^2(\phi) \Rightarrow \cos^2(\phi) = \sin^2(\phi) \Rightarrow \tan^2(\phi) \\
&= 1 \Rightarrow \tan(\phi) = 1 \Rightarrow \phi_{\text{max}} = \tan^{-1}(1) \Rightarrow \phi_{\text{max}} = \frac{\pi}{4}
\end{aligned}$$

$$0 \leq \phi \leq \frac{\pi}{4}$$

***θ* Limits**: Since the projection (Fig. 3.33) is a disk quadrant in the *xy xy*-plane:

$$0 \leq \theta \leq \frac{\pi}{2}$$

The integral transforms to:

$$\int_0^{\sqrt{2}} \int_0^{\sqrt{2-x^2}} \int_{\sqrt{x^2+y^2}}^{\sqrt{4-x^2-y^2}} zdzdydx = \int_0^{\frac{\pi}{2}} \int_0^{\frac{\pi}{4}} \int_0^{2} \rho \cos(\phi) \times \rho^2 \sin(\phi) d\rho d\phi d\theta$$

$$= \int_0^{\frac{\pi}{2}} \int_0^{\frac{\pi}{4}} \int_0^{2} \rho^3 \sin(\phi) \cos(\phi) d\rho d\phi d\theta = \int_0^{2} \rho^3 d\rho \int_0^{\frac{\pi}{4}} \sin(\phi) \cos(\phi) d\phi \int_0^{\frac{\pi}{2}} d\theta$$

Let's calculate each integral separately.

$$\int_0^{2} \rho^3 d\rho = \left. \frac{\rho^4}{4} \right|_0^2 = \frac{2^4}{4} - \frac{0^4}{4} = 4$$

$$\int_0^{\frac{\pi}{4}} \sin(\phi) \cos(\phi) d\phi = \left. \frac{\sin^2(x)}{2} \right|_0^{\frac{\pi}{4}} = \frac{\sin^2\left(\frac{\pi}{4}\right)}{2} - \frac{\sin^2(0)}{2}$$

$$= \frac{\left(\frac{\sqrt{2}}{2}\right)^2}{2} - \frac{0^2}{2} = \frac{\frac{1}{2}}{2} - \frac{0}{2} = \frac{1}{4} - 0 = \frac{1}{4} = 0.25$$

$$\int_0^{\frac{\pi}{2}} d\theta = \frac{\pi}{2} - 0 = \frac{\pi}{2} = 1.5708$$

Therefore,

$$\int_0^{\sqrt{2}} \int_0^{\sqrt{2-x^2}} \int_{\sqrt{x^2+y^2}}^{\sqrt{4-x^2-y^2}} zdzdydx$$

$$= \int_0^{2} \rho^3 d\rho \int_0^{\frac{\pi}{4}} \sin(\phi) \cos(\phi) d\phi \int_0^{\frac{\pi}{2}} d\theta = 4 \times \frac{1}{4} \times \frac{\pi}{2} = \frac{\pi}{2}$$

Example 3.50 Evaluate $\int_0^1 \int_0^{\sqrt{1-x^2}} \int_0^{\sqrt{1-x^2-y^2}} \frac{1}{\sqrt{1-(x^2+y^2+z^2)}} dzdydx.$

Solution Let's use spherical coordinates(ρ, ϕ, θ), where $x^2 + y^2 + z^2 = \rho^2$ and $dV = \rho^2 sin(\phi) d\rho d\phi d\theta.$

The Cartesian limits define the region E:

***y* Limits**: $0 \leq x \leq 1.$

***x* Limits**: $0 \leq y \leq \sqrt{1 - x^2}$. Together with $x \geq 0$, this is the portion of the disk $x^2 + y^2 \leq 1$ in the first quadrant of the xy-plane (Fig. 3.34).

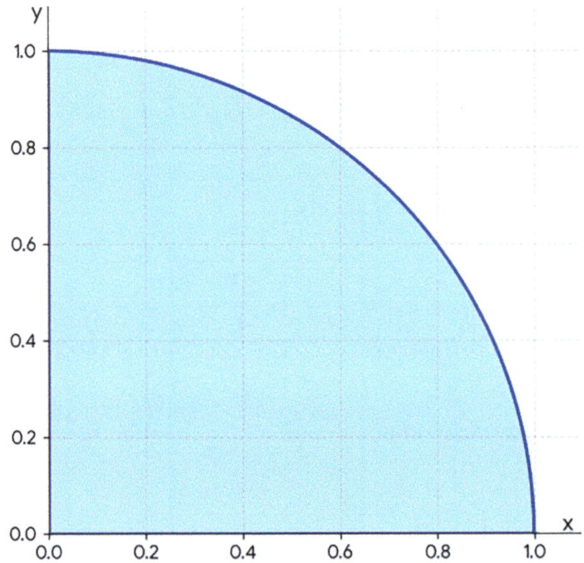

Fig. 3.34 The region defined by $\{(x, y)|x \geq 0, y \geq 0, x^2 + y^2 \leq 1\}$

***z* Limits**: $0 \leq z \leq \sqrt{1 - x^2 - y^2}.$

$z = 0$ is the xy-plane.

$z = \sqrt{1 - x^2 - y^2}$ is the upper hemisphere of the sphere $x^2 + y^2 + z^2 = 1.$

The region of integration is the portion of the unit ball $(x^2 + y^2 + z^2 \leq 1)$ lying in the first octant $(x \geq 0, y \geq 0$ and $z \geq 0)$.

Let's convert to spherical coordinates:

Integrand: $\frac{1}{\sqrt{1-(x^2+y^2+z^2)}} = \frac{1}{\sqrt{1-\rho^2}}.$

***ρ* Limits**: Bounded by the origin $(\rho = 0)$ and the sphere $\rho = 1.$

$$0 \leq \rho \leq 1$$

ϕ **Limits**: Bounded by the positive z-axis ($\phi = 0$) and the xy-plane ($\phi = \frac{\pi}{2}$).

$$0 \leq \phi \leq \frac{\pi}{2}$$

$\boldsymbol{\theta}$ **Limits**: Since the projection (Fig. 3.34) is a disk quadrant in the xy-plane:

$$0 \leq \theta \leq \frac{\pi}{2}$$

The integral transforms to:

$$\int_0^1 \int_0^{\sqrt{1-x^2}} \int_0^{\sqrt{1-x^2-y^2}} \frac{1}{\sqrt{1-\left(x^2+y^2+z^2\right)}} dzdydx$$

$$= \int_0^{\frac{\pi}{2}} \int_0^{\frac{\pi}{2}} \int_0^1 \frac{1}{\sqrt{1-\rho^2}} \times \rho^2 \sin(\phi) d\rho d\phi d\theta$$

$$= \int_0^{\frac{\pi}{2}} \int_0^{\frac{\pi}{2}} \int_{0\sqrt{1-\rho^2}}^1 \frac{\rho^2}{\sqrt{1}} \sin(\phi) d\rho d\phi d\theta = \int_0^1 \frac{\rho^2}{\sqrt{1-\rho^2}} d\rho \int_0^{\frac{\pi}{2}} \sin(\phi) d\phi \int_0^{\frac{\pi}{2}} d\theta$$

Let's calculate each integral separately.

$$\int_0^1 \frac{\rho^2}{\sqrt{1-\rho^2}} d\rho \overset{\begin{cases} \rho = \sin(u) \\ d\rho = \cos(u)du \end{cases}}{\Rightarrow} \int_{\sin^{-1}(0)}^{\sin^{-1}(1)} \frac{\sin^2(u)}{\sqrt{1-\sin^2(u)}} \cos(u)du$$

$$= \int_0^{\frac{\pi}{2}} \frac{\sin^2(u)}{\sqrt{1-\sin^2(u)}} \cos(u)du$$

$$= \int_0^{\frac{\pi}{2}} \frac{\sin^2(u)}{\sqrt{\cos^2(u)}} \cos(u)du$$

$$= \int_0^{\frac{\pi}{2}} \frac{\sin^2(u)}{|\cos(u)|} \cos(u)du \overset{0 \leq u \leq \frac{\pi}{2} \Rightarrow \cos(u) \geq 0}{\Rightarrow} \int_0^{\frac{\pi}{2}} \frac{\sin^2(u)}{\cos(u)} \cos(u)du$$

$$= \int_0^{\frac{\pi}{2}} \sin^2(u)du$$

$$= \int_0^{\frac{\pi}{2}} \frac{1 - \cos(2u)}{2} du = \frac{u}{2} - \frac{\sin(2u)}{4} \Bigg|_0^{\frac{\pi}{2}}$$

$$= \frac{\pi}{4} - \frac{\sin\left(2 \times \frac{\pi}{2}\right)}{4} - \left(\frac{0}{2} - \frac{\sin(2 \times 0)}{4}\right) = \frac{\pi}{4}$$

$$\int_0^{\frac{\pi}{2}} sin(\phi)d\phi = -\cos(\phi)\Bigg|_0^{\frac{\pi}{2}} = \cos(\phi)\Bigg|_{\frac{\pi}{2}}^{0} = \cos(0) - \cos\left(\frac{\pi}{2}\right) = 1 - 0 = 1$$

$$\int_0^{\frac{\pi}{2}} d\theta = \frac{\pi}{2} - 0 = \frac{\pi}{2}$$

Therefore,

$$\int_0^1 \int_0^{\sqrt{1-x^2}} \int_0^{\sqrt{1-x^2-y^2}} \frac{1}{\sqrt{1 - \left(x^2 + y^2 + z^2\right)}} dzdydx$$

$$= \int_0^1 \frac{\rho^2}{\sqrt{1 - \rho^2}} d\rho \int_0^{\frac{\pi}{2}} \sin(\phi)d\phi \int_0^{\frac{\pi}{2}} d\theta = \frac{\pi}{4} \times 1 \times \frac{\pi}{2} = \frac{\pi^2}{8} \approx 1.2337$$

Example 3.51 Evaluate $\iiint\limits_E e^{(x^2+y^2+z^2)^{\frac{3}{2}}} dV$ where E is the solid sphere $x^2+y^2+z^2 =$ 1.

Solution The solid sphere $x^2 + y^2 + z^2 \leq 1$ translates to the following limits for spherical coordinates:

Radial Distance (ρ):$1 \leq \rho \leq 1$.

Polar Angle (ϕ): $0 \leq \phi \leq \pi$.

Azimuthal Angle (θ):$0 \leq \theta \leq 2\pi$.

Therefore,

$$\iiint\limits_E e^{(x^2+y^2+z^2)^{\frac{3}{2}}} dV$$

$$= \int_0^{2\pi} \int_0^{\pi} \int_0^{1} e^{(\rho^2)^{\frac{3}{2}}} \times \rho^2 \sin(\phi)d\rho d\phi d\theta$$

$$= \int_0^{2\pi}\int_0^{\pi}\int_0^{1} e^{\rho^3} \times \rho^2 \sin(\phi) d\rho d\phi d\theta$$

$$= \int_0^{2\pi}\int_0^{\pi}\int_0^{1} \rho^2 e^{\rho^3} \sin(\phi) d\rho d\phi d\theta$$

$$= \int_0^{1} \rho^2 e^{\rho^3} d\rho \int_0^{\pi} \sin(\phi) d\phi \int_0^{2\pi} d\theta$$

Let's calculate each integral separately.

$$\int_0^1 \rho^2 e^{\rho^3} d\rho = \frac{1}{3}\int_0^1 3\rho^2 e^{\rho^3} d\rho = \left.\frac{e^{\rho^3}}{3}\right|_0^1$$

$$= \frac{e^{1^3}}{3} - \frac{e^{0^3}}{3} = \frac{e}{3} - \frac{1}{3} = \frac{e-1}{3}$$

$$\int_0^{\pi} \sin(\phi) d\phi = -\cos(\phi)|_0^{\pi} = \cos(\phi)|_{\pi}^{0}$$

$$= \cos(0) - \cos(\pi) = 1 - (-1) = 2$$

$$\int_0^{2\pi} d\theta = 2\pi - 0 = 2\pi$$

Therefore,

$$\iiint_E e^{(x^2+y^2+z^2)^{\frac{3}{2}}} dV = \int_0^{1} \rho^2 e^{\rho^3} d\rho \int_0^{\pi} \sin(\phi) d\phi \int_0^{2\pi} d\theta$$

$$= \frac{e-1}{3} \times 2 \times 2\pi = \frac{4(e-1)}{3}\pi \approx 7.1975$$

Exercise 3.13 Evaluate $\iiint_E e^{(x^2+y^2+z^2)^{\frac{3}{2}}} dV$ where E is the region of the solid sphere $x^2+y^2+z^2 \leq 4$ that lies in the first octant. (Ans. $\frac{(e^8-1)\pi}{6} \approx 1560.3$).

Example 3.52 Use MATLAB to evaluate $\int_0^1 \int_0^z \int_0^{x+z} (x+y) dydxdz$.

Solution The required code is shown in Fig. 3.35.

```
Command Window
>> syms x y z
>> f=x+y;
>> I=int(int(int(f,y,0,x+z),x,0,z),z,0,1)

I =

1/2

fx >>
```

Fig. 3.35 Evaluation of $\int_0^1 \int_0^z \int_0^{x+z} (x+y)dydxdz$ using nested int command

Example 3.53 Use MATLAB to evaluate $\int_{-1}^{1} \int_{-\sqrt{1-x^2}}^{\sqrt{1-x^2}} \int_0^2 z^2 dzdydx$.

Solution The required code is shown in Fig. 3.36.

```
Command Window
>> syms x y z
>> f=z^2;
>> I=int(int(int(f,z,0,2),y,-sqrt(1-x^2),sqrt(1-x^2)),x,-1,1)

I =

(8*pi)/3

>> eval(I)

ans =

    8.3776

fx >> |
```

Fig. 3.36 Evaluation of $\int_{-1}^{1} \int_{-\sqrt{1-x^2}}^{\sqrt{1-x^2}} \int_0^2 z^2 dzdydx$ using nested int command

Example 3.54 Use MATLAB to evaluate $\int_0^{2\pi} \int_0^{\pi} \int_0^1 \rho^2 e^{\rho^3} \sin(\phi) d\rho d\phi d\theta$.

Solution The required code is shown in Fig. 3.37.

```
Command Window
>> syms rho phi theta
>> f=rho^2*exp(rho^3)*sin(phi);
>> I=int(int(int(f,rho,0,1),phi,0,pi),theta,0,2*pi)

I =

(4*pi*(exp(1) - 1))/3

>> eval(I)

ans =

    7.1975

fx >>
```

Fig. 3.37 Evaluation of $\int_0^{2\pi}\int_0^{\pi}\int_0^1 \rho^2 e^{\rho^3}\sin(\phi)d\rho d\phi d\theta$ using nested int command

Example 3.55 Use MATLAB to evaluate $\int_0^{\pi}\int_0^{1+\cos^2(y)}\int_0^{\sqrt{yz}}\sin\left(x^2+y^2\right)e^{-z}dxdzdy$.

Solution While most triple integrals can be evaluated analytically using the int command, some cases result in complex symbolic expressions rather than numerical values (Fig. 3.38). For this specific problem, applying nested int commands produces an unwieldy result; therefore, using the integral3 command is more efficient for obtaining a direct numerical solution.

```
Command Window
>> syms x y z
>> f=sin(x^2+y^2)*exp(-z);
>> I=int(int(int(f,x,0,sqrt(y*z)),z,0,1+cos(y)^2),y,0,pi)

I =

int(int((2^(1/2)*pi^(1/2)*cos(y^2)*exp(-z)*fresnels((2^(1/2)*(y*z)^

fx >>
```

Fig. 3.38 Nested int commands can't evaluate the integral $\int_0^{\pi}\int_0^{1+\cos^2(y)}\int_0^{\sqrt{yz}}\sin\left(x^2+y^2\right)e^{-z}dxdzdy$

The script in Fig. 3.39 utilizes the `integral3` command to efficiently compute the numerical value of the given integral.

```
Command Window
>> f = @(y,z,x) sin(x.^2+y.^2).*exp(-z);
>> xmax= @(z,y) sqrt(y.*z);
>> zmax= @(y) 1+cos(y).^2;
>> I=integral3(f,0,pi,0,zmax,0,xmax)

I =

    0.3516

fx >>
```

Fig. 3.39 Evaluation of $\int_0^{\pi}\int_0^{1+\cos^2(y)}\int_0^{\sqrt{yz}}\sin\left(x^2+y^2\right)e^{-z}dxdzdy$ using the integral3 command

In MATLAB's integral3 command, the variable ordering in the function handle determines the integration sequence. Table 3.1 summarizes the required MATLAB function handle syntax for each triplet of integration variables.

Example 3.56 Use MATLAB to evaluate $\int_0^1\int_0^x\int_0^{x+z}(x+y)dydzdx$.

Solution Either the script presented in Fig. 3.40 or Fig. 3.41 may be used. Since the integration order is *dydzdx*, the function handle for the integral3 command is defined as @(x, z, y).

Table 3.1 Function handle definitions for triple integration orders

Integral	Integration order	Function Handle
$\int_{z0}^{z1}\int_{i(z)}^{j(z)}\int_{g(y,z)}^{h(y,z)}f(x,y,z)dxdydz$	*dxdydz*	@(z,y,x)
$\int_{y0}^{y1}\int_{i(y)}^{j(y)}\int_{g(z,y)}^{h(z,y)}f(x,y,z)dxdzdy$	*dxdzdy*	@(y,z,x)
$\int_{z0}^{z1}\int_{i(z)}^{j(z)}\int_{g(x,z)}^{h(x,z)}f(x,y,z)dydxdz$	*dydxdz*	@(z,x,y)
$\int_{x0}^{x1}\int_{i(x)}^{j(x)}\int_{g(z,x)}^{h(z,x)}f(x,y,z)dydzdx$	*dydzdx*	@(x,z,y)
$\int_{y0}^{y1}\int_{i(y)}^{j(y)}\int_{g(x,y)}^{h(x,y)}f(x,y,z)dzdxdy$	*dzdxdy*	@(y,x,z)
$\int_{x0}^{x1}\int_{i(x)}^{j(x)}\int_{g(y,x)}^{h(y,x)}f(x,y,z)dzdydx$	*dzdydx*	@(x,y,z)

```
Command Window
>> syms x y z
>> f=x+y;
>> I=int(int(int(f,y,0,x+z),z,0,x),x,0,1)

I =

2/3

fx >>
```

Fig. 3.40 Evaluation of $\int_0^1 \int_0^x \int_0^{x+z} (x+y)dydzdx$ using nested int commands

```
Command Window
>> f = @(x,z,y) x+y;
>> ymax = @(x,z) x+z;
>> zmax= @(x) x;
>> I = integral3(f,0,1,0,zmax,0,ymax)

I =

    0.6667

fx >>
```

Fig. 3.41 Evaluation of $\int_0^1 \int_0^x \int_0^{x+z} (x+y)dydzdx$ using the integral3 command

Example 3.57 Use MATLAB to evaluate $\int_0^2 \int_0^z \int_0^{y+z} (x+y+z)dxdydz$.

Solution Either the script presented in Fig. 3.42 or Fig. 3.43 may be used. Since the integration order is *dxdydz*, the function handle for the integral3 command is defined as @(z, y, x).

```
Command Window
>> syms x y z
>> f = x + y + z;
>> I=int(int(int(f,x,0,y+z),y,0,z),z,0,2)

I =

14

fx >>
```

Fig. 3.42 Evaluation of $\int_0^2 \int_0^z \int_0^{y+z} (x+y+z)dxdydz$ using nested int commands

```
Command Window
>> f = @(z,y,x) x+y+z;
>> xmax= @(y,z) y+z;
>> ymax= @(z) z;
>> I=integral3(f,0,2,0,ymax,0,xmax)

I =

   14.0000

fx >>
```

Fig. 3.43 Evaluation of $\int_0^2 \int_0^z \int_0^{y+z} (x+y+z)dxdydz$ using the integral3 command

Example 3.58 Use MATLAB to evaluate $\int_0^1 \int_0^{1-x} \int_0^{1-x-y} 1dzdydx$.

Solution Either the script presented in Fig. 3.44 or Fig. 3.45 may be used. Since the integration order is *dzdydx*, the function handle for the integral3 command is defined as @(x, y, z).

```
Command Window
>> syms x y z
>> f=1;
>> I=int(int(int(f,z,0,1-x-y),y,0,1-x),x,0,1)

I =

1/6

fx >>
```

Fig. 3.44 Evaluation of $\int_0^1 \int_0^{1-x} \int_0^{1-x-y} 1dzdydx$ using nested int commands

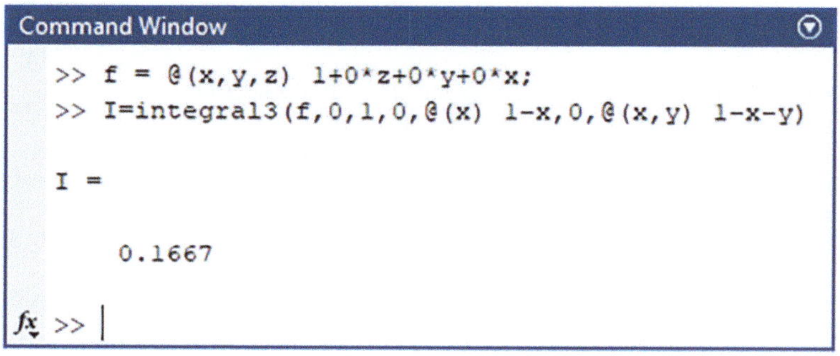

Fig. 3.45 Evaluation of $\int_0^1 \int_0^{1-x} \int_0^{1-x-y} 1 dzdydx$ using the integral3 command

Example 3.59 Use MATLAB to evaluate $\int_0^1 \int_0^2 \int_0^3 xdxdydz$.

Solution Either the script presented in Fig. 3.46 or Fig. 3.47 may be used. Since the integration order is *dxdydz*, the function handle for the integral3 command is defined as @(z, y, x).

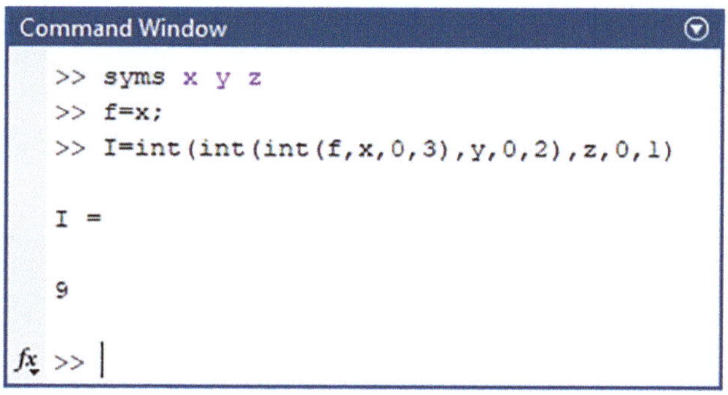

Fig. 3.46 Evaluation of $\int_0^1 \int_0^2 \int_0^3 xdxdydz$ using nested int commands

Fig. 3.47 Evaluation of $\int_0^1 \int_0^2 \int_0^3 xdxdydz$ using the integral3 command

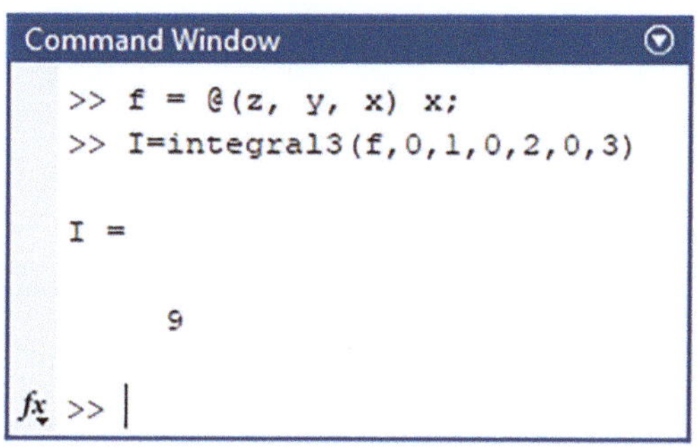

References

1. Calculus, George Thomas, Maurice Weir, Joel Hass, Pearson, 2014.
2. Calculus, James Stewart, Cengage Learning, 2015.
3. Engineering Mathematics with MATLAB® and Simulink®, Farzin Asadi, Springer, 2025.
4. Applied Numerical Analysis with MATLAB®/Simulink®: For Engineers and Scientists, Farzin Asadi, Springer, 2023.
5. A Problem-Solving Approach to Single-Variable Calculus (Volume I), Farzin Asadi, Springer, 2026.
6. A Problem-Solving Approach to Single-Variable Calculus (Volume II), Farzin Asadi, Springer, 2026.
7. A Problem-Solving Approach to Multi-Variable Calculus (Volume I), Farzin Asadi, Springer, 2026.
8. A Problem-Solving Approach to Multi-Variable Calculus (Volume III), Farzin Asadi, Springer, 2026.
9. A Problem-Solving Approach to Ordinary Differential Equations (Volume I), Farzin Asadi, Springer, 2026.
10. A Problem-Solving Approach to Ordinary Differential Equations (Volume II), Farzin Asadi, Springer, 2026.
11. A Problem-Solving Approach to Ordinary Differential Equations (Volume III), Farzin Asadi, Springer, 2026.
12. A Problem-Solving Approach to Ordinary Differential Equations (Volume IV), Farzin Asadi, Springer, 2026.

Chapter 4
Essentials of MATLAB®

4.1 Introduction

MATLAB® (short for MATrix LABoratory) is a programming and numerical computing platform created specifically for engineers and scientists. It is used to analyze and design the systems and products that are transforming our world.

At the core of MATLAB is the MATLAB language, a matrix-based language that provides the most natural and expressive way to write computational mathematics.

The keywords of the MATLAB language are listed in Table 4.1.

Millions of engineers and scientists worldwide use MATLAB for a wide range of applications in both industry and academia. MATLAB serves as both an analysis and a design tool. This chapter will introduce the essentials of MATLAB and its common daily uses.

4.2 MATLAB Environment

The MATLAB environment is shown in Fig. 4.1. It is composed of four parts. First part is a collection of icons that are required frequently. The second section (current folder browser) enables you to interactively manage files and folders in MATLAB. Use the current folder browser to view, create, open, move, and rename files and folders in the current folder. Use the icon shown in Fig. 4.2 to open the desired folder. The third section of MATLAB is command window. The MATLAB commands are entered here and their results are shown here. You need to press Enter key of your key board to run the commands written in command window. The fourth section of MATLAB (the workspace) contains variables that you create within or import into MATLAB from data files or other programs.

F. Asadi, *A Problem-Solving Approach to Multi-Variable Calculus*,
https://doi.org/10.1007/978-3-032-21115-6_4

Table 4.1 Keywords of the MATLAB language

break	case	catch	classdef
continue	else	elseif	end
for	function	global	if
otherwise	parfor	persistent	return
spmd	switch	try	while

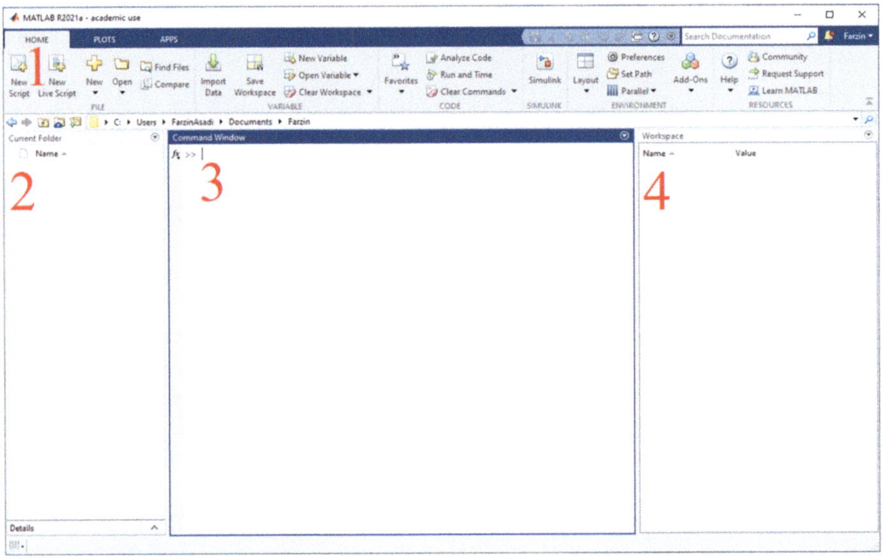

Fig. 4.1 MATLAB environment

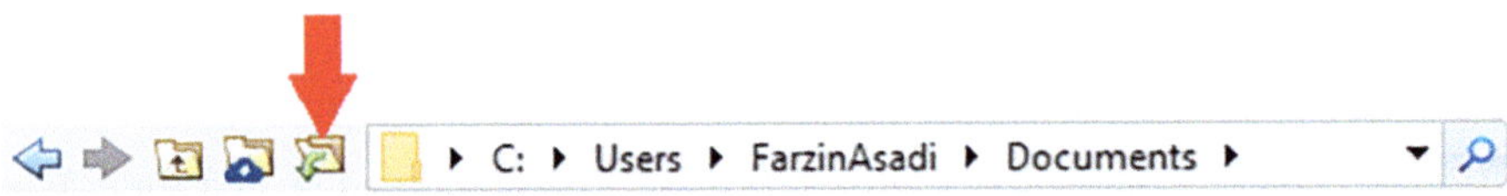

Fig. 4.2 Brows for folder icon

The layout of MATLAB can be changed using the layout icon (Fig. 4.3). This book uses the default layout.

Fig. 4.3 Layout icon

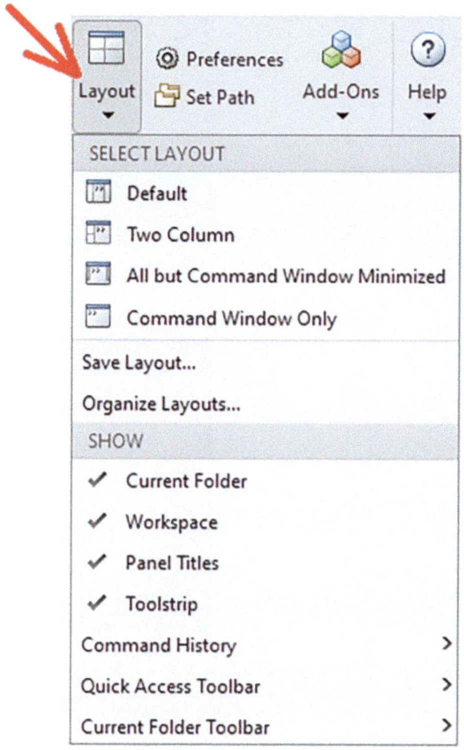

4.3 Basic Operations with MATLAB

Let's define a complex variable a with value of $1 + 2i$. This can be done with the aid of commands shown in Figs. 4.4, 4.5 or 4.6. In MATLAB, both of i and j can be used to enter the imaginary part of a complex number.

Fig. 4.4 Defining variable a $= 1 + 2$i

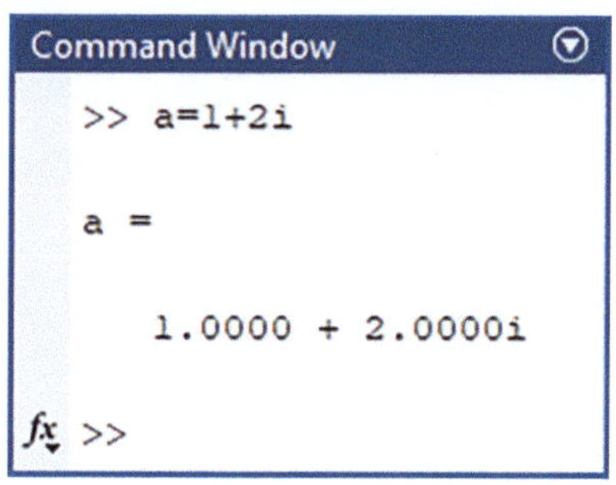

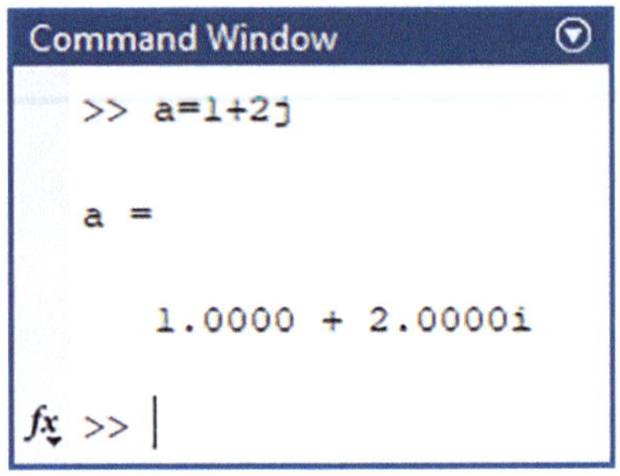

Fig. 4.5 Defining variable a = 1 + 2i

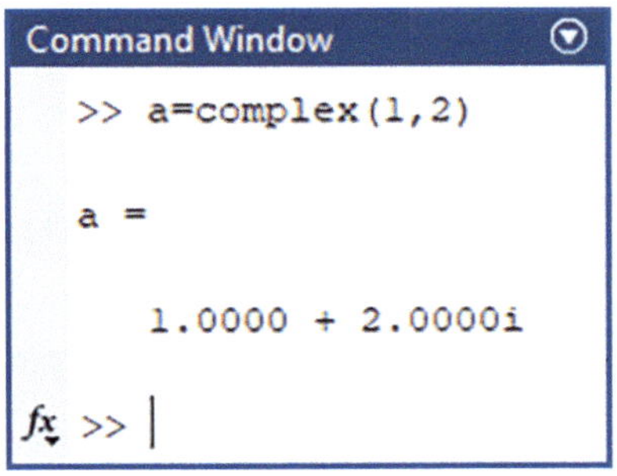

Fig. 4.6 Defining variable a = 1 + 2i

After running the commands of Figs. 4.4, 4.5 or 4.6, a new variable will be added to the workspace window (Fig. 4.7).

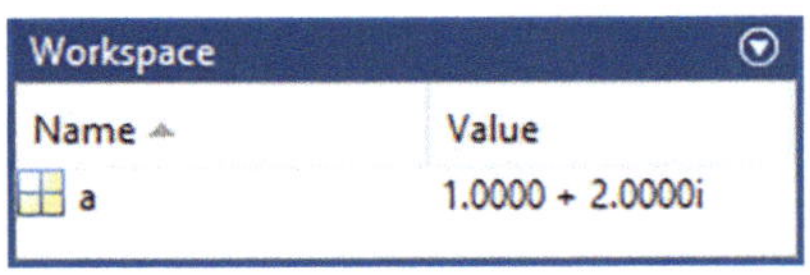

Fig. 4.7 Variable a is added to the Workspace

To suppress the output of a command (i.e., prevent the result from being displayed in the command window), simply end the line with a semicolon (Fig. 4.8).

```
Command Window
>> a=1+2*i

a =

   1.0000 + 2.0000i

>> 3+5

ans =

     8
```

```
Command Window
>> a=1+2*i;
>> 3+5;
fx >>
```

Fig. 4.8 The semicolon suppresses the output of a command

You can use the built-in functions real and imag to extract the real and imaginary parts of any complex number, respectively (Fig. 4.9).

Fig. 4.9 Calculation of real and imaginary parts with real and imag commands

```
Command Window
>> a=1+2*i;
>> real(a)

ans =

     1

>> imag(a)

ans =

     2

fx >>
```

You can use the conj function to calculate the complex conjugate of a number (Fig. 4.10).

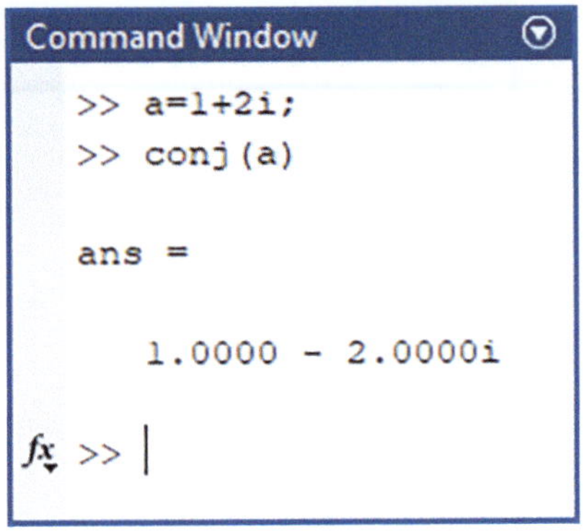

Fig. 4.10 Calculation of the complex conjugate with conj command

The point $a = 1 + 2i$ is shown in Fig. 4.11. The polar form of this point is $\sqrt{1^2 + 2^2}e^{i\tan^{-1}\left(\frac{2}{1}\right)}$. Value of $\sqrt{1^2 + 2^2}$ and $\tan^{-1}\left(\frac{2}{1}\right)$ is calculated with the aid commands shown in Fig. 4.12. The magnitude and principal argument of a complex number can be calculated using the commands shown in Fig. 4.13.

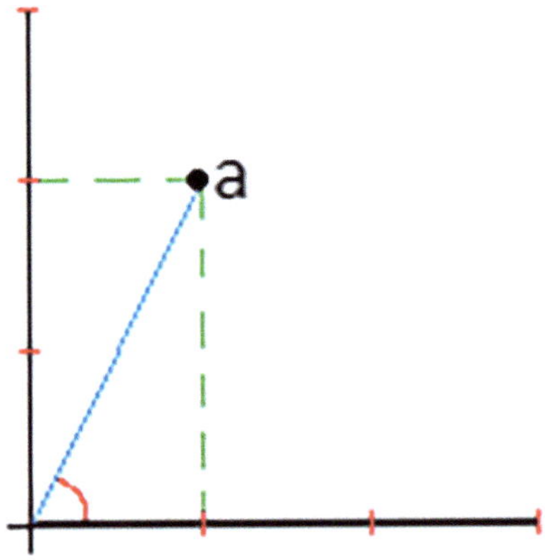

Fig. 4.11 Representation of 1 + 2i in the complex plane

Fig. 4.12 Calculation of magnitude and angle of $1 + 2i$ $1 + 2i$

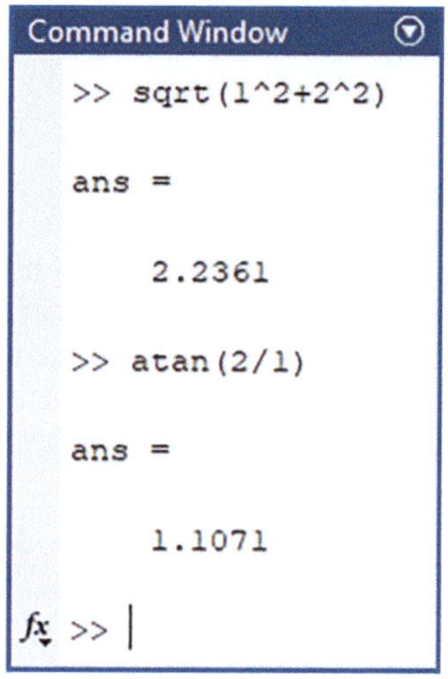

Fig. 4.13 Calculation of magnitude and angle of $1 + 2i$

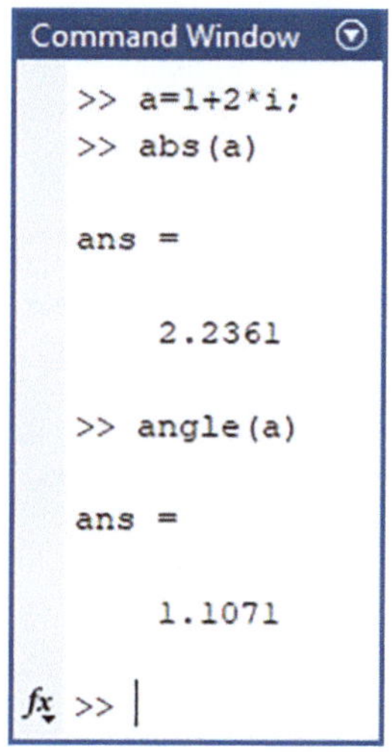

Case sensitivity dictates whether a programming language treats uppercase and lowercase letters as distinct (case-sensitive) or equivalent (case-insensitive). MATLAB is a case-sensitive language; this characteristic is demonstrated in Figs. 4.14 and 4.15.

Fig. 4.14 MATLAB is a case sensitive language

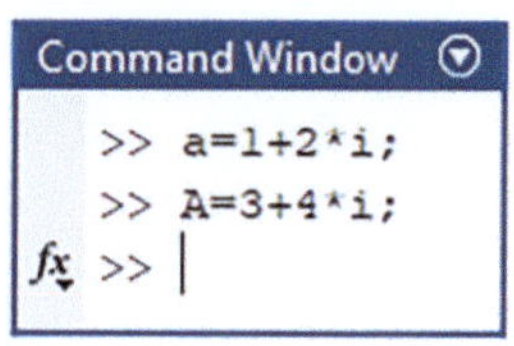

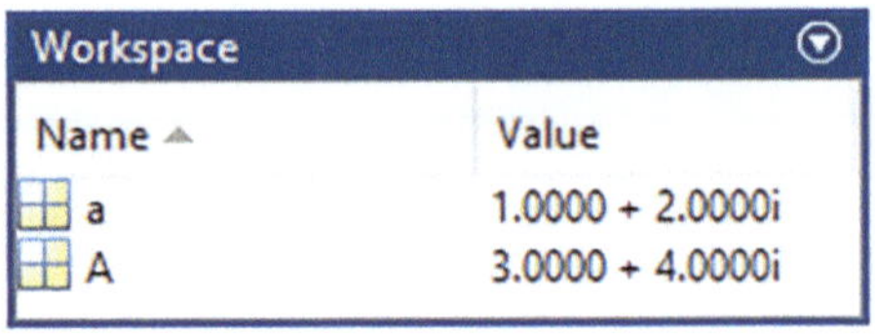

Fig. 4.15 A and A are two different variables

MATLAB automatically uses the default variable `ans` (short for answer) to store the result of a command when no output variable is explicitly defined. For example, in Fig. 4.16, the result of the multiplication is stored in the user-defined variable `c`. Conversely, in Fig. 4.17, the result is saved in `ans` because the user did not define any variable.

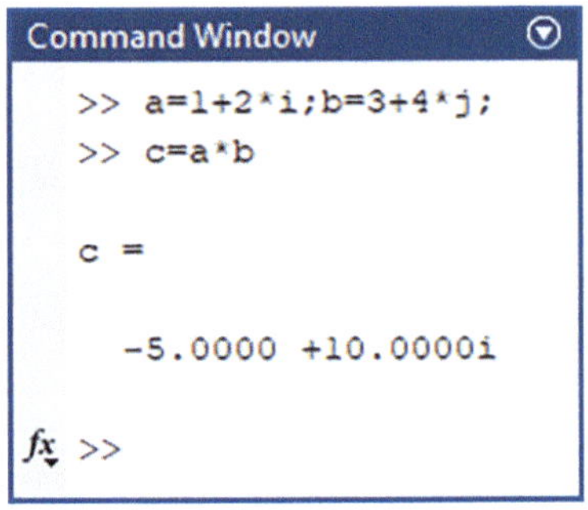

Fig. 4.16 Variable c equals to product of a and b

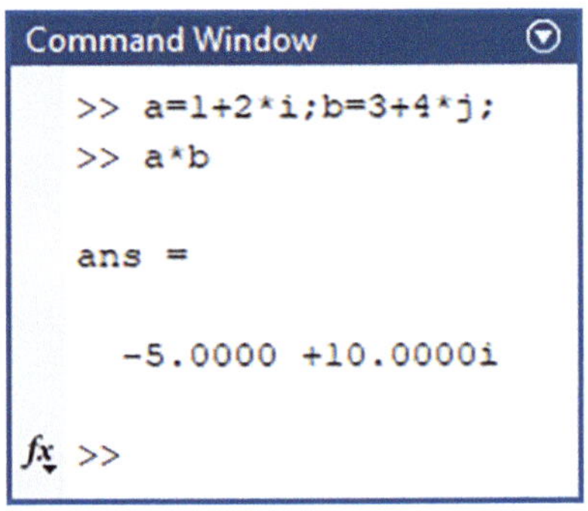

Fig. 4.17 Default variable ans saves the calculation result

Commands in Fig. 4.18 do some basic operations on two complex numbers. The percent symbol (%) is used for indicating a comment line.

Fig. 4.18 Basic operations in MATLAB

```
Command Window
>> a=1+2*i;b=3+4*j;
>> c=a*b

c =

  -5.0000 +10.0000i

>> d=a+b

d =

   4.0000 + 6.0000i

>> e=a/b

e =

   0.4400 + 0.0800i

>> f=a-b

f =

  -2.0000 - 2.0000i

>> g=a^b %^ shows power

g =

   0.1290 + 0.0339i

fx >>
```

The reminder of division of two real numbers can be found using the mod or rem functions (Fig. 4.19).

Fig. 4.19 Rem and mod commands

```
Command Window

>> rem(17,3)

ans =

     2

>> mod(17,3)

ans =

     2

>> rem(17,-3)

ans =

     2

>> mod(17,-3)

ans =

    -1

fx >>
```

The concept of remainder after division is not uniquely defined, and the two functions mod and rem each compute a different variation. The mod function produces a result that is either zero or has the same sign as the divisor. The rem function produces a result that is either zero or has the same sign as the dividend.

Another difference is the convention when the divisor is zero. The mod function follows the convention that mod(a,0) returns a, whereas the rem function follows the convention that rem(a,0) returns NaN. NaN stands for "Not a Number" and is used to represents values that are not real or complex numbers. Expressions like 0/0 and inf/inf result in NaN.

Both mod and rem functions have their uses. For example, in signal processing, the mod function is useful in the context of periodic signals because its output is periodic (with period equal to the divisor).

4.4 Clearing the Command Window and Workspace

To clear the contents of the command window, use the `clc` command (Fig. 4.20). Figure 4.21 shows the appearance of the command window after `clc` has been executed.

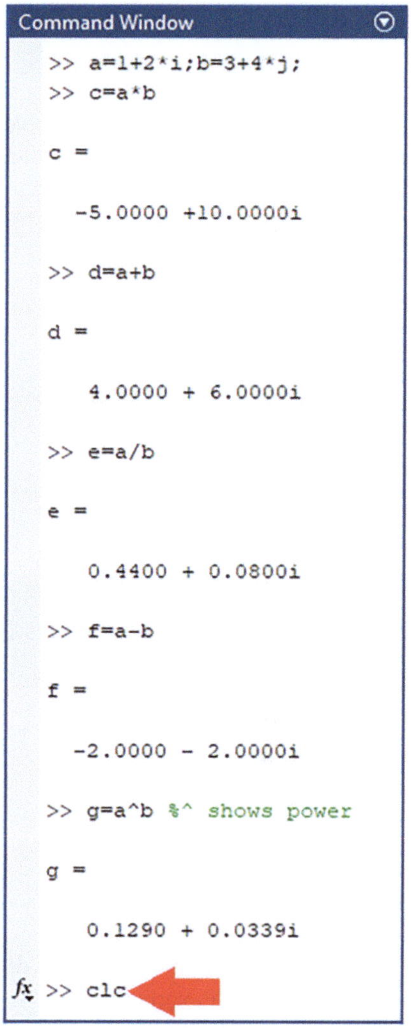

Fig. 4.20 Clc command clears the Command Window

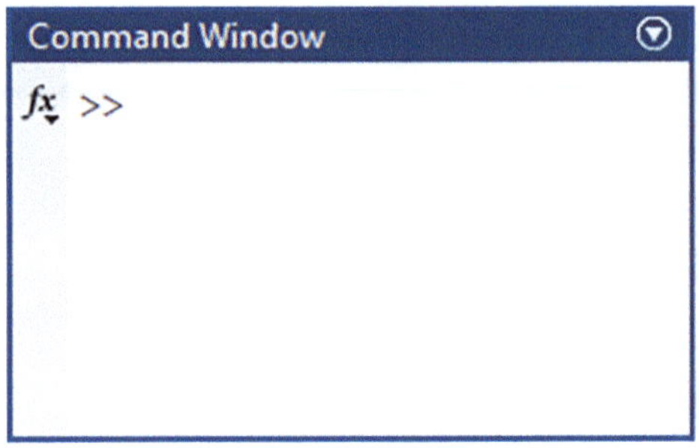

Fig. 4.21 The Command Window is cleared using the clc command

Alternatively, you can clear the command window by right-clicking anywhere within it and selecting Clear Command Window (Fig. 4.22) from the context menu that appears.

Fig. 4.22 Clicking the clear command window is another way to clear the command window

Clearing the command window does not affect the variables (Fig. 4.23). If you want to clear all the variables, you need to use the command shown in Fig. 4.24 or 4.25. If you want to clear a specific variable, then you need to write its name after the clear command. For instance, the command shown in Fig. 4.26, clears the variable g only. There is another way to remove the variable g as well. You can right click on it and select delete from the appeared menu (Fig. 4.27).

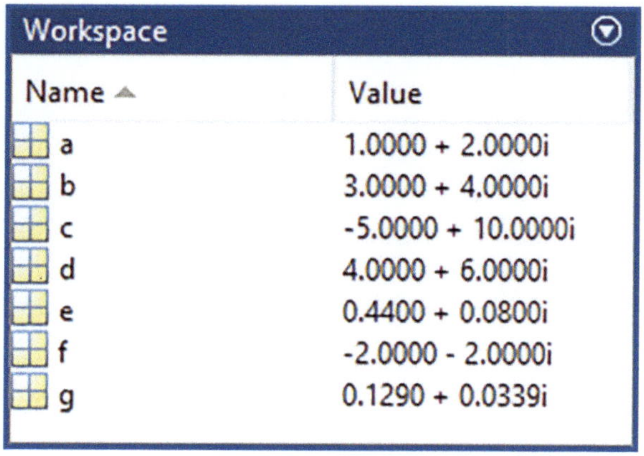

Fig. 4.23 Clearing the command window has no effect on the variables

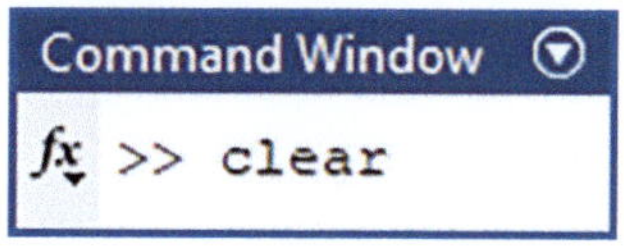

Fig. 4.24 Deleting all the Workspace variables

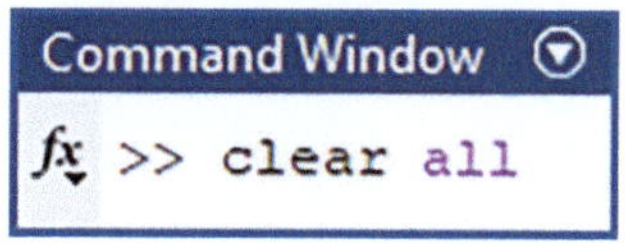

Fig. 4.25 Deleting all the Workspace variables

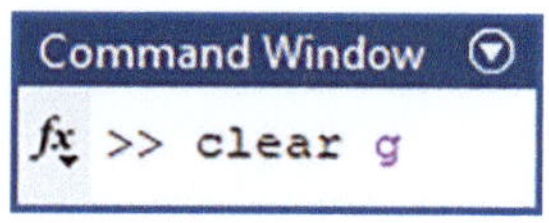

Fig. 4.26 Deleting the variable g of Workspace

Fig. 4.27 Deleting the variable g of Workspace

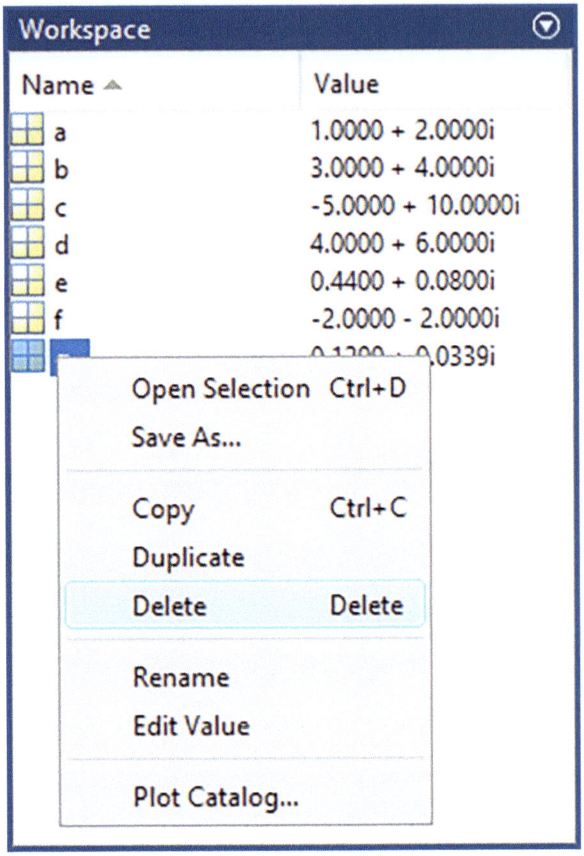

You can use the up and down arrows of the keyboard to access the previous entered commands (Fig. 4.28).

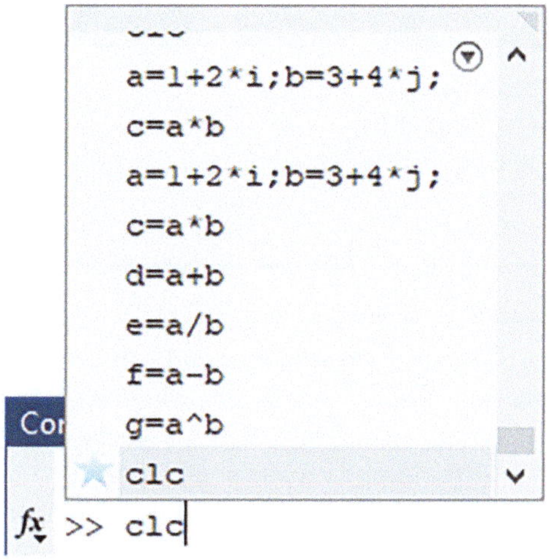

Fig. 4.28 Accessing to the previously entered commands

4.5 Basic Matrix Operations

Anything in MATLAB is a matrix. Even a number is considered as a 1×1 matrix. Assume that you want to enter $A = \begin{bmatrix} 1 & 2 & 3 \end{bmatrix}$, $B = \begin{bmatrix} 4 \\ 5 \\ 6 \end{bmatrix}$ and $C = \begin{bmatrix} 7 & 8 & 9 \\ 10 & 11 & 12+2i \\ 13 & 14 & 16 \end{bmatrix}$ to MATLAB. The commands shown in Figs. 4.29 or 4.30 do this job.

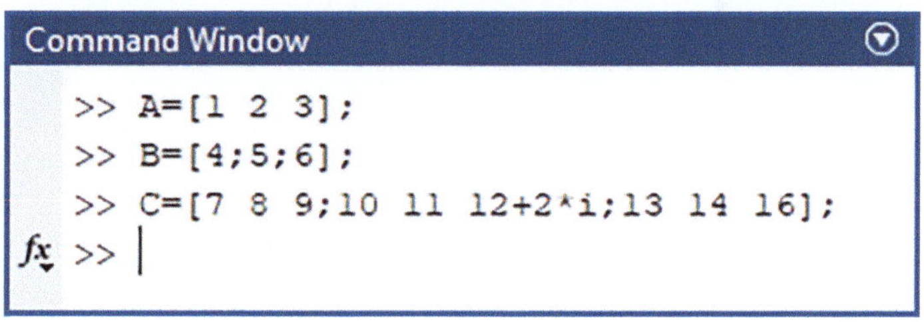

Fig. 4.29 Defining the matrices A, B and C

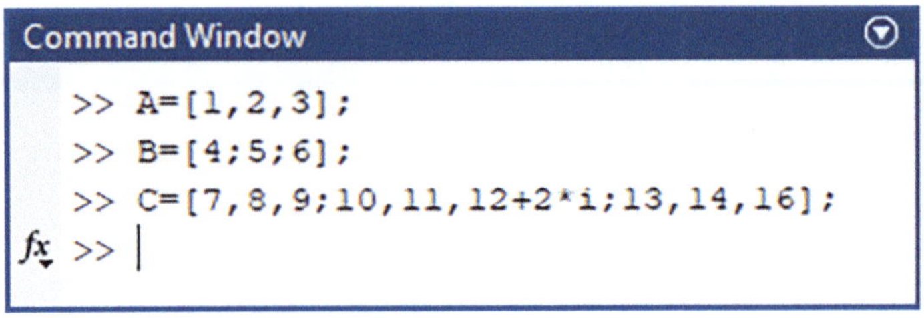

Fig. 4.30 Defining the matrices A, B and C

You can access the element in the i-th row and j-th column of a matrix x using the syntax x(i,j). for matrices that are strictly one-dimensional (vectors), a shorthand notation is available: if x is a 1 × n (row) vector, you can use x(i) instead of x(1,i). Similarly, if x is an n × 1 (column) vector, you can use x(i) instead of x(i,1). The term vector is generally applied to both 1 × n and n × 1 matrices.

Figure 4.31 presents examples of commands used to read the elements of the matrices defined in Figs. 4.29 and 4.30.

Fig. 4.31 Accessing the elements of the matrices (Matrices A, B and C are defined in Fig. 4.29)

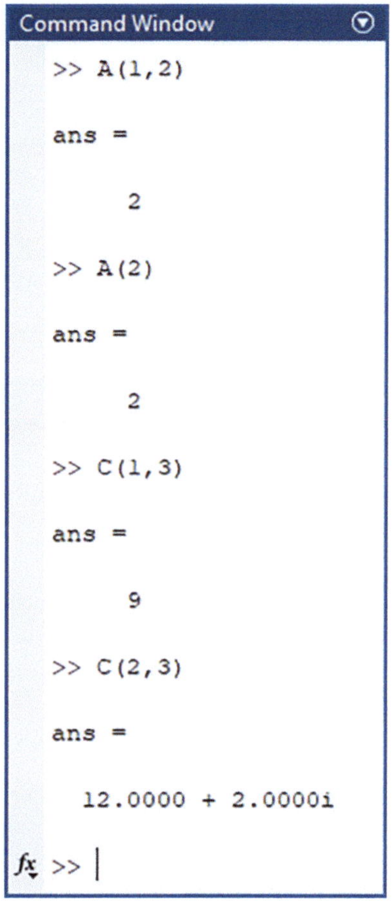

The command shown in Fig. 4.32, changes the element in the second row and third column of matrix C to 12 + 12i.

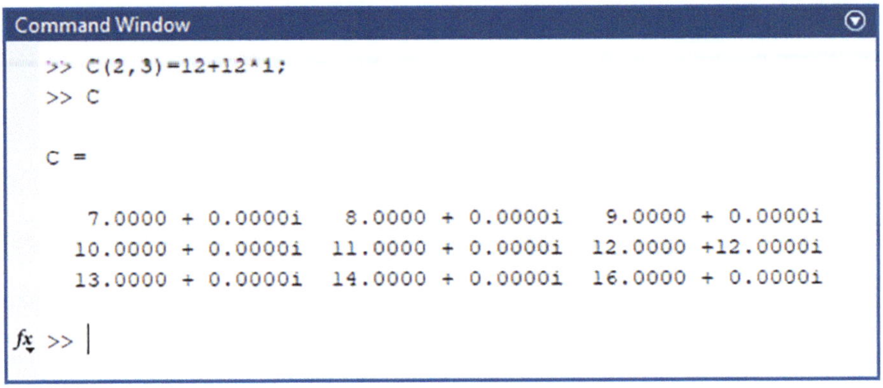

Fig. 4.32 Changing the element on the second row and third column of C (Matrix C is defined in Fig. 4.29)

To obtain the dimensions (number of rows and columns) of a matrix, use the size command (Fig. 4.33).

Fig. 4.33 Calculation of number of rows and columns of a matrix (Matrices A, B and C are defined in Fig. 4.29)

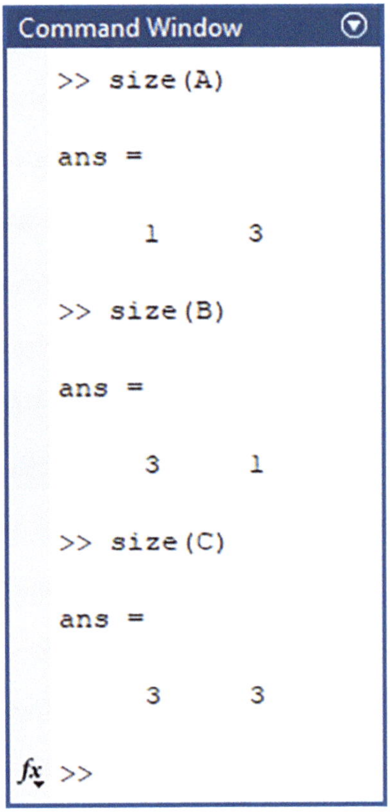

You can save the number of rows and the number of columns in separate variables. The command shown in Fig. 4.34 saves the number of rows and the number of columns of matrix A in the variables RowA and ColumnA, respectively.

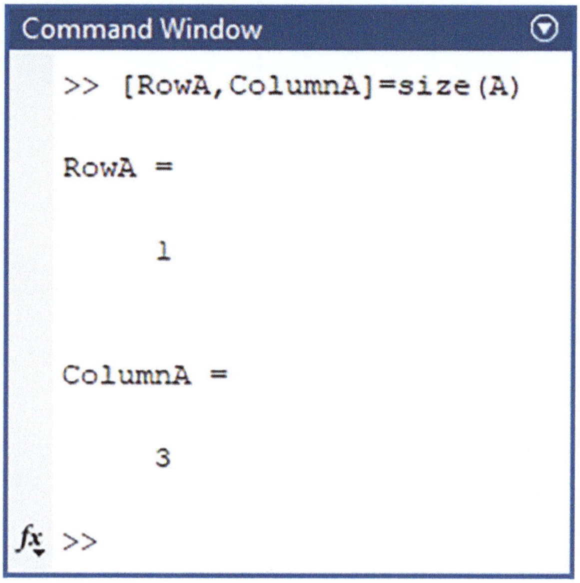

Fig. 4.34 Number of rows and columns of matrix A

The multiplication, summation and subtraction of two matrices can be done with the aid of *, + and − operators, respectively. Note that the size of two matrices must be the same for summation and subtraction and the number of columns of the first matrix must be the same as the number of rows of the second matrix, otherwise the operation will not be done.

MATLAB has elementwise operators as well. The element wise operator does the operation on the corresponding elements. For instance, in Fig. 4.35, [1 2 3].^[4 5 6] = [1^4 2^5 3^6] = [1 32 729] or [1 2 3].*[4 5 6] = [1*4 2*5 3*6] = [4 10 18].

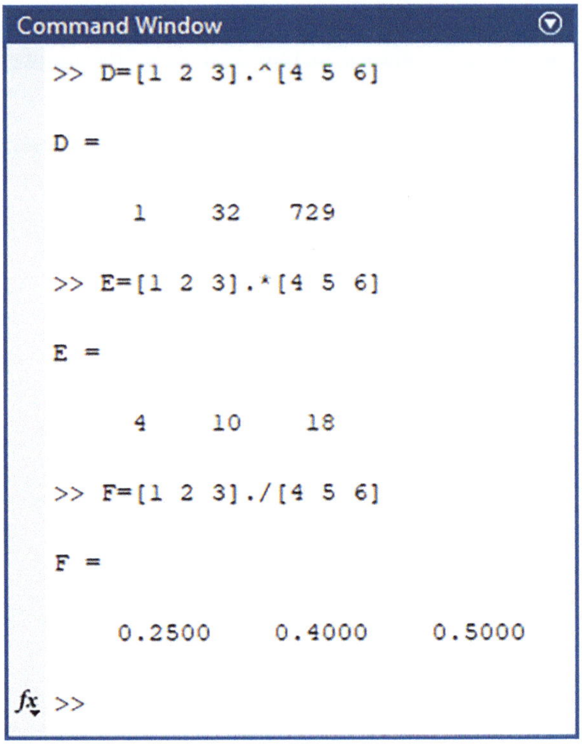

Fig. 4.35 Elementwise commands

You can find the minimum and maximum of a vector by using the min and max commands, respectively (Fig. 4.36).

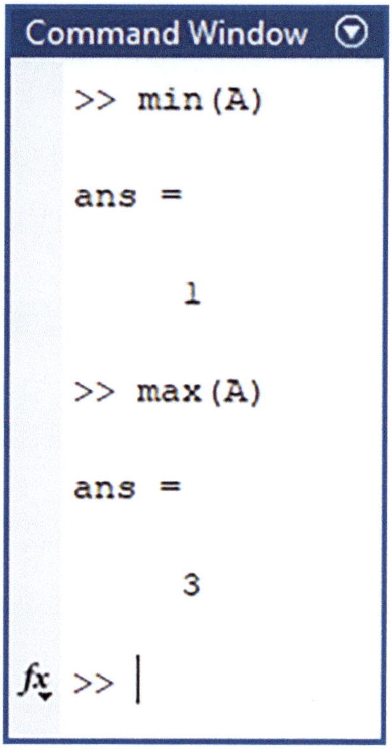

Fig. 4.36 Calculation of minimum and maximum of vector A

The min and max functions can be applied to matrices as well. If the input of these functions is a m × n matrix, they return an n × 1 vector containing the minimum/maximum element from each column (Fig. 4.37).

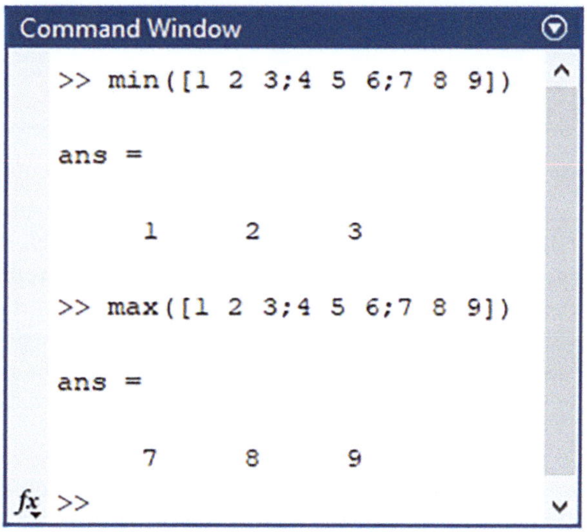

Fig. 4.37 The min and max commands can also be applied to matrices

When a matrix contains complex elements, the min and max functions operate based on the magnitude of each element. Recall that the magnitude of a complex number $a + bi$ is $\sqrt{a^2 + b^2}$. In Fig. 4.38, the corresponding magnitude matrix is $\begin{bmatrix} 7 & 8 & 9 \\ 10 & 11 & 16.971 \\ 13 & 14 & 16 \end{bmatrix}$. Therefore, the max function returns 12 + 12i as the maximum of the third column, since 12 + 12i has the largest magnitude among the elements in that column.

```
Command Window

>> C

C =

   7.0000 + 0.0000i   8.0000 + 0.0000i   9.0000 + 0.0000i
  10.0000 + 0.0000i  11.0000 + 0.0000i  12.0000 +12.0000i
  13.0000 + 0.0000i  14.0000 + 0.0000i  16.0000 + 0.0000i

>> min(C)

ans =

     7     8     9

>> max(C)

ans =

  13.0000 + 0.0000i  14.0000 + 0.0000i  12.0000 +12.0000i

fx >>
```

Fig. 4.38 Calculation of minimum and maximum element of matrix C

The transpose of a matrix can be calculated using the single quote operator or transpose function (Fig. 4.39). If the matrix is real, then two methods produce the same results. However, when the matrix is complex, the results are not the same. The transpose function simply flips the matrix over its diagonal (i.e., switches the row and column indices of the matrix), whereas the single quote function calculates the complex conjugate of elements and then change the row and column indices.

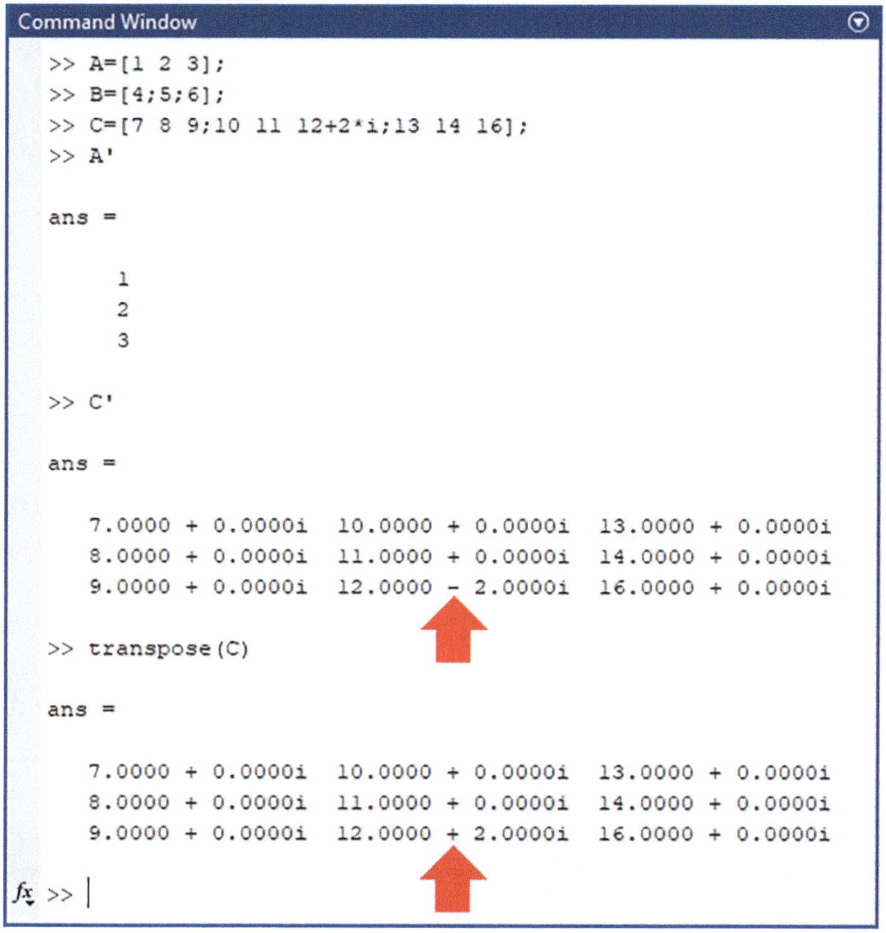

Fig. 4.39 Different methods of calculation of transpose of a matrix

You can use the det function to calculate the determinant of square matrix. The inverse of a matrix X can be calculated using the inv(X) or X^−1 commands (Fig. 4.40).

```
Command Window

>> det(C)

ans =

  -3.0000 +12.0000i

>> inv(C)

ans =

  -2.3529 - 0.0784i   0.0392 + 0.1569i   1.3137 - 0.0784i
   2.1176 - 0.1961i   0.0980 + 0.3922i  -1.2157 - 0.1961i
   0.0588 + 0.2353i  -0.1176 - 0.4706i   0.0588 + 0.2353i

>> C^-1

ans =

  -2.3529 - 0.0784i   0.0392 + 0.1569i   1.3137 - 0.0784i
   2.1176 - 0.1961i   0.0980 + 0.3922i  -1.2157 - 0.1961i
   0.0588 + 0.2353i  -0.1176 - 0.4706i   0.0588 + 0.2353i

fx >>
```

Fig. 4.40 Calculation of the determinant and inverse of matrix C

Assume you want to ensure the two matrices calculated in Fig. 4.40—using inv(C) and C^−1—are identical. You can verify this by computing the difference between them. As shown in Fig. 4.41, the difference is zero, which confirms that the results obtained from both commands are the same.

```
Command Window

>> C^-1-inv(C)

ans =

     0     0     0
     0     0     0
     0     0     0

fx >>
```

Fig. 4.41 Calculation of the difference between the C^−1 and inv(C) results

Assume that you want to solve the linear system $\begin{bmatrix} 7 & 8 & 9 \\ 10 & 11 & 12+2i \\ 13 & 14 & 16 \end{bmatrix} x = \begin{bmatrix} 4 \\ 5 \\ 6 \end{bmatrix}$. The commands shown in Fig. 4.42 can be used to calculate the vector x.

```
Command Window

>> C=[7 8 9;10 11 12+2i;13 14 16];
>> B=[4;5;6];
>> C\B

ans =

  -1.3333 + 0.0000i
   1.6667 + 0.0000i
  -0.0000 - 0.0000i

fx >>
```

```
Command Window

>> C=[7 8 9;10 11 12+2i;13 14 16];
>> B=[4;5;6];
>> x=C^-1*B

x =

  -1.3333 + 0.0000i
   1.6667 + 0.0000i
  -0.0000 - 0.0000i

fx >>
```

```
Command Window

>> C=[7 8 9;10 11 12+2i;13 14 16];
>> B=[4;5;6];
>> inv(C)*B

ans =

  -1.3333 + 0.0000i
   1.6667 + 0.0000i
  -0.0000 - 0.0000i

fx >>
```

Fig. 4.42 Calculation of vector x

4.6 Trigonometric Functions

Trigonometric functions of MATLAB are listed in Table 4.2. The listed functions can also take complex arguments.

For instance, assume we want to calculate $\sin(x)^4 - 3\cos(x)$ for $x = 30$ °. The commands shown in Figs. 4.43 or 4.44 perform this calculation.

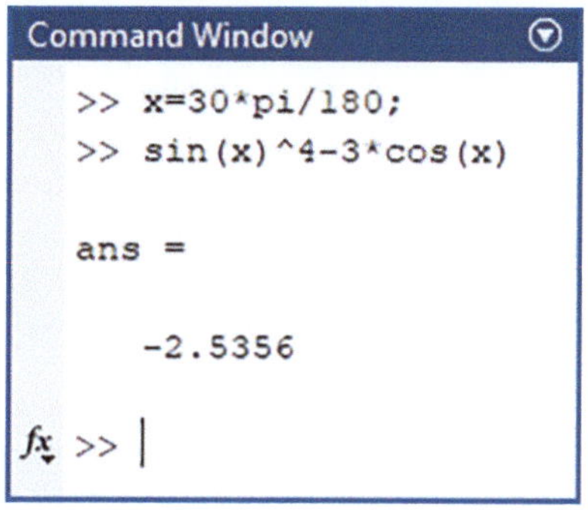

Fig. 4.43 Calculation of $\sin(x)^4 - 3\cos(x)$ for $x = 30$ °

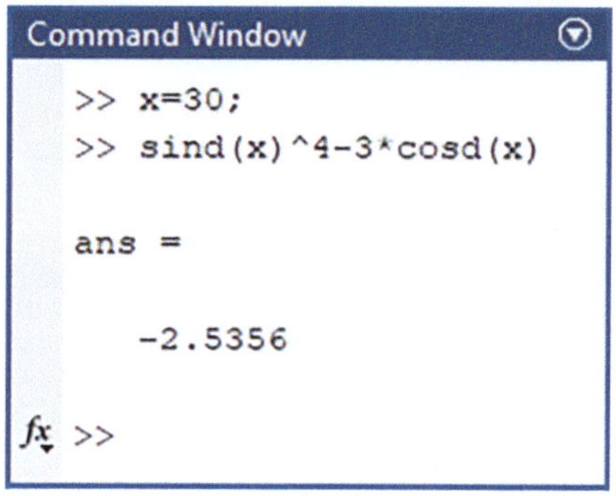

Fig. 4.44 Calculation of $\sin(x)^4 - 3\cos(x)$ for $x = 30$ °

4.7 Hyperbolic Functions

Hyperbolic functions of MATLAB are listed in Table 4.3. The listed functions can also take complex arguments.

4.8 Logarithmic and Exponential Function

Logarithmic and exponential functions of MATLAB are listed in Table 4.4. The listed functions can also take complex arguments.

Table 4.2 Trigonometric functions of MATLAB

MATLAB Function	Description	Example

`sin`	sin(x). Takes the input in radians	`>> sin(pi/2)`
		`ans =`
		`1`

`sind`	sin(x). Takes the input in degrees	`>> sind(30)`
		`ans =`
		`0.5000`

`asin`	$\sin^{-1}(x)$. Returns the output in radians	`>> asin(0.5)`
		`ans =`
		`0.5236`

`asind`	$\sin^{-1}(x)$. Returns the output in degrees	`>> asind(0.5)`
		`ans =`
		`30.0000`

`cos`	cos(x). Takes the input in radians	`>> cos(1)`
		`ans =`
		`0.5403`

`cosd`	cos(x). Takes the input in degrees	`>> cosd(60)`
		`ans =`
		`0.5000`

`acos`	$\cos^{-1}(x)$. Returns the output in radians	`>> acos(0.5)`
		`ans =`
		`1.0472`

`acosd`	$\cos^{-1}(x)$. Returns the output in degrees	`>> acosd(0.5)`
		`ans =`
		`60.0000`

`tan`	tan(x). Takes the input in radians	`>> tan(1)`
		`ans =`
		`1.5574`

`tand`	tan(x). Takes the input in degrees	`>> tand(60)`
		`ans =`
		`1.7321`

`atan`	$\tan^{-1}(x)$. Returns the output in radians	`>> atan(1)`
		`ans =`
		`0.7854`

`atand`	$\tan^{-1}(x)$. Returns the output in degrees	`>> atand(1)`
		`ans =`
		`45`

`sec`	sec(x). Takes the input in radians	`>> sec(1)`
		`ans =`
		`1.8508`

`secd`	sec(x). Takes the input in degrees	`>> secd(10)`
		`ans =`
		`1.0154`

`asec`	$\sec^{-1}(x)$. Returns the output in radians	`>> asec(30)`
		`ans =`
		`1.5375`

(continued)

Table 4.2 (continued)

MATLAB Function	Description	Example
`asecd`	$\sec^{-1}(x)$. Returns the output in degrees	`>> asecd(30)` `ans =` `88.0898`
`csc`	$\csc(x)$. Takes the input in radians	`>> csc(1)` `ans =` `1.1884`
`cscd`	$\csc(x)$. Takes the input in degrees	`>> cscd(1)` `ans =` `57.2987`
`acsc`	$\csc^{-1}(x)$. Returns the output in radians	`>> acsc(1)` `ans =` `1.5708`
`acscd`	$\csc^{-1}(x)$. Returns the output in degrees	`>> acscd(1)` `ans =` `90`
`cot`	$\cot(x)$. Takes the input in radians	`>> cot(1)` `ans =` `0.6421`
`cotd`	$\cot(x)$. Takes the input in degrees	`>> cotd(30)` `ans =` `1.7321`
`acot`	$\cot^{-1}(x)$. Returns the output in radians	`>> acot(1)` `ans =` `0.7854`
`acotd`	$\cot^{-1}(x)$. Returns the output in degrees	`>> acotd(1)` `ans =` `45`

Table 4.3 Hyperbolic functions of MATLAB

Function	MATLAB Representation
$\sinh(x)$	`sinh(x)`
$\cosh(x)$	`cosh(x)`
$\tanh(x)$	`tanh(x)`
$\coth(x)$	`coth(x)`
$\mathrm{sech}(x)$	`sech(x)`
$\mathrm{csch}^{-1}(x)$	`csch(x)`
$\sinh^{-1}(x)$	`asinh(x)`
$\cosh^{-1}(x)$	`acosh(x)`
$\tanh^{-1}(x)$	`atanh(x)`
$\coth^{-1}(x)$	`acoth(x)`
$\mathrm{sech}^{-1}(x)$	`asech(x)`
$\mathrm{csch}^{-1}(x)$	`acsch(x)`

Table 4.4 Logarithmic and exponential functions of MATLAB

MATLAB Function	Description	Example
`log`	Calculates the natural logarithm (with base of $e = 2.7182$)	`>> log(2.7182)` `ans =` `1.0000`
`log10`	Calculates the common logarithm (with base of 10)	`>> log10(100)` `ans =` `2`
`log2`	Calculates the binary logarithm with (with base of 2)	`>> log2(4)` `ans =` `2`
`exp`	Calculates e^x	`>> exp(2)` `ans =` `7.3891`
`sqrt`	Calculates the square root	`>> sqrt(16)` `ans =` `4`
`power`	Calculates the power function	`>> power(2,3)` `ans =` `8`
`^`	Calculates the power function	`>> 2^3` `ans =` `8`

4.9 Rounding Functions

Rounding functions of MATLAB are shown in Table 4.5.

Table 4.5 Rounding function of MATLAB

MATLAB function	Expression	Example
`fix`	It truncates the decimal part	`fix([-4.6 4.6])` `ans =` `[-4 4]`
`floor`	It rounds down to the nearest integer	`>> floor(8.4687)` `ans =` `8`
`ceil`	It rounds up to the nearest integer	`>> ceil(4.4)` `ans =` `5`
`round`	Rounds to the nearest integer	`>> round(4.55)` `ans =` `5`

4.10 Colon Operator

The colon operator can be used for creating regularly spaced vectors, index into arrays, and define the bounds of a `for` loop. For instance, assume that you want to make a vector t from 0 to 1 with 0.1 steps. The command shown in Fig. 4.45 makes this vector.

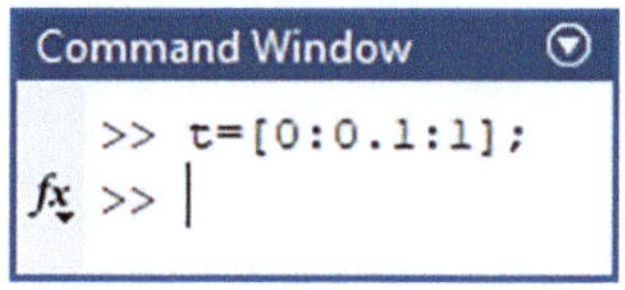

Fig. 4.45 Defining the vector t with colon operator

The values of vector t are shown in Fig. 4.46.

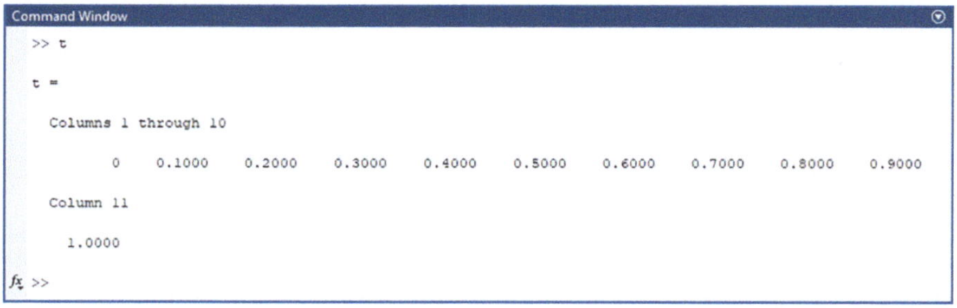

Fig. 4.46 Values of vector t defined in Fig. 4.45

The default step of colon operator equals to one. So, x = [3:10] and x = [3:1:10] are the same. Both of these commands define x = [3 4 5 6 7 8 9 10]. The command shown in Fig. 4.47 defines a random matrix with 8 rows and 6 columns. The A(5:8,:) command in Fig. 4.47, selects the elements in the 5th, 6th, 7th and 8th rows.

```
Command Window

>> A=rand(8,6)

A =

    0.9619    0.8001    0.5797    0.0760    0.9448    0.3897
    0.0046    0.4314    0.5499    0.2399    0.4909    0.2417
    0.7749    0.9106    0.1450    0.1233    0.4893    0.4039
    0.8173    0.1818    0.8530    0.1839    0.3377    0.0965
    0.8687    0.2638    0.6221    0.2400    0.9001    0.1320
    0.0844    0.1455    0.3510    0.4173    0.3692    0.9421
    0.3998    0.1361    0.5132    0.0497    0.1112    0.9561
    0.2599    0.8693    0.4018    0.9027    0.7803    0.5752

>> A(5:8,:)

ans =

    0.8687    0.2638    0.6221    0.2400    0.9001    0.1320
    0.0844    0.1455    0.3510    0.4173    0.3692    0.9421
    0.3998    0.1361    0.5132    0.0497    0.1112    0.9561
    0.2599    0.8693    0.4018    0.9027    0.7803    0.5752

fx >>
```

Fig. 4.47 Defining a random matrix and selecting rows 5, 6, 7, and 8

The command A(:,4:6) in Fig. 4.48 selects all the elements in the 4th, 5th, 6th columns.

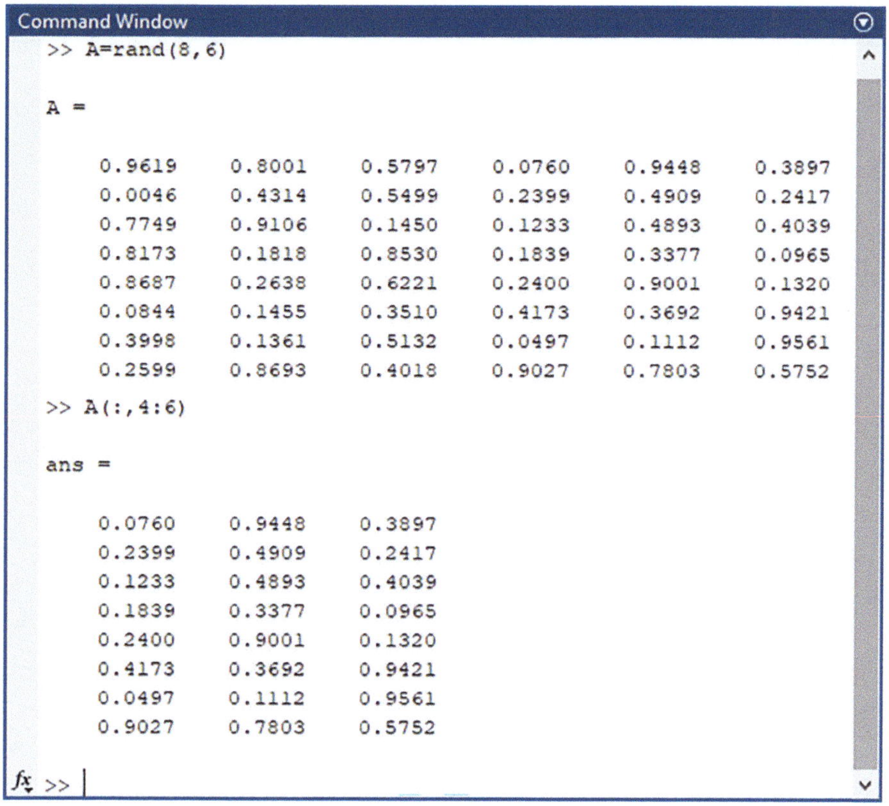

```
Command Window
>> A=rand(8,6)

A =

    0.9619    0.8001    0.5797    0.0760    0.9448    0.3897
    0.0046    0.4314    0.5499    0.2399    0.4909    0.2417
    0.7749    0.9106    0.1450    0.1233    0.4893    0.4039
    0.8173    0.1818    0.8530    0.1839    0.3377    0.0965
    0.8687    0.2638    0.6221    0.2400    0.9001    0.1320
    0.0844    0.1455    0.3510    0.4173    0.3692    0.9421
    0.3998    0.1361    0.5132    0.0497    0.1112    0.9561
    0.2599    0.8693    0.4018    0.9027    0.7803    0.5752

>> A(:,4:6)

ans =

    0.0760    0.9448    0.3897
    0.2399    0.4909    0.2417
    0.1233    0.4893    0.4039
    0.1839    0.3377    0.0965
    0.2400    0.9001    0.1320
    0.4173    0.3692    0.9421
    0.0497    0.1112    0.9561
    0.9027    0.7803    0.5752

fx >>
```

Fig. 4.48 Defining a random matrix and selecting columns 4, 5 and 6

The command A(2:5,3:4) in Fig. 4.49 selects the elements A_{ij} where $2 \leq i \leq 5$ and $3 \leq j \leq 4$.

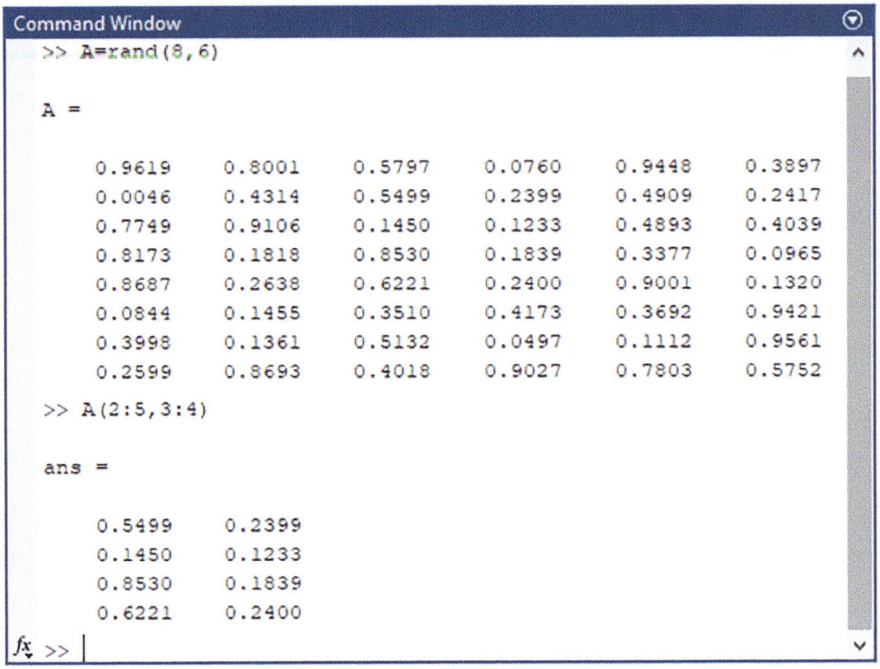

Fig. 4.49 Defining a random matrix and selecting the elements A_{ij} where $2 \le i \le 5$ and $3 \le j \le 4$

The colon operator is used in for loops as well. For instance, the code in Fig. 4.50 counts the number of elements that are greater than or equal to 0.5.

```
Command Window
>> A=rand(8,6)

A =

    0.9619    0.8001    0.5797    0.0760    0.9448    0.3897
    0.0046    0.4314    0.5499    0.2399    0.4909    0.2417
    0.7749    0.9106    0.1450    0.1233    0.4893    0.4039
    0.8173    0.1818    0.8530    0.1839    0.3377    0.0965
    0.8687    0.2638    0.6221    0.2400    0.9001    0.1320
    0.0844    0.1455    0.3510    0.4173    0.3692    0.9421
    0.3998    0.1361    0.5132    0.0497    0.1112    0.9561
    0.2599    0.8693    0.4018    0.9027    0.7803    0.5752

>> [m,n]=size(A);
>> s=0;
>> for i=1:m
for j=1:n
if (A(i,j)>=0.5)
s=s+1;
end
end
end
>> s

s =

    19

fx >>
```

Fig. 4.50 Counting the number of elements which are equal or bigger than 0.5

4.11 Linspace and Logspace Commands

The linspace(X1, X2, N) generates N linearly spaced points between X1 and X2. MATLAB assumes N = 100 when it is not entered. For instance, the command in Fig. 4.51 generates a linear space from 1 to 10 with 6 members. We expect the difference between two consecutive elements to be $\frac{10-1}{6-1} = 1.8$.

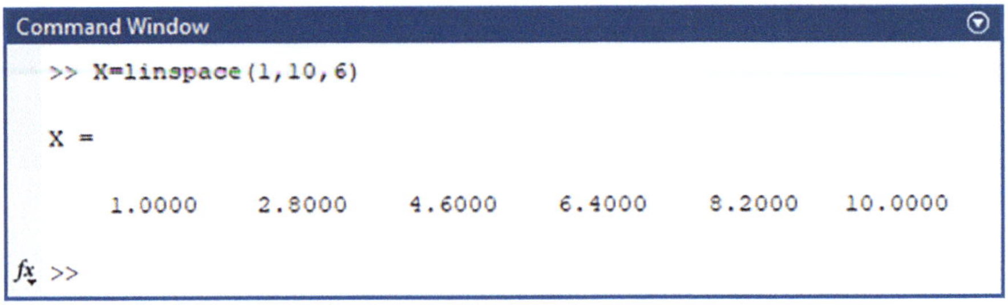

Fig. 4.51 Defining the variable X with linspace command

The difference between two consecutive elements can be calculated using the diff command (diff command can be used to calculate the derivative of a function as well). According to Fig. 4.52, the difference between two consecutive elements is 1.8.

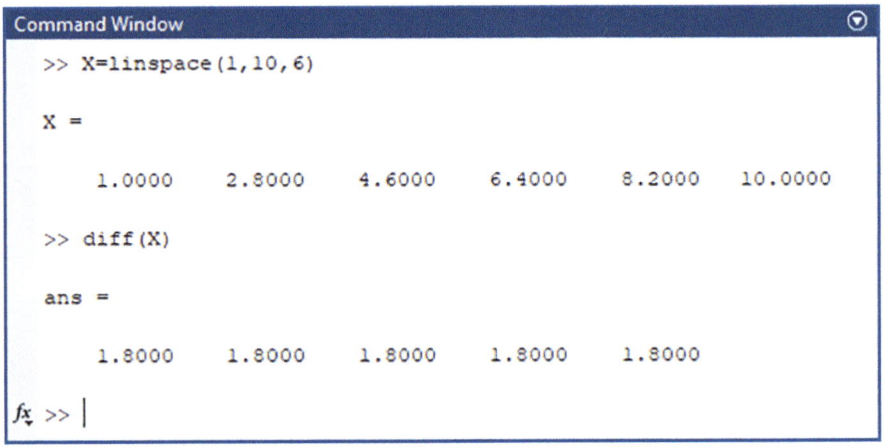

Fig. 4.52 Difference between the consecutive elements of vector X

logspace(X1, X2,N) generates a row vector of N logarithmically equally spaced points between 10^{X1} and 10^{X2}. MATLAB assumes N = 100 when it is not entered. For instance, the command in Fig. 4.53, makes a logarithmically equally spaced from 100 to 1000 with 5 members.

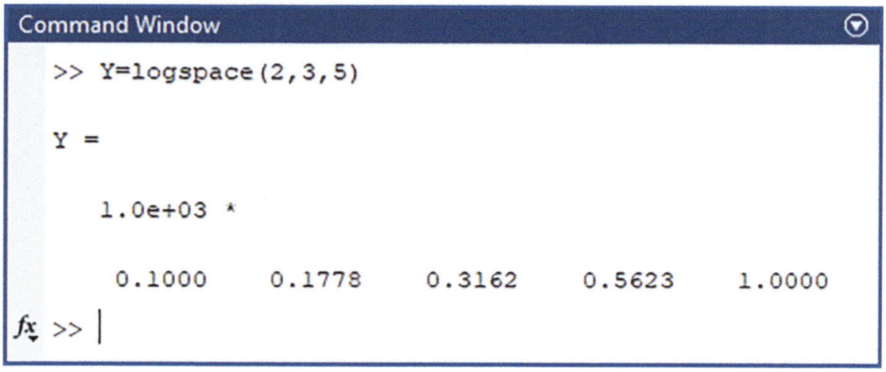

Fig. 4.53 Defining the variable Y with logspace command

The ratio between any two consecutive elements is constant (Fig. 4.54). This ratio can be calculated from the $10^2 \times x^{5-1} = 10^3$ equation. Figure 4.55 solves this equation. According to Fig. 4.55, the only positive answer of this equation is 1.7783. This value matches the one shown in Fig. 4.54.

```
Command Window

>> Y=logspace(2,3,5)

Y =

   1.0e+03 *

    0.1000    0.1778    0.3162    0.5623    1.0000

>> Y(2)/Y(1)

ans =

    1.7783

>> Y(3)/Y(2)

ans =

    1.7783

>> Y(4)/Y(3)

ans =

    1.7783

>> Y(5)/Y(4)

ans =

    1.7783

fx >>
```

Fig. 4.54 Ratio between the consecutive elements of vector Y

Fig. 4.55 Solution of $10^2 \times x^{5-1} = 10^3$

```
Command Window
>> syms x
>> solve(100*x^4==1000)

ans =

    10^(1/4)
   -10^(1/4)
-10^(1/4)*1i
 10^(1/4)*1i

>> eval(ans)

ans =

   1.7783 + 0.0000i
  -1.7783 + 0.0000i
   0.0000 - 1.7783i
   0.0000 + 1.7783i

fx >>
```

4.12 Ones, Zeros and Eye Commands

The ones(m,n) command makes an m × n matrix which all of its elements are one. The zeros (m,n) makes an m × n matrix which all of its elements are zero. The eye(n) command makes an n × n identity matrix. The eye(m,n) command makes an m × n matrix with 1's on the diagonal and 0 elsewhere (Fig. 4.56).

Fig. 4.56 Example for ones, zeros and eye commands

```
Command Window
>> ones(3,4)

ans =

     1     1     1     1
     1     1     1     1
     1     1     1     1

>> zeros(2,3)

ans =

     0     0     0
     0     0     0

>> eye(5)

ans =

     1     0     0     0     0
     0     1     0     0     0
     0     0     1     0     0
     0     0     0     1     0
     0     0     0     0     1

>> eye(2, 3)

ans =

     1     0     0
     0     1     0

fx >>
```

4.13 Format Command

The format command permits you to set the output display format for command window. The long format displays results with greater precision in the decimal places (Fig. 4.57).

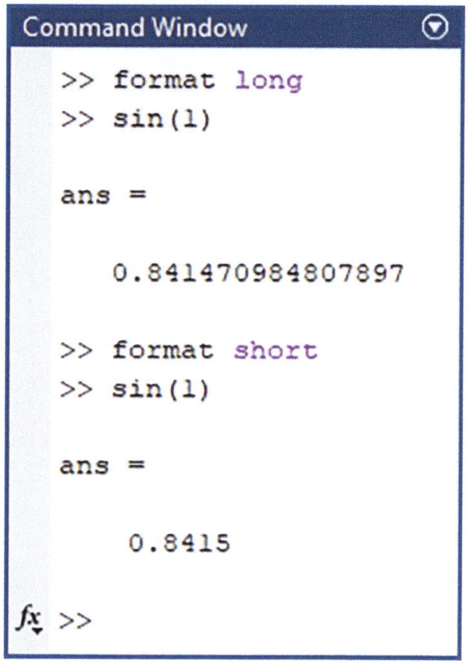

Fig. 4.57 The format command

4.14 Polynomial Functions

Assume that you want to solve the $x^2 - 4x + 3 = 0$. The roots command takes the coefficients and return the roots (Fig. 4.58).

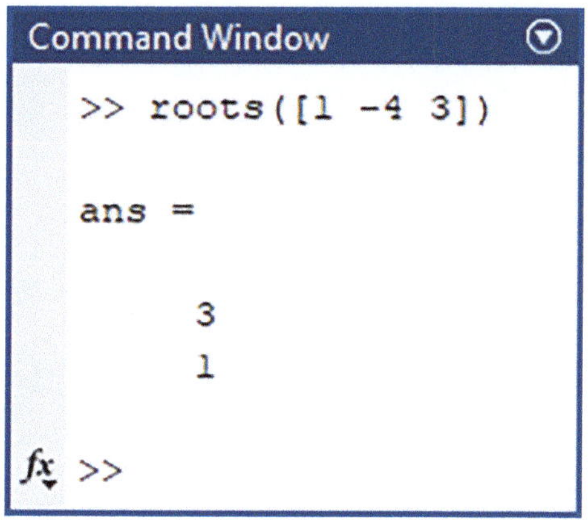

Fig. 4.58 The roots command calculates the roots of a polynomial

Assume that you want to find the polynomial whose its roots are $-1 + 2i$, $-1-2i$ and 6. The poly function gives the coefficients of the polynomial which has that roots. According to Fig. 4.59, the polynomial is $p(x) = x^3 - 4x^2 - 7x - 30$.

Fig. 4.59 Poly command takes the roots of a polynomial function and gives its coefficients

```
Command Window
>> poly([-1+2i -1-2i 6])

ans =

     1    -4    -7   -30

fx >>
```

Assume that we want to calculate the value of $p(x) = x^3 - 4x^2 - 7x - 30$ at $x = 5$. The polyval function can be used for this purpose (Fig. 4.60).

Fig. 4.60 Calculation of a polynomial function's value at a point with polyval command

```
Command Window
>> polyval([1 -4 -7 -30],5)

ans =

   -40

fx >>
```

Assume that you want to calculate the product of $p_1(x) = x + 3$ and $p_2(x) = x^2 + 5x + 6$. The commands shown in Fig. 4.61 calculates the product of these two polynomials. According to Fig. 4.61, the product is $p(x) = x^3 + 8x^2 + 21x + 18$.

Fig. 4.61 Calculation of the product of two polynomials with the conv command

```
Command Window
>> conv([1 3],[1 5 6])

ans =

     1     8    21    18

fx >>
```

4.15 Partial Fraction Expansion

You can use the residue command to calculate the partial fraction expansions. For instance, assume that we want to find the partial fraction expansion of $\frac{s+1}{s^2+5s+6}$. The commands shown in Fig. 4.62 calculate the residues at the poles.

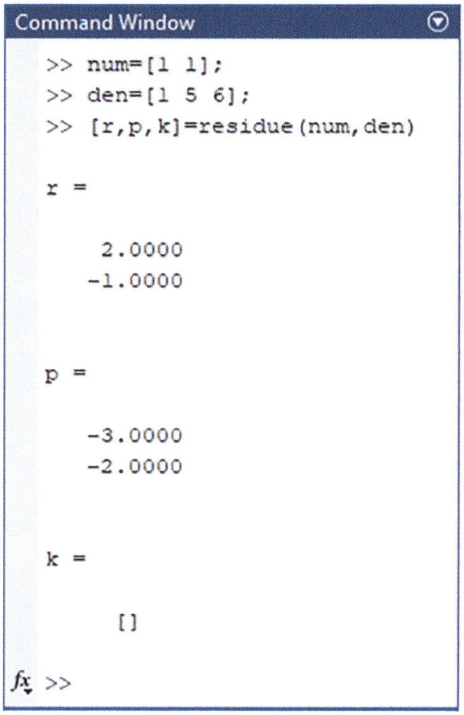

Fig. 4.62 Residue command output for the function $\frac{s+1}{s^2+5s+6}$

The poles of a rational function are the values of the variable that make the denominator equal to zero. Residues are the coefficients of the terms in a partial fraction expansion that correspond to the poles of the function.

According to Fig. 4.62, the partial fraction expansion of $\frac{s+1}{s^2+5s+6}$ is equal to $\frac{2}{s+3} + \frac{-1}{s+2}$. The residue at pole -3 is 2, and the residue at pole -2 is -1.

Let's consider another example: Finding the partial fraction expansion of the expression $\frac{1}{s^3(s-0.5)}$. By applying the partial fraction expansion technique, it is found that $\frac{1}{s^3(s-0.5)} = \frac{-2}{s^3} + \frac{-4}{s^2} + \frac{-8}{s} + \frac{8}{s-0.5}$. Figure 4.63 demonstrates that this result is correct.

```
Command Window
>> syms s
>> simplify(-2/s^3-4/s^2-8/s+8/(s-0.5))

ans =

2/(s^3*(2*s - 1))

fx >>
```

Fig. 4.63 Verification of $\frac{-2}{s^3}+\frac{-4}{s^2}+\frac{-8}{s}+\frac{8}{s-0.5}=\frac{1}{s^3(s-0.5)}$

The commands shown in Fig. 4.64 calculate the residues. The calculated residues agree with the hand calculations.

Fig. 4.64 Residue command output for the function $\frac{1}{s^3(s-0.5)}$

```
Command Window
>> num=[1];
>> den=[1 -.5 0 0 0];
>> [r,p,k]=residue(num,den)

r =

     8
    -8
    -4
    -2


p =

    0.5000
         0
         0
         0


k =

     []

fx >> |
```

In both of the above examples, the matrix **k** was a null matrix. The matrix **k** is always null when the degree of the numerator is less than the degree of the denominator. However, when the degree of the numerator is equal to or greater than the degree of the denominator, it will not be null.

For instance, consider the expression $\frac{s^3+3s^2+7s+4}{s^2+2s+1}$. The commands shown in Fig. 4.65 calculate the partial fraction of this function. Note that matrix k is not null in this case, since the degree of the numerator is greater than that of denominator. According to Fig. 4.65, $\frac{s^3+3s^2+7s+4}{s^2+2s+1} = s + 1 + \frac{4}{(s+1)} + \frac{-1}{(s+1)^2}$.

```
Command Window
>> [r,p,k]=residue([1 3 7 4],[1 2 1])

r =

     4
    -1


p =

    -1
    -1


k =

     1     1

fx >>
```

Fig. 4.65 Residue command output for the function $\frac{s^3+3s^2+7s+4}{s^2+2s+1}$

4.16 Calculation of Limit, Derivative and Integral

Table 4.6 shows the calculation of limits, derivatives, and integrals using MATLAB. the syms command is used to create the necessary symbolic variables for these operations.

Table 4.6 Calculating limits, derivatives, and integrals in MATLAB

Mathematical expression	MATLAB representation
$\lim\limits_{x\to 0}\frac{sin(x)}{x}=?$	`>> syms x` `>> limit(sin(x)/x,0)`
$\frac{d}{dx}\left(\frac{x^3-x}{x^2+8}\right)=?$	`>> syms x` `>> diff((x^3-x)/(x^2 + 8))`
$\int\frac{x^3-x}{x^2+8}dx=?$	`>> syms x` `>> int((x^3-x)/(x^2 + 8))`
$\int_0^1\frac{x^3-x}{x^2+8}dx=?$	`>> syms x` `>> int((x^3-x)/(x^2 + 8),x,0,1)`
$\int_0^1\frac{x^3-x}{x^2+8}dx=?$	`>> f = @(x)(x.^3-x)./(x.^2 + 8);` `>> integral(f,0,1)`

The commands in Fig. 4.66 calculate $\frac{d^2f(x)}{dx^2}$ for the function $f(x)=\sin(x)+\cos\left(x^2\right)$ and evaluate it at $x=\frac{\pi}{6}$.

```
Command Window

>> syms x
>> f=sin(x)+cos(x^2);
>> d2f_dx2=simplify(diff(f,x,2))

d2f_dx2 =

- 2*sin(x^2) - sin(x) - 4*x^2*cos(x^2)

>> eval(subs(d2f_dx2,x,pi/6))

ans =

   -2.0971

fx >>
```

Fig. 4.66 Calculation of $\frac{d^2}{dx^2}\left(\sin(x)+\cos\left(x^2\right)\right)$ and its evaluation at $x=\frac{\pi}{6}$

For calculating indefinite integrals in MATLAB, the primary tool is the int command from the symbolic math toolbox. This command is used to find the general antiderivative of a symbolic expression. To use it, one must first define the variable(s) as symbolic using the syms command. For example, Fig. 4.67 shows the code that calculates the integral $\int\sin(3x+7)dx$.

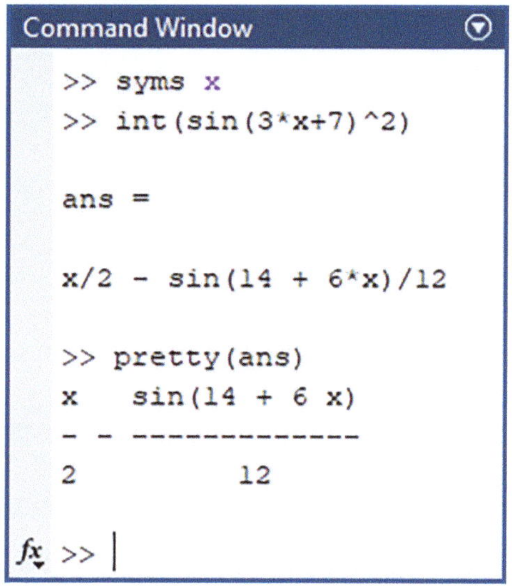

Fig. 4.67 Calculation of $\int \sin^2(3x+7)dx$ using the int command

The MATLAB int and integral commands both calculate definite integrals, but they differ fundamentally in their approach and the type of result they return.

The int command, part of the symbolic math toolbox, performs symbolic integration, meaning it first finds the exact mathematical antiderivative of the function and then evaluates it at the specified limits; this method yields the exact mathematical value of the definite integral.

In contrast, the integral command performs numerical integration, employing adaptive algorithms to approximate the area under the curve; this method returns a double-precision floating-point number that is a high-accuracy approximation of the definite integral.

Consequently, int is used when an exact, formulaic solution is required, while integral is used for faster computation or when integrating a function for which no simple antiderivative exists, or when the function is defined by data points rather than a symbolic expression.

The code in Figs. 4.68 or 4.69 calculates the definite integral $\int_0^1 \frac{x^3-x}{x^2+8}dx$. Both the int and integral commands successfully calculate the value of this integral.

Fig. 4.68 Calculation of $\int_0^1 \frac{x^3-x}{x^2+8}dx$ using the int command

```
Command Window
>> syms x
>> int((x^3-x)/(x^2+8),x,0,1)

ans =

log((8192*2^(1/2))/19683) + 1/2

>> eval(ans)

ans =

   -0.0300

fx >>
```

Fig. 4.69 Calculation of $\int_0^1 \frac{x^3-x}{x^2+8}dx$ using the integral command

```
Command Window
>> f=@(x) (x.^3-x)./(x.^2+8);
>> integral(f,0,1)

ans =

   -0.0300

fx >>
```

The code in Figs. 4.70 or 4.71 calculates the definite integral $\int_0^1 e^{-x^2}dx$. Both the int and integral commands successfully calculate the value of this integral.

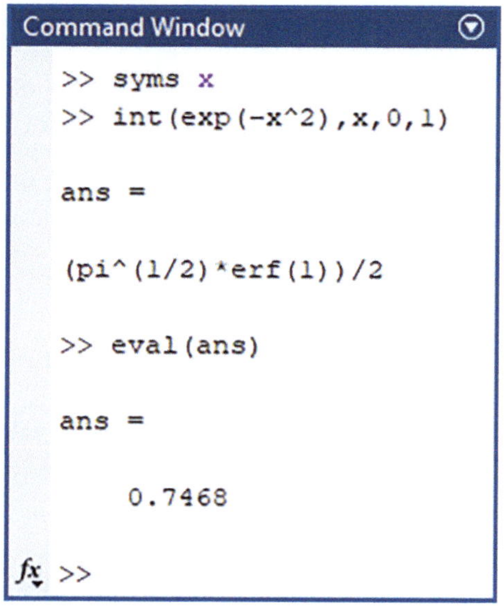

Fig. 4.70 Calculation of $\int_0^1 e^{-x^2}dx$ using the int command

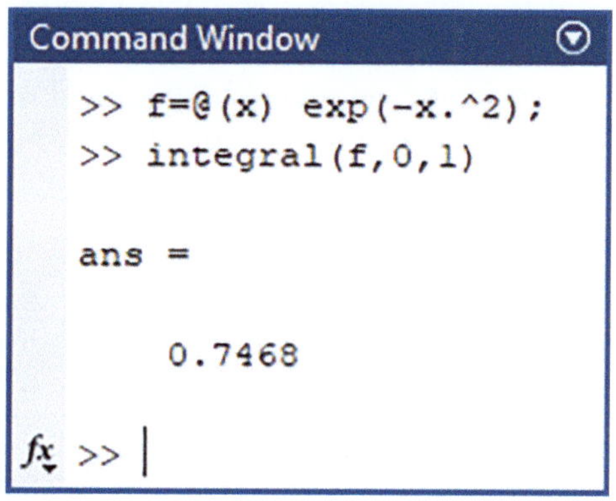

Fig. 4.71 Calculation of $\int_0^1 e^{-x^2}dx$ using the integral command

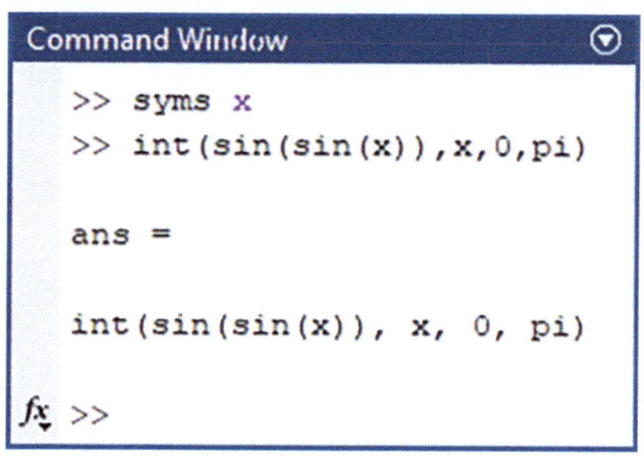

Fig. 4.72 The int command cannot calculate $\int_0^{\pi} \sin(\sin(x))dx$

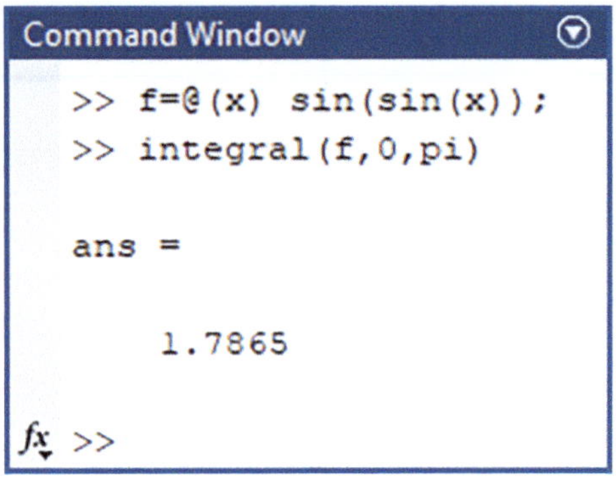

Fig. 4.73 The integral command successfully calculates $\int_0^{\pi} \sin(\sin(x))dx$

The code in Fig. 4.72 used the int command and failed to calculate the integral $\int_0^{\pi} \sin(\sin(x))dx$, while the code in Fig. 4.73 used the integral command to successfully calculate its value.

The trapz function is specifically designed to perform numerical integration on a set of discrete numeric data points, where the mathematical function that generated the data is unknown or unavailable. Let's examine an example using the data provided in Table 4.7.

To compute the numerical integral of the data presented in Table 4.7, you would use the commands illustrated in Fig. 4.74.

Table 4.7 Sample data points

x	y
0	0.0000
0.1	0.0998
0.2	0.1987
0.3	0.2955
0.4	0.3894
0.5	0.4794

```
Command Window
>> x=[0:0.1:0.5];
>> y=[0 0.0998 0.1987 0.2955 0.3894 0.4794];
>> trapz(x,y)

ans =

    0.1223

fx >>
```

Fig. 4.74 Integrating Table 4.7's data using the quad command

MATLAB results can sometimes be difficult to read when displayed on a single line. In these instances, you can use the pretty function to display the result in a more readable format, as shown in Fig. 4.75.

Fig. 4.75 The pretty command improves the output's readability

```
Command Window
>> syms x
>> int((x^3-x)/(x^2+8))

ans =

x^2/2 - (9*log(x^2 + 8))/2

>> pretty(ans)
 2           2
x     9 log(x   + 8)
-- - -------------
 2          2

fx >> |
```

The simplify command, can be used to simplify the result of symbolic calculation (Fig. 4.76).

Fig. 4.76 Simplification of $\frac{1}{x+1} + \frac{1}{x-1}$

```
Command Window

>> syms x
>> simplify(1/(x+1)+1/(x-1))

ans =

(2*x)/(x^2 - 1)

>> pretty(ans)
  2 x
------
 2
x  - 1

fx >>
```

4.17 Plotting the Graph of Data

Plotting a function's graph in MATLAB is easily achieved using the plot command. The commands illustrated in Fig. 4.77 generate a plot of the function $y = \sin(x)$ over the interval $0 \le x \le 2\pi$, with the resulting graphic output shown in Fig. 4.78.

```
Command Window

>> x=[0:0.01:2*pi];
>> y=sin(x);
>> plot(x,y)
fx >>
```

Fig. 4.77 Plotting the function $\sin(x)$ over the interval $[0, 2\pi]$

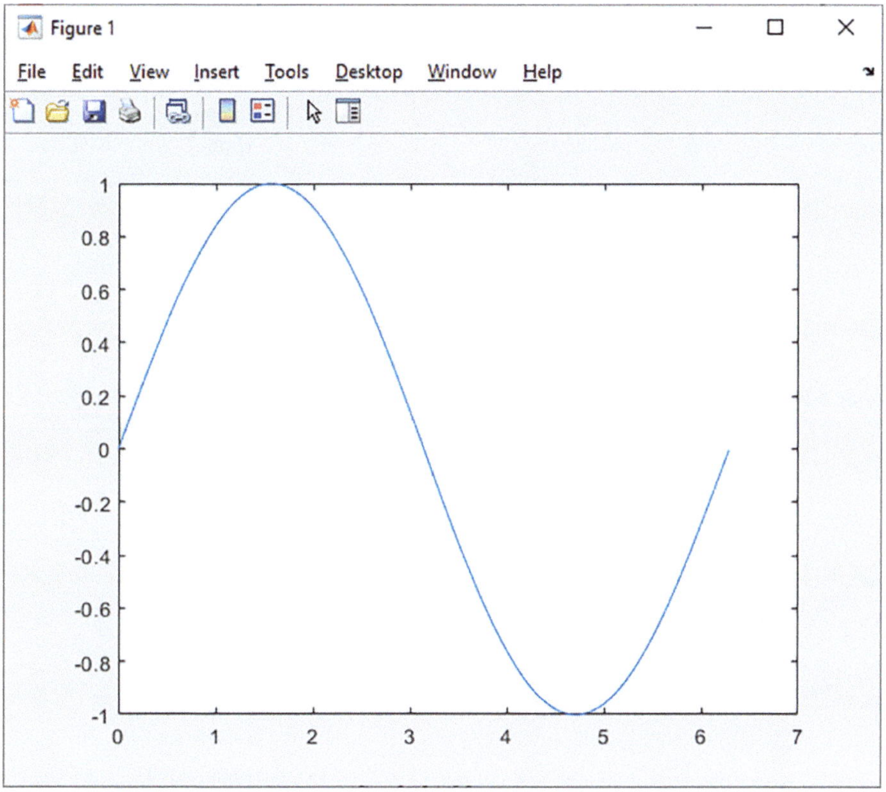

Fig. 4.78 The output generated by the code in Fig. 4.77

To create more customized plots, you can utilize the operators detailed in Tables 4.8, 4.9, and 4.10.

Let's study an example. Table 4.11 presents voltage and current values for a resistor.

We aim to plot this data, using circles to mark the data points and connecting them with a black dashed line. The plot must be appropriately labeled, using "Voltage (V)" for the horizontal axis, "Current (A)" for the vertical axis, and "I-V for a resistor" as the title. The necessary MATLAB commands are shown in Fig. 4.79, and the resulting graph is displayed in Fig. 4.80.

Table 4.8 Line types

MATLAB command	Line type
-	Solid
:	Dotted
-..	Dashdot
--	Dashed

Table 4.9 Colors

MATLAB command	Color
R	Red
g	Green
b	Blue
c	Cyan
m	Magenta
y	Yellow
k	Black
w	White

Table 4.10 Plot symbols

MATLAB command	Plot symbol
.	Point
+	Plus
*	Star
O	Circle
x	x-mark
s	Square
d	Diamond
v	Triangle (down)
^	Triangle (up)
<	Triangle (left)
>	Triangle (right)

Table 4.11 Voltage current measurements for resistor R1

V (Volt)	I (Amper)
0.499	0.10
0.985	0.20
1.508	0.31
1.969	0.41
2.528	0.53
2.935	0.61
3.481	0.73
3.971	0.83
4.486	0.94
4.960	1.04
5.502	1.15
6.007	1.26
6.60	1.38

```
Command Window
>> V=[0.499 0.985 1.508 1.969 2.528 2.935 3.481 3.971 4.486 4.960 5.502 6.007 6.60];
>> I=[0.1 0.2 0.31 0.41 0.53 0.61 0.73 0.83 0.94 1.04 1.15 1.26 1.38];
>> plot(V,I,'--ko')
>> title('I-V for a resistor')
>> xlabel('Voltage(V)')
>> ylabel('Current(A)')
>> grid on
fx >>
```

Fig. 4.79 The commands for generating the plot of Table 4.11

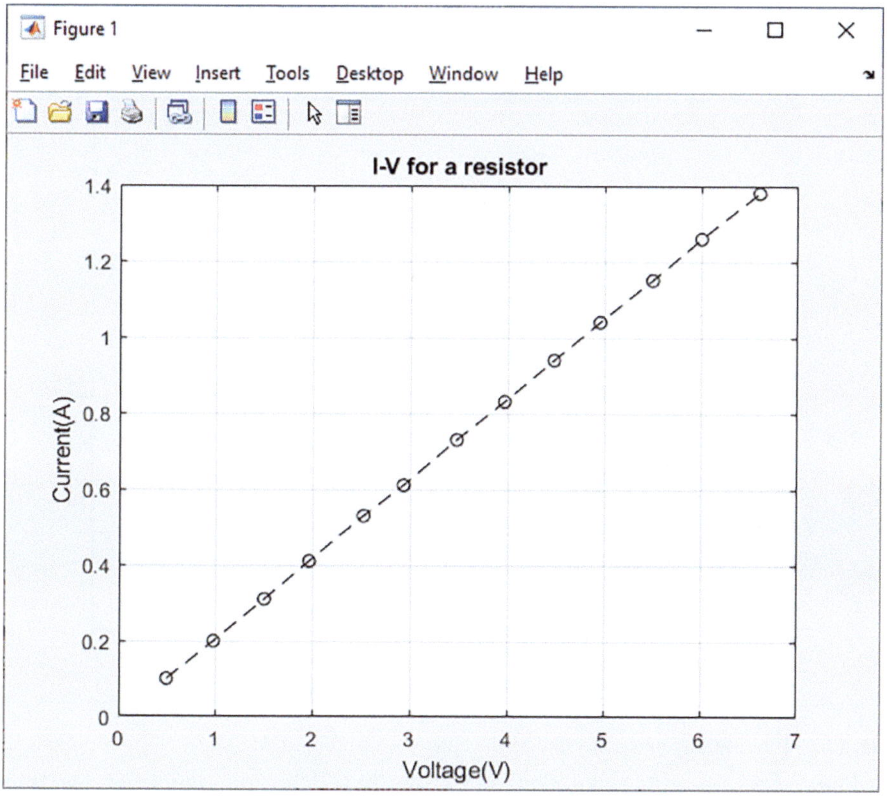

Fig. 4.80 The resulting plot generated by the code shown in Fig. 4.79

Figure 4.81 shows the commands to plot the data contained in Table 4.11. Unlike the previous plot, this version uses red stars for data points and connects them with a solid black line (the output is shown in Fig. 4.82).

```
Command Window
>> V=[0.499 0.985 1.508 1.969 2.528 2.935 3.481 3.971 4.486 4.960 5.502 6.007 6.60];
>> I=[0.1 0.2 0.31 0.41 0.53 0.61 0.73 0.83 0.94 1.04 1.15 1.26 1.38];
>> plot(V,I,'k',V,I,'r*'),grid minor
fx >>
```

Fig. 4.81 The commands for generating the plot of Table 4.11

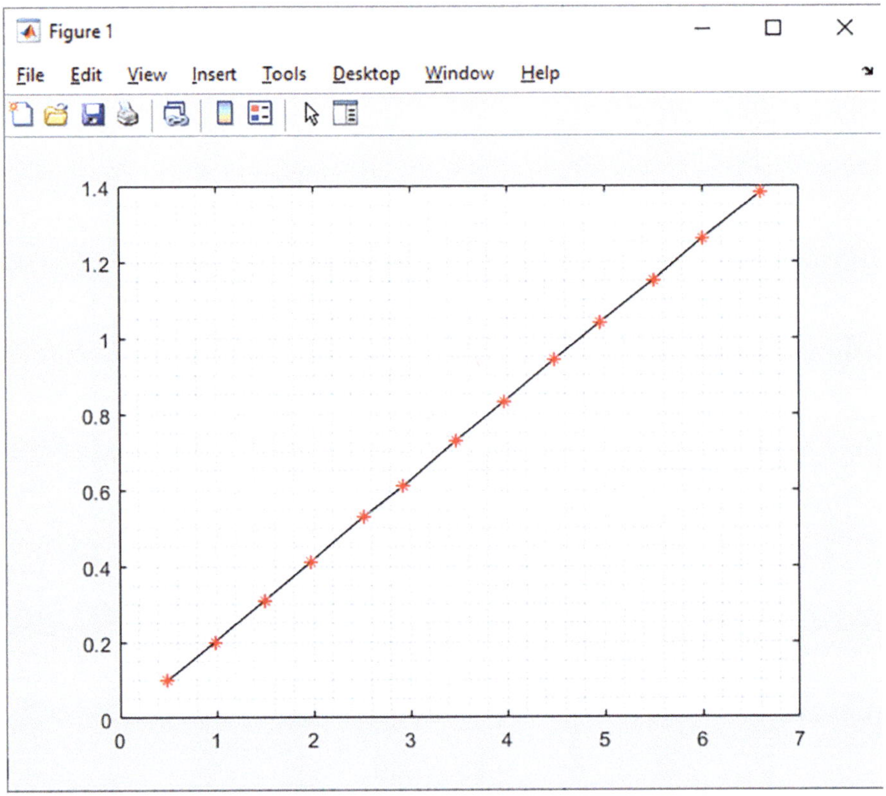

Fig. 4.82 The resulting plot generated by the code shown in Fig. 4.81

Figure 4.82 currently lacks labels and a title. To add the necessary labels and title, you can use the Insert menu (Fig. 4.83).

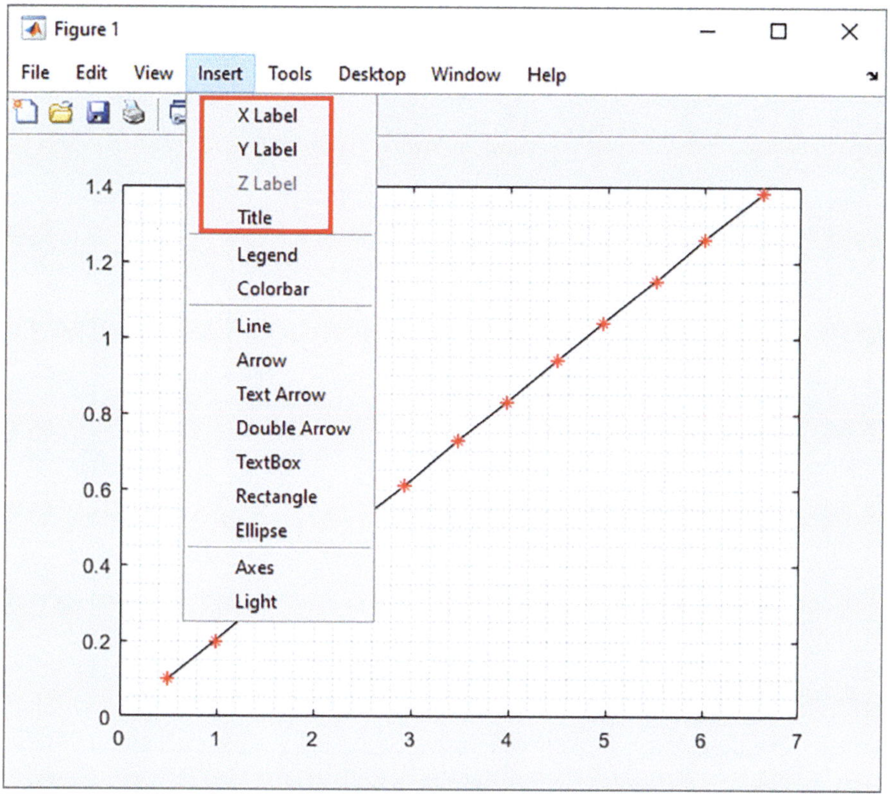

Fig. 4.83 The insert menu can be used for adding a title and axis labels

To transfer the resulting graph to the clipboard, select the Edit > Copy Figure command (Fig. 4.84). Once on the clipboard, the image can be pasted into other applications, such as MS Word®, by using the Ctrl + V keyboard shortcut.

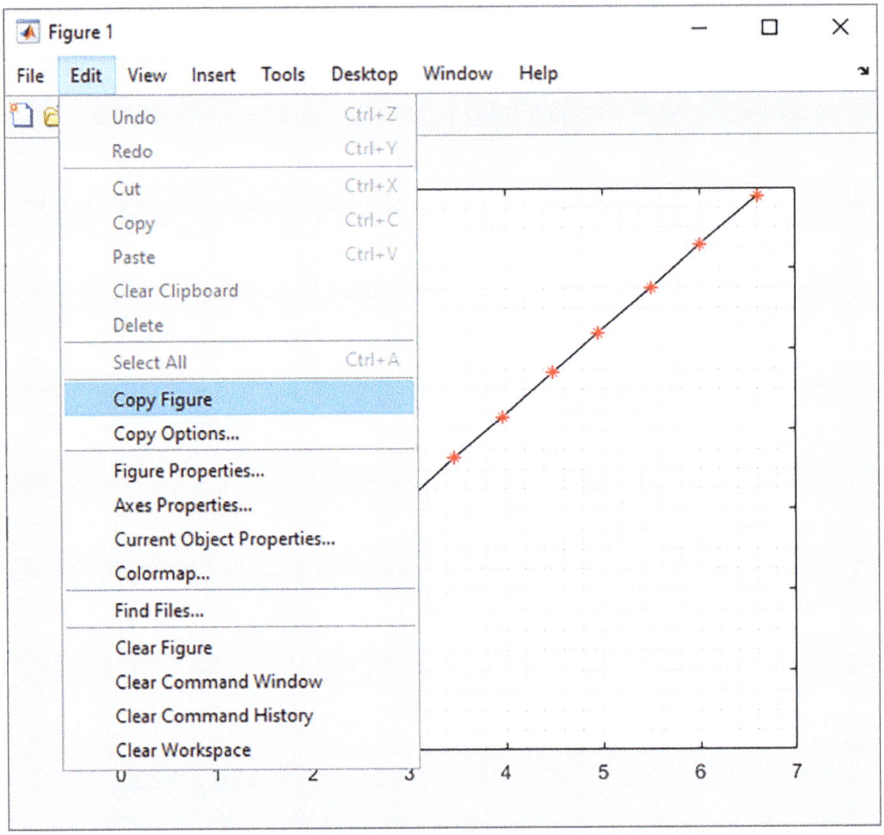

Fig. 4.84 Copying the figure to the clipboard

You can also save the generated graph as a graphical file. To do this, use File > Save as (Fig. 4.85). When the Save As window appears, select the desired output file format from the Save as type drop-down list (Fig. 4.86).

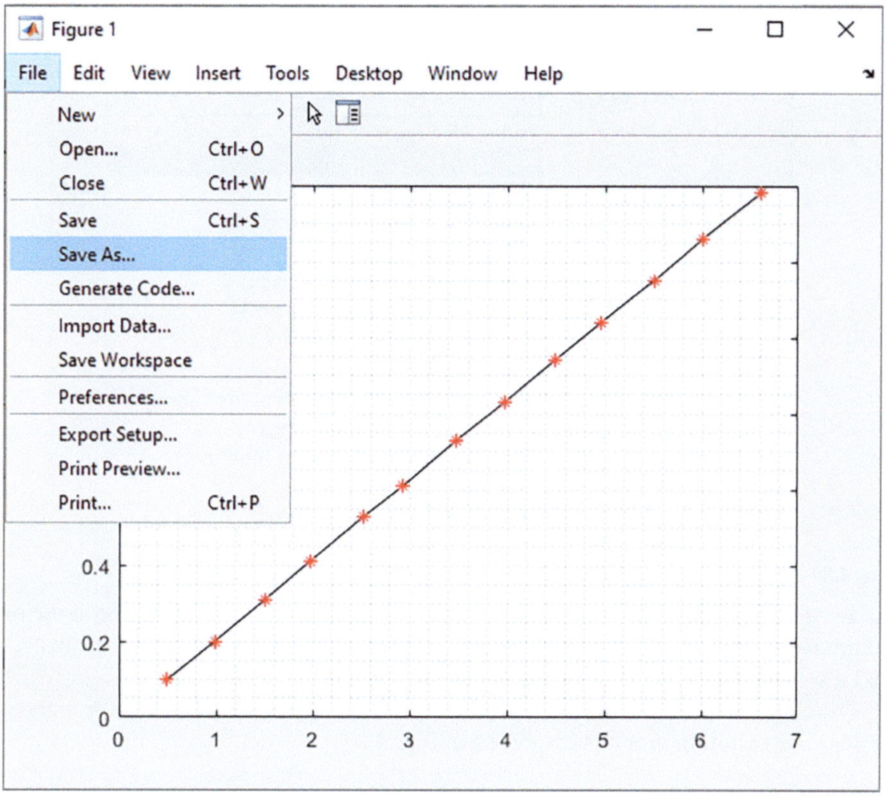

Fig. 4.85 Exporting the figure to a graphical file format

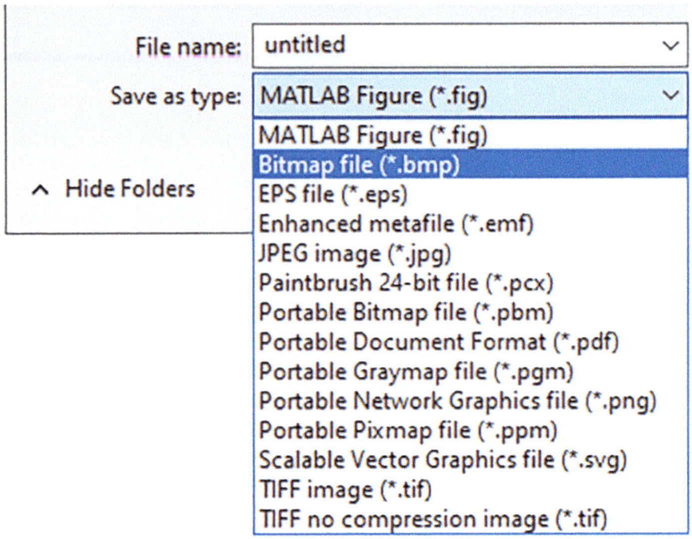

Fig. 4.86 Choice of the desired file type

To show multiple datasets concurrently on a single set of axes, the hold on command is required. Consider an exercise where we plot the data from Table 4.11 and a second dataset in Table 4.12 on the same figure.

The commands shown in Fig. 4.87 generate a combined plot of the data from both tables. The resulting graph is displayed in Fig. 4.88.

Table 4.12 Voltage current measurements for resistor R2

V (Volt)	I (Amper)
0.579	0.10
0.978	0.17
1.598	0.28
1.976	0.34
2.496	0.43
2.953	0.51
3.458	0.60
4.068	0.71
4.450	0.78
4.917	0.86
5.35	0.93
5.75	1.01
6.37	1.11
6.60	1.15

```
Command Window
>> V1=[0.499 0.985 1.508 1.969 2.528 2.935 3.481 3.971 4.486 4.960 5.502 6.007 6.60];
>> I1=[0.1 0.2 0.31 0.41 0.53 0.61 0.73 0.83 0.94 1.04 1.15 1.26 1.38];
>> V2=[0.579 0.978 1.598 1.976 2.496 2.953 3.458 4.068 4.450 4.917 5.35 5.75 6.37 6.60];
>> I2=[0.1 0.17 0.28 0.34 0.43 0.51 0.60 0.71 0.78 0.86 0.93 1.01 1.11 1.15];
>> plot(V1,I1,'b',V1,I1,'r*')
>> hold on
>> plot(V2,I2,'k',V2,I2,'r+')
>> grid minor
>> xlabel('Voltage (V)')
>> ylabel('Current (A)')
>> title('Comparison of I-V plot of R1 and R2')
fx >>
```

Fig. 4.87 Plotting data from Tables 4.11 and 4.12 on a single graph

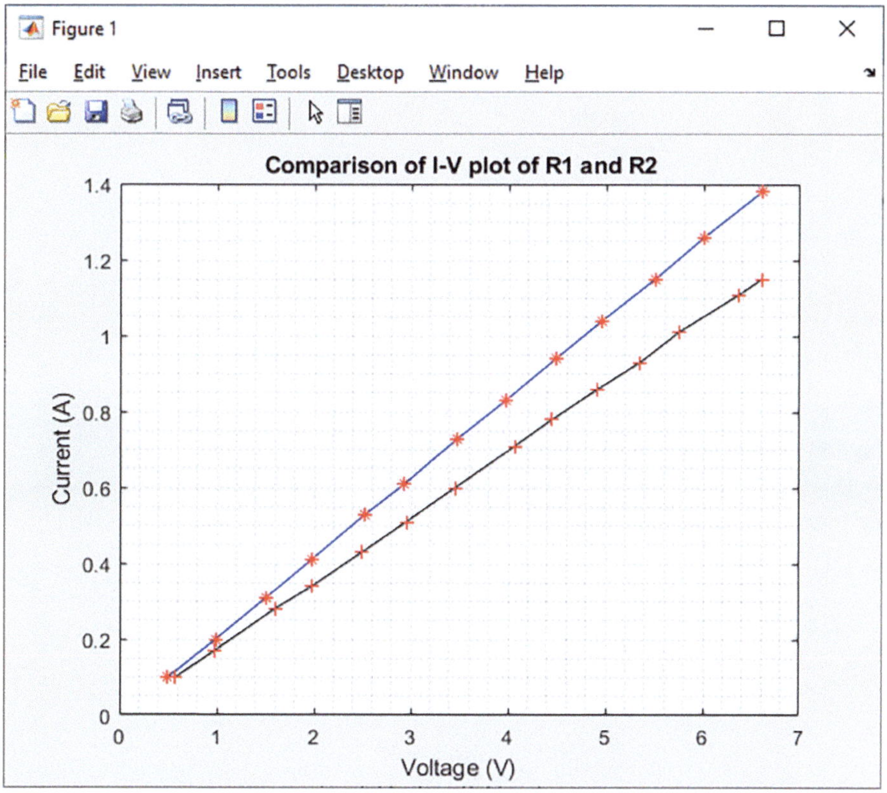

Fig. 4.88 The combined plot generated by the code shown in Fig. 4.87

You can add a legend to the graph by selecting Insert > Legend (Fig. 4.89).

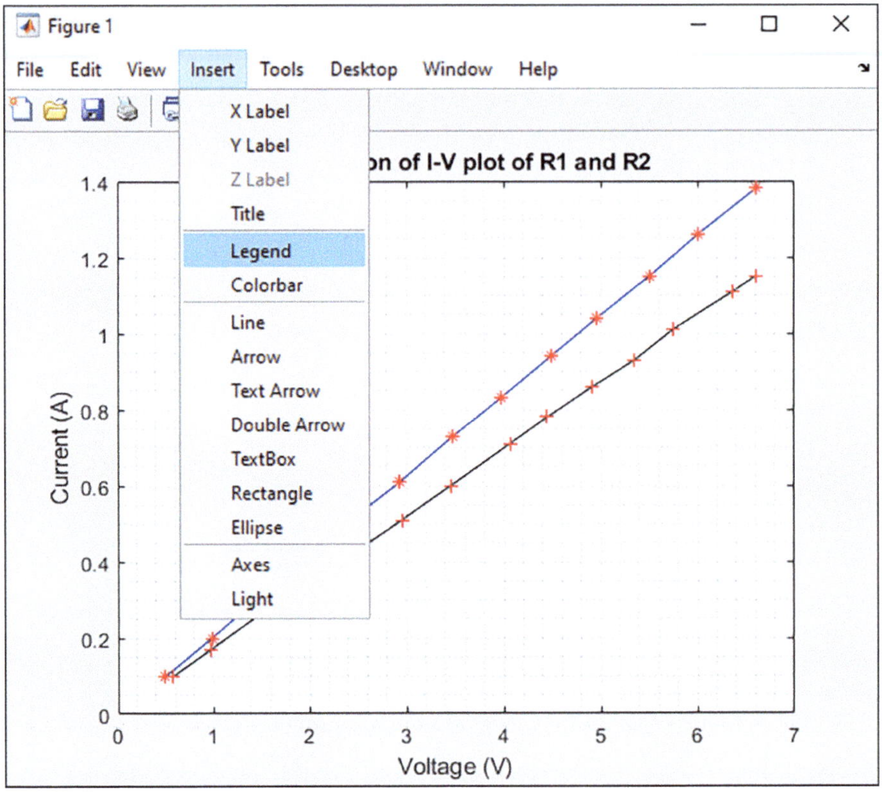

Fig. 4.89 Adding a legend to the graph

The legend shown in Fig. 4.90 is added to the graph immediately after executing the Insert > Legend command.

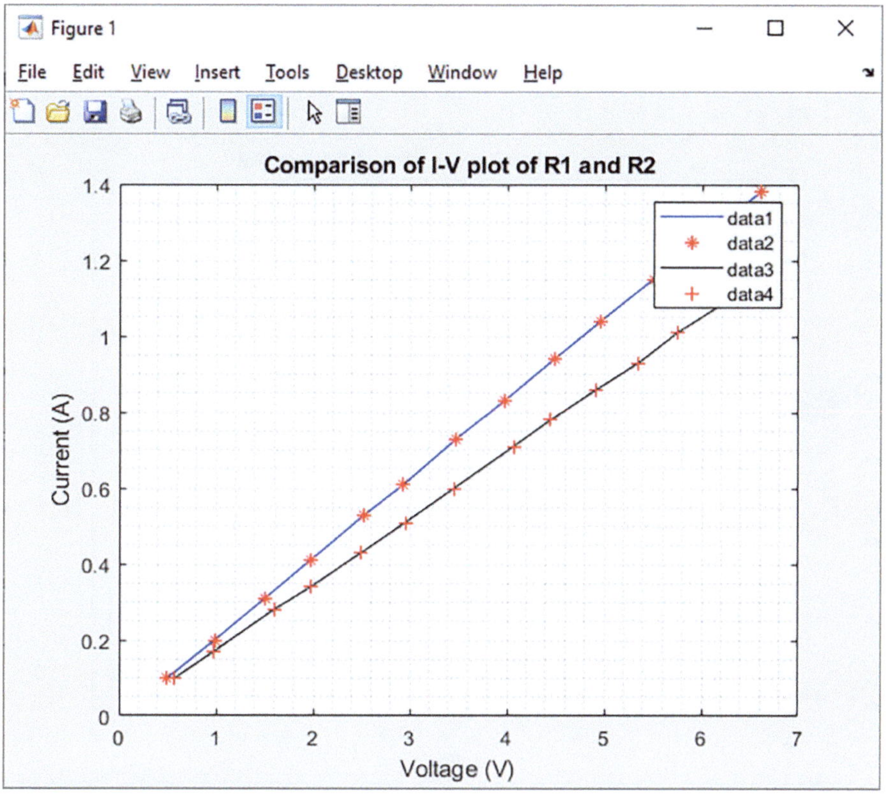

Fig. 4.90 The legend added to the graph

To customize the legend labels shown in Fig. 4.90, double-click the default text (e.g., "data1") and enter the appropriate description. Apply this editing process to data 2, data3, and data4. the legend's position can be adjusted manually by dragging it (Fig. 4.91), or you may right-click the box to choose from a list of predefined placement options (Fig. 4.92).

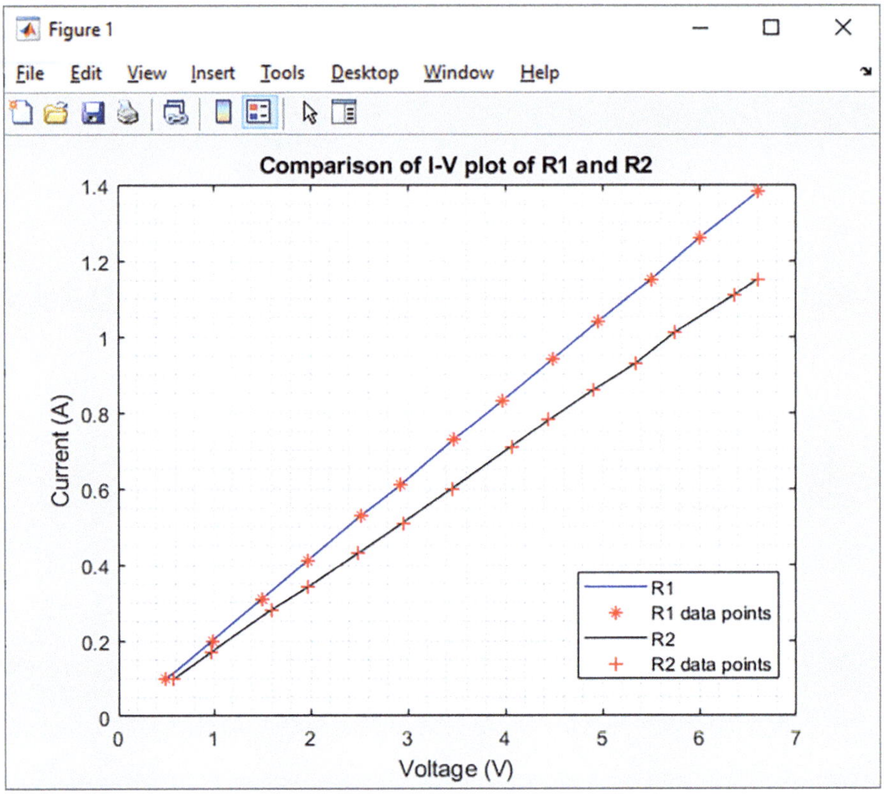

Fig. 4.91 Customizing the plot legend

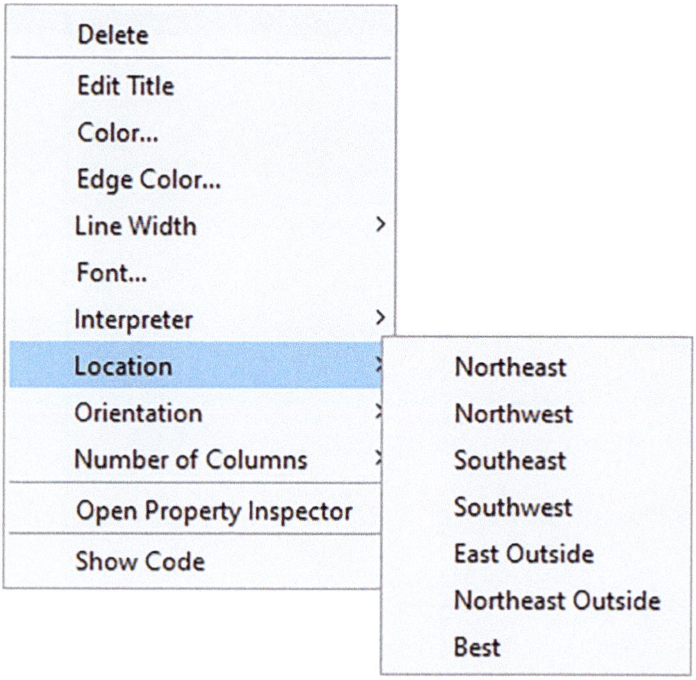

Fig. 4.92 Predefined locations for legend insertion

4.18 Accessing Help in MATLAB

To retrieve more information about the commands covered in this chapter, use the help command. For instance, entering help roots will display the documentation for the roots command (Fig. 4.93).

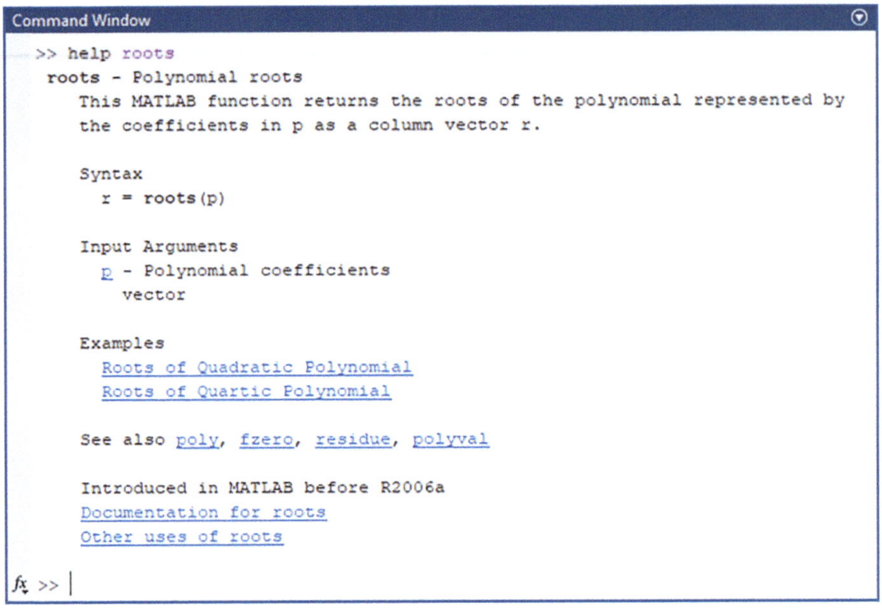

Fig. 4.93 Documentation for the roots command

Use the doc command (Fig. 4.94) to access the MATLAB Help Center.

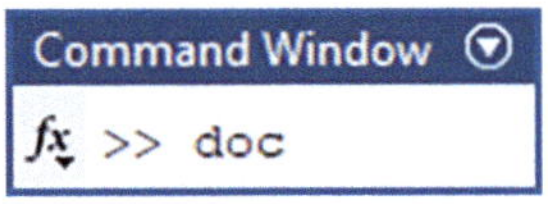

Fig. 4.94 Doc command

Let's examine an alternative method for seeking assistance. On the left side of the command window, there is a small f_x icon (Fig. 4.95). Clicking this icon (or pressing Shift + F1) opens a drop-down menu (Fig. 4.96). This menu lists your installed MATLAB products. You can then click any product name to view a complete list of its related functions (Fig. 4.97). Clicking a function will display its specific help documentation (Fig. 4.98).

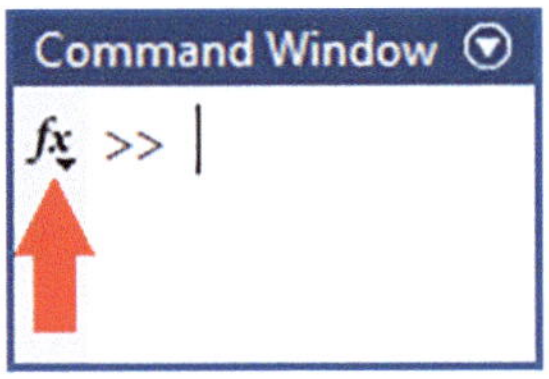

Fig. 4.95 The f_x button

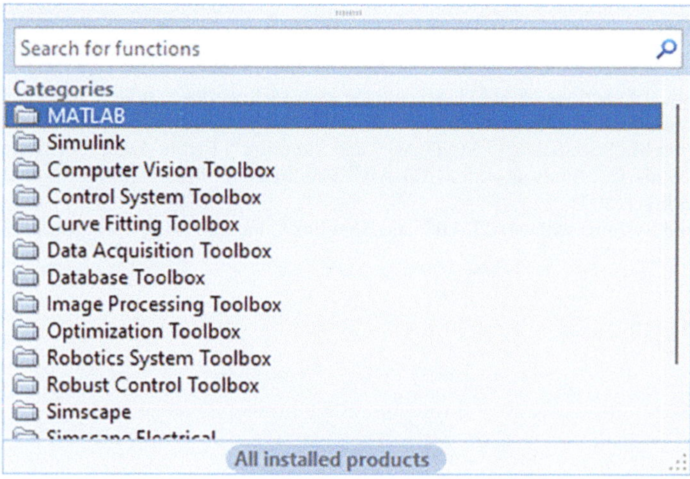

Fig. 4.96 The window displayed after clicking the f_x button

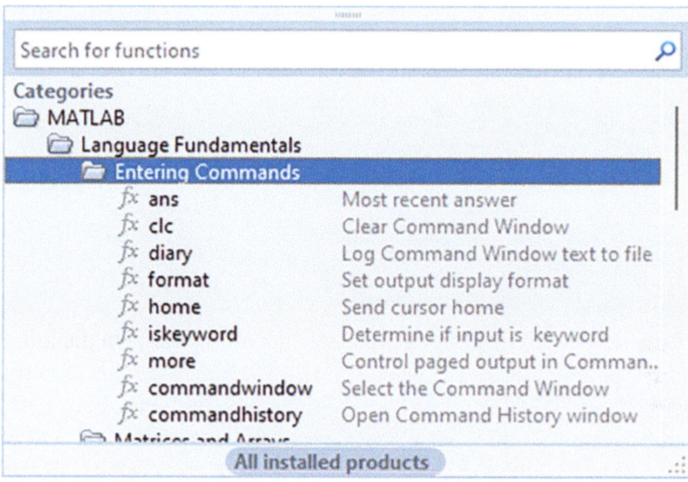

Fig. 4.97 List of commands under the MATLAB Language Fundamentals category

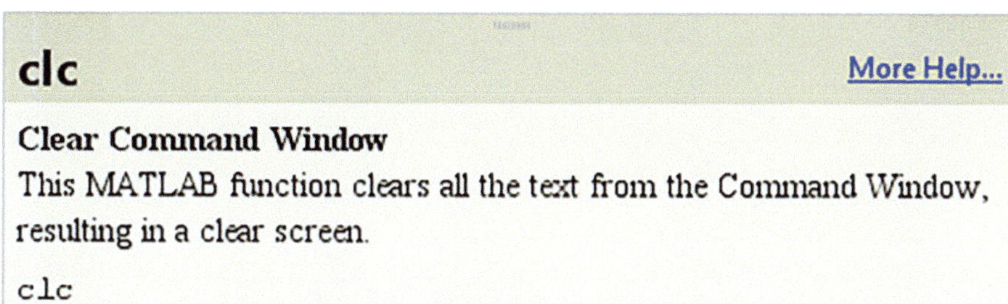

Fig. 4.98 Documentation for the clc command

References

1. Mathematical Functions of MATLAB: https://www.mathworks.com/help/symbolic/mathematical-functions.html
2. Engineering Mathematics with MATLAB® and Simulink®, Farzin Asadi, Springer, 2025.
3. Applied Numerical Analysis with MATLAB®/Simulink®: For Engineers and Scientists, Farzin Asadi, Springer, 2023.
4. Signals and Systems with MATLAB® and Simulink®, Farzin Asadi, Springer, 2023.

Chapter 5
Calculator Skills and Formula Writing in Microsoft® Word®

5.1 Introduction

Calculators have become an essential tool for engineers, significantly enhancing their efficiency and accuracy in a wide range of tasks. Calculators play an important role in university exams as well. So, it is important to learn how to work with your calculator correctly. By mastering your calculator, you can significantly improve your problem-solving skills and achieve better results in your engineering studies.

A common mistake in calculator usage is setting the wrong angle mode. If your calculator is in radian mode but you input an angle in degrees, the result will be incorrect. Always double-check your calculator's mode setting to ensure it's in degrees or radians, as needed, before performing calculations.

First part of this chapter offers reference operations to aid in your calculator usage. Endeavor to replicate the provided correct results. Depending on your calculator's model, some operations may not be possible. Second part of this chapter shows how to write a formula in Microsoft® Word®.

5.1.1 Basic Operations

The following are some basic operations and their values.

1. $12345679 \times 8 = 98765432$
2. $54.63 \times 632.58 = 34557.8454$
3. $12.653 \times 98.74 = 1249.3572$
4. $\frac{542.68}{6.24} = 86.9679$
5. $\frac{364.56}{5.25} = 69.44$
6. $(1 + 2j) + (3 - 8j) = 4 - 6j$
7. $(1 + 2j) - (3 - 8j) = -2 + 10j$
8. $(1 + 2j) \times (3 - 8j) = 19 - 2j$

F. Asadi, *A Problem-Solving Approach to Multi-Variable Calculus*,
https://doi.org/10.1007/978-3-032-21115-6_5

9. $\frac{(1+2j)}{3-8j} = -0.1781 + 0.1918j$

5.2 Exponents and Radicals

The following are some example calculations for exponents and radicals.

1. $14^{2.5} = 733.3648$
2. $3.2^{4.3} = 148.6432$
3. $12.96^{2.24} = 310.6241$
4. $\sqrt{17.63} = 4.1988$
5. $\sqrt{965.22} = 31.0680$
6. $\sqrt[3]{752.53} = 9.0958$
7. $\sqrt[5]{12036} = 6.5478$

5.3 Logarithm

The following are some example calculations for logarithms.

1. $\log_{10}56.3 = 1.7505$
2. $\log_{10} 1200.698 = 3.0794$
3. $\ln(56.3) = 4.0307$
4. $\ln(1200.698) = 7.0907$
5. $\ln(0.33) = -1.1087$

5.4 Conversion of Complex Numbers to Polar Form

The following are some examples of converting complex numbers between Cartesian and polar forms.

1. $1 + 6j = 6.0828e^{1.4056j}$
2. $-3 + 5j = 5.8310e^{j2.1112}$
3. $-16 - 3j = 16.2788e^{-2.9562j}$
4. $10 - 9j = 13.4536e^{-j0.7328}$
5. $12.56e^{-j1.8} = -2.8537 - 12.2315j$
6. $7.23e^{j.8} = 5.0372 + 5.1865j$

5.5 Roots of Polynomial and Non-Linear Equations

The following are some examples of finding the roots of polynomials and non-linear equations.

1. $x^2 - 3x + 2 = 0 \Rightarrow x_1 = 1, x_2 = 2$
2. $x^3 - 11.77x^2 + 41.985x - 47.025 = 0 \Rightarrow x_1 = 2.5, x_2 = 3, x_3 = 6.27$
3. $x^2 + 2x + 10 = 0 \Rightarrow x_1 = -1 + 3j, x_2 = -1 - 3j$
4. $x^3 + 2x^2 - 51x - 232 = 0 \Rightarrow x_1 = -5 - 2j, x_2 = -5 + 2j, x_3 = 8$
5. $x^2 - 2\cos(x) = 0 \Rightarrow x_1 = -1.02162, x_2 = 1.02162$
6. $x + e^{-x} - 2 \Rightarrow x_1 = -1.146189, x_2 = +1.841402$

5.6 Derivative and Integrals

The following are some examples of evaluating integrals and derivatives.

1. $\int_0^3 \sin(x)dx = 1.9900$
2. $\int_0^1 x^2 dx = 0.3333$
3. $\int_0^2 x^2 sin(2x)dx = 0.1371$
4. $\int_0^{0.5} e^{-6x}dx = 0.1584$
5. $\int_0^2 \frac{3}{x+1}dx = 3.2958$
6. $\frac{d}{dx}(\cos(x))\Big|_{\frac{\pi}{3}} = -0.8660$
7. $\frac{d}{dx}\left(1 + tan^2(x)\right)\Big|_{0.25} = 0.5440$
8. $\frac{d}{dx}\left(\frac{1+x}{\sin(x)}\right)\Big|_{0.5} = -3.6413$
9. $\frac{d}{dx}\left(\frac{x^2+9}{(x^3+6x-4)}\right)\Big|_{1.5} = -1.6868$

5.7 Trigonometric Ratios

Table 5.1 presents trigonometric ratios for several angles.

5.8 Inverse Trigonometric Ratios

Table 5.2 shows the inverse trigonometric functions for a sample of trigonometric ratios.

Table 5.1 Trigonometric ratios for some sample angles

Radian	Degrees
$\sin(0.2563) = 0.2535$	$\sin(10°) = 0.1736$
$\sin(0.5236) = 0.5000$	$\sin(30°) = 0.5000$
$\sin(0.7238) = 0.6622$	$\sin(50°) = 0.7660$
$\sin(1.5) = 0.9975$	$\sin(70°) = 0.9397$
$\sin(2) = 0.9093$	$\sin(123°) = 0.8387$
$\cos(0.2563) = 0.9673$	$\cos(10°) = 0.9848$
$\cos(0.5236) = 0.8660$	$\cos(30°) = 0.866$
$\cos(0.7238) = 0.7493$	$\cos(50°) = 0.6428$
$\cos(1.5) = 0.0707$	$\cos(70°) = 0.3420$
$\cos(2) = -0.4161$	$\cos(123°) = -0.5446$
$\tan(0.2563) = 0.2621$	$\tan(10°) = 0.1763$
$\tan(0.5236) = 0.5774$	$\tan(30°) = 0.5774$
$\tan(0.7238) = 0.8838$	$\tan(50°) = 1.1918$
$\tan(1.5) = 14.1014$	$\tan(70°) = 2.7475$
$\tan(2) = -2.1850$	$\tan(123°) = -1.5399$

Table 5.2 Inverse trigonometric functions for a sample of trigonometric ratios

$\sin^{-1}(0.310) = 0.3152 Rad = 18.0596°$	$\cos^{-1}(0.52) = 1.0239 Rad = 58.6677°$
$\sin^{-1}(0.652) = 0.7102 Rad = 40.6915°$	$\cos^{-1}(0.73) = 0.7525 Rad = 43.1136°$
$\sin^{-1}(0.711) = 0.7909 Rad = 45.3163°$	$\cos^{-1}(0.95) = 0.3176 Rad = 18.1949°$
$\tan^{-1}(1.45) = 0.967 Rad = 55.4077°$	$\cot^{-1}(0.65) = 0.9944 Rad = 56.9761°$
$\tan^{-1}(3.35) = 1.2807 Rad = 73.3792°$	$\cot^{-1}(7.35) = 0.1352 Rad = 7.7477°$
$\tan^{-1}(8.62) = 1.4553 Rad = 83.3827°$	$\cot^{-1}(10.87) = 0.0917 Rad = 5.2562°$

5.9 Microsoft Word Equation Editor

Click the Insert button (Fig. 5.1) to open the Insert ribbon (Fig. 5.2).

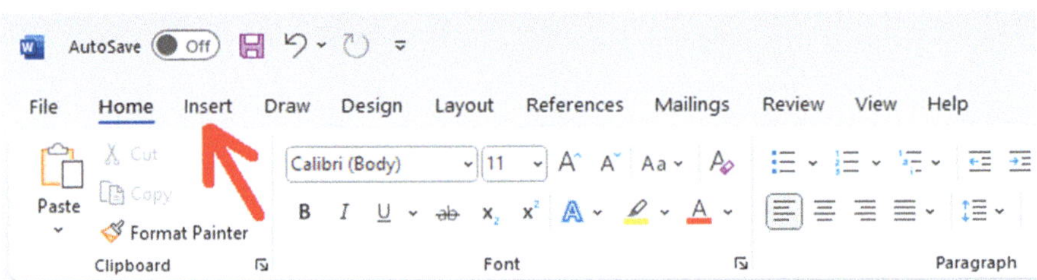

Fig. 5.1 Insert button

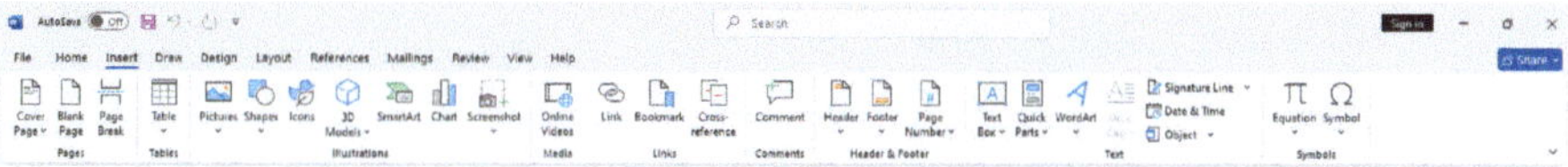

Fig. 5.2 Insert ribbon

Click on the Equations button (Fig. 5.3).

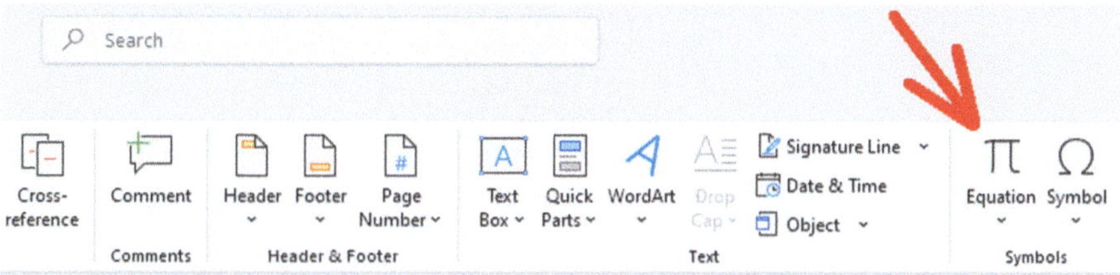

Fig. 5.3 Equations button

Clicking the Equation button will insert an equation box (Fig. 5.4) into your Word document. You can then use the tools provided in the toolbar to input your mathematical expressions (Fig. 5.5).

Fig. 5.4 Equation box

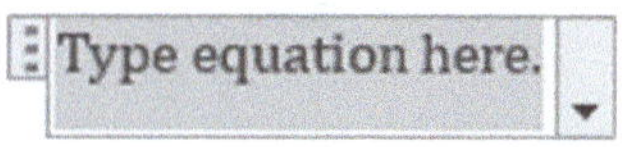

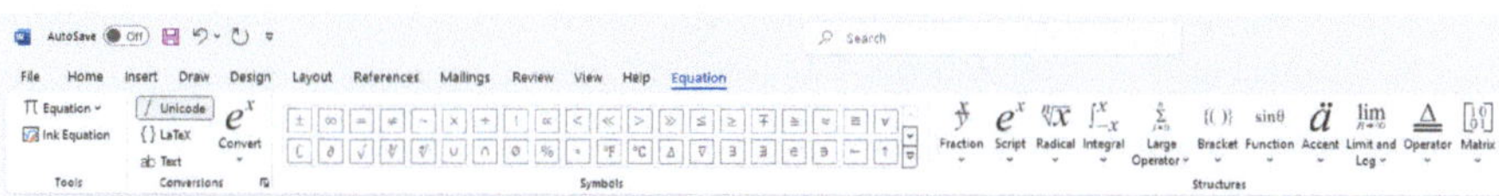

Fig. 5.5 Tools for writing mathematical expressions

Clicking the "Other" button (Fig. 5.6) will open a new window (Fig. 5.7), providing access to additional tools. From there, expand the "Basic Math" drop-down menu (Fig. 5.8) to explore a wider range of mathematical symbols and functions.

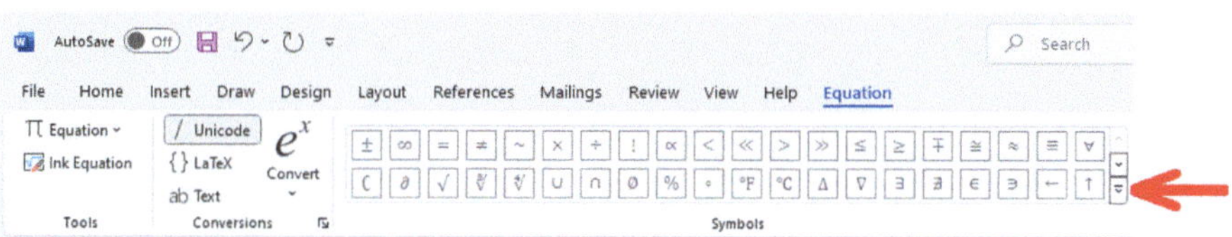

Fig. 5.6 Other button

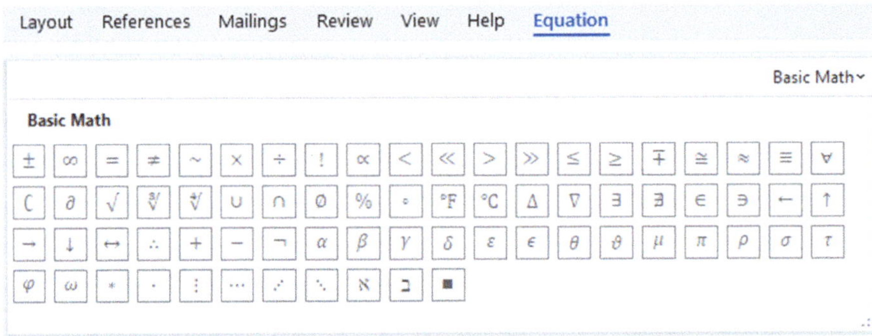

Fig. 5.7 Opened window

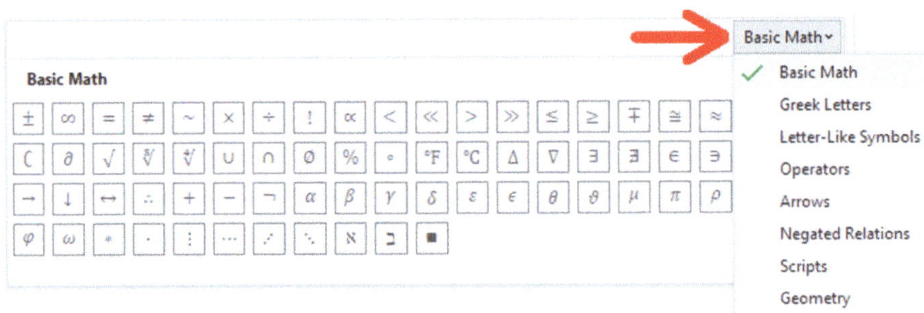

Fig. 5.8 Provided mathematical symbols and functions

References

1. Engineering Mathematics with MATLAB®and Simulink®, Farzin Asadi, Springer, 2025.
2. Applied Numerical Analysis with MATLAB®/Simulink®: For Engineers and Scientists, Farzin Asadi, Springer, 2023.

Index

F. Asadi, *A Problem-Solving Approach to Multi-Variable Calculus*,
https://doi.org/10.1007/978-3-032-21115-6

V

Z